D1230337

Metal Contamination of Food

Metal Contamination of Food

Its Significance for Food Quality and Human Health

Third edition

Conor Reilly
BSc, BPhil, PhD, FAIFST
Emeritus Professor of Public Health
Queensland University of Technology, Brisbane, Australia

Visiting Professor of Nutrition
Oxford Brookes University, Oxford, UK

Blackwell
Science

Editorial Offices:
Osney Mead, Oxford OX2 0EL, UK
Tel: +44 (0)1865 206206
Blackwell Science, Inc., 350 Main Street,
Malden, MA 02148-5018, USA
Tel: +1 781 388 8250
Iowa Street Press, a Blackwell Publishing Company,
2121 State Avenue, Ames, Iowa 50014-8300, USA
Tel: +1 515 292 0140
Blackwell Publishing Asia Pty Ltd, 550 Swanston
Street, Carlton South, Melbourne, Victoria 3053,
Australia
Tel: +61 (0)3 9347 0300
Blackwell Wissenschafts Verlag,
Kurfürstendamm 57, 10707 Berlin, Germany
Tel: +49 (0)30 32 79 060

First edition published 1980 by Elsevier Science
Publishers
Second edition published 1991
Third edition published 2002 by Blackwell
Science Ltd

Library of Congress
Cataloging-in-Publication Data
Reilly, Conor.
Metal contamination of food: its significance
for food quality and human health/Conor
Reilly. – 3rd ed.
p. cm.
Includes bibliographical references and index.
ISBN 0-632-05927-3 (alk. paper)
1. Food contamination. 2. Food – Anlaysis.
3. Metals – Analysis. I. Title.

TX571.M48 R45 2003
363.19′2 – dc21

2002026281

ISBN 0-632-05927-3

A catalogue record for this title is available from the
British Library

Set in 10/11pt Sabon
by DP Photosetting, Aylesbury, Bucks
Printed and bound in Great Britain by
MPG Books Ltd, Bodmin, Cornwall

For further information on
Blackwell Science, visit our website:
www.blackwell-science.com

Contents

Preface to the third edition

The first edition of this book was written in the late 1970s at the invitation of a commissioning editor who had seen a short paper I had published on copper and zinc uptake by foods cooked in tinned-copper utensils. I was happy to respond to the invitation since in the course of my research I had come to see the need for a text which, as I was to write in the *Preface*, would 'present a brief and useful summary of available information [on metals in food, and] make available in one place what otherwise might require time-consuming literature searches'.

The book was well received. It was favourably reviewed internationally, especially in food-related journals. These ranged from *Food Science* (Japan) to the UK *Journal of the Association of Public Analysts*, and included similar journals published in Italy, Germany, Spain, Belgium, USA, Canada, South Africa, Japan, Australia and New Zealand. It was described by one reviewer as 'a welcome addition to the library of any laboratory interested in food analysis and related topics... useful for students, consumers, exporters, regulators and even executives of the food industry'. The book's wide appeal was underlined by its translation into Russian and publication in Moscow in 1985.

Some ten years after *Metal Contamination of Food* first appeared I was asked to prepare a second edition. I agreed on the grounds that, as I said in the *Preface*, there was a continuing need for 'a useful, compact source of information that otherwise would not be readily available to the hard-pressed scientist and to others who do not have easy access to technical databases'. There had, moreover, been several significant developments in the field and the continuing flow of new data on metals in food and on their technical and health implications had made some of the material in the first edition obsolescent. Again, the book, in its revised form, was given a welcome by reviewers. One of them noted that it was 'an ideal handbook for students in the chemistry, food and medical fields [and] can be recommended not only for the medical and biological scientist, but as an invaluable tool to the food scientist'. A German reviewer described it as 'a rich source of information that is often lacking in German scientific literature'.

Advances have continued to be made in the field of food science and metal analysis in the decade since the second edition was published. During that time certain aspects of the book have become dated and there is a need to replace some of the earlier data with new information obtained with the use of more precise and reliable analytical procedures. Today we know considerably more about a number of metals that were only briefly considered, if at all, in earlier editions. Several of these, once studied only by the specialist metallurgist, are now widely used in the electronic and chemical industries. These must now be added to the list of possible food contaminants, and information about them needs to be readily available to the food scientist. From the point of view of the analyst there have been several significant developments, such as the ready availability of multielement analysis, improved sample preparation procedures and a growing interest in chemical species and not simply total concentrations of metals in foods. There have also been changes in legislative requirements and in consumer expectations in relation to food quality, not least its metal content, with major implications for the food industry.

This new edition of *Metal Contamination of Food* has been written with these developments in mind. Data on metal levels in foods and diets have been updated with information gathered from recent international literature. More than 80% of the text has been rewritten, and, as the addition of the subtitle *Its Significance for Food Quality and Human Health* indicates, greater account is taken than in earlier editions of the importance of the nutritional properties of many of the metals we consume. While information is still provided which may be useful for the less experienced, the new edition also reviews in some detail recent pertinent topics that are currently at the forefront of research. No attempt is made, however, to cover all topics in depth or to compete with specialist texts. References are provided for those who wish to proceed beyond the limits of the book. It is hoped that in this way it will continue to meet the needs of a wide range of readers.

My gratitude is owed to many people for the help they have given to me in writing this work: to my wife Ann for her encouragement and patience; to the librarians at Oxford Brookes University and the Bodleian Library, Oxford University; to many academic and professional colleagues who assisted in several ways, especially Professor J Henry, Dr U Tinggi, Dr J Arthur, Dr H Crews and others who remain anonymous.

Conor Reilly

Preface to the second edition

It is a brave publisher who will launch a new technical journal or a new book on to what must appear to be, to all but the most optimistic, an already saturated market. Not only do scientists groan at the continuous flood of publications with which they feel obliged to keep abreast, but they must pay attention to shrinking budgets which demand that a new periodical for the library be paid for by the cancellation of one already on the shelves. Prasad, writing in the editorial of the first issue of a new trace element journal[1], sought to justify the venture on the grounds that there was a need for a single publication that would bring together data currently scattered in many journals, making it difficult for researchers to keep trace of significant advances which were taking place so rapidly. He expressed the hope that his readers would benefit greatly from the journal's comprehensive overview of the many facets of trace element research.

I used a somewhat similar justification in the Preface of the first edition of this book. My aim, I wrote, was to present a brief and useful summary of available information gathered from many sources and thus to make available in one place what otherwise might require time-consuming literature searches.

Now, 10 years later, I make the same claim, that this is a useful, compact source of information that otherwise would not be readily available to the hard-pressed scientist and to others who do not have easy access to technical databases. The need for a new edition is reinforced by the fact that, during that decade, there have been several significant developments which have made the first edition no longer as useful as it was meant to be to its users.

The continuing flow of new data on metals in food and on their technical and health implications has made some of the earlier information obsolescent. Of major importance have been a variety of improvements in analytical equipment and techniques, notably the introduction of inductively coupled plasma mass spectrometry (ICP-MS) as well as of Zeeman mode atomic absorption spectrophotometry and the availability of second-generation certified reference materials (CRMs). Perhaps no less significant than the technical developments has been a wider acceptance of the need for greater care and the practise of immaculate laboratory hygiene when dealing with trace levels of metals in food.

Another development of significance has already been referred to, namely the appearance of a number of new periodicals dealing with aspects of metals in foods and their relation to health. Several of these, such as *Food Additives and Contaminants* (from 1984), *Journal of Micronutrient Analysis* (from 1985), *Journal of Trace Elements and Electrolytes in Health and Disease* (from 1987) and *Journal of Trace Elements in Experimental Medicine* (from 1988), have been of considerable use in the preparation of this volume.

Reviews of the first edition showed a surprisingly wide spread of interest in the subject of metals in food among scientists in different fields. They included physicians, environmental health officers, food manufacturers, technologists and public analysts. The book was recommended by one reviewer to food scientists, quality control professionals and regulatory officials and, as a library reference in schools of public health and environmental science. Another reviewer noted that it would be of value also to the interested lay reader.

It was pleasing for the author to find that the book was welcomed in many countries, because of the international nature of its subject and the worldwide problems that metal contamination can cause. A New Zealand reviewer noted that the material was presented in a way to which his fellow countrymen could easily relate. In Scotland it was recommended for adoption as a textbook for students preparing for the Mastership in Chemical Analysis. A Russian translation, published in 1985, testified to its perceived usefulness in the USSR.

It is hoped that the new edition will also appeal to a wide, international audience. To some extent the style used and selection of material for inclusion in the text have been determined by an expectation of a wide readership spectrum. Apologies are made to specialist readers for the over-simplifications and the generalisations which resulted from this approach. It is to be hoped that the disadvantages will be outweighed by the advantage of presenting a thorough overview of the field in a concise, readable and not too technical manner for less specialised readers from many areas of interest and from different backgrounds.

My gratitude is due to many people for help in writing this book. My wife, Ann, my secretaries, the ever helpful information librarians at QUT, as well as my research associates and graduate students who shared in my enthusiasm for investigating trace metals in the diet, are all thanked for their contributions.

Conor Reilly
Brisbane, Queensland, 23 March 1990

References

1. Prasad, A.S. (1988). Editorial: Why a new trace element journal? *J. Trace Elem. Exp. Med.* 1, 1–2.

Preface to the first edition

Some eighty of the hundred-odd elements of the Periodic Table are metals. Several of the metals are known to be essential for human life; others are toxic, even in small amounts. These metals occur in all foodstuffs, in greater or lesser amounts depending on various circumstances. The presence of metals in food can have both good and bad consequences. It can be of interest to food processors, nutritionists, toxicologists and a wide variety of other scientists. This book is concerned entirely with the presence of these metals in food. It concentrates on metals which are generally considered undesirable, at least in more than trace quantities, in food. In a general way it follows the pathways by which those metals get into food, and examines the significance – from the point of view of the manufacturer as well as of the scientist – of that contamination. It considers how man has sought to protect himself from the undesirable effects of metal contamination of food by passing laws and regulations. The international consequences of such laws are evidenced by the efforts of the United Nations and other organisations to develop a uniform and harmonised universal code of regulations concerning food.

In the second part of the study attention is given to individual metals which occur as food contaminants. A great deal of investigation has been carried out on many of these metals from biochemical, medical and other points of view. The findings of these studies have been published in many different journals and some are still only available in restricted form. An attempt has been made in this book to present a brief and useful summary of available information gathered from many sources. The aim has been to make available in one place what otherwise might require time-consuming literature searches.

Metals have served man well since he started on the long road of technological progress. We chart human progress by the metals we have used: thus we refer to the Bronze and Iron Ages. Only relatively recently have we begun to appreciate the significance of metals in a facet of human life besides that of technology. We now know a great deal about the part played by metals in the structure and function of the human body. We classify some metals among the essential nutrients. We know that others are destructive of human life and development. But we are also aware that certain metals can have both good and bad effects on the human organism, depending on the amount of metal present and other factors. Apart from the biological and medical scientist, the food chemist and technologist also have a major interest in the metals. They know the importance of metals in food not only with regard to health and toxicity, but also from the point of view of food quality, processing and commercial characteristics.

The general public is more aware today than ever before of the part metals in food play in people's lives. They are exposed to advertising extolling the nutrient value of minerals in food and in health supplements of many kinds. They are alert to the dangers of heavy metal contamination of what they eat and look to legislation to protect themselves against an excess of undesirable metals in their diet.

It is with these diverse groups of people in mind that this book was written. They all have a commendable interest in the topic, but information has not always been readily available to meet their needs. True, the specialist will frequently want to delve deeply into the question and seek out original reports, but often the intelligent layman, as well as the technician and specialist, will want a comprehensive, general treatment of the subject such as is given here.

Metal contamination of food is by no means confined to any one nation. Data given here on metals in Israeli canned food, Australian fish, Polish margarine and Zambian gin show the international nature of the problem. Water supplies in Boston as well as in Glasgow, as will be seen, can carry the same contaminating metals as do illicit spirits in Kentucky. For this reason summaries of food laws relating to metals from the UK, the US, Australia and other English-speaking countries, as well as the current international codes, are given in the text. It is hoped that, in this way, the basic data on metals in food and their significance for human health will be related to the practical context of the day-to-day user and the food producer.

Gratitude is owed to many for help given in writing this work: to my wife Ann, especially, for encouragement and practical help with references and bibliography; to Drs George and Olga Berg of the University of Rochester, New York, for setting me right on US food legislation, and for other advice and help; to my brother Brian who, from the point of view of his own profession of engineering, contributed to sections on metal contamination during food processing; and to many others who will remain anonymous. Finally, a word of gratitude to the hard-working secretaries who deciphered my writing and endured my re-editing with so much patience.

Conor Reilly
Brisbane, 4 March 1980

Part I
The Metals We Consume

Chapter 1
Introduction

1.1 Ash

'The ashes of an Oak in the Chimney, are no epitaph of that Oak ... the dust of great persons' graves is speechless too, it says nothing'

[John Donne 1571?–1631]

Everyone knows what ash is – the greyish-white fluffy powder that is left after something is burned in a fire. The Oxford Dictionary gives some further meanings of the word, including 'the remains of the human body after cremation ... a cricket trophy competed for regularly by Australia and England ... material thrown out by a volcano'[1]. None of these seem particularly pertinent to the subject of this book. However, the Dictionary adds another meaning which not only challenges the claim of the poet that ash is speechless, but also shows why it is appropriate to discuss it in the context of metals in food. It is also, we are told, a 'residue used in chemical analysis, e.g. to assess mineral content'. The oak tree's ash is not, then, speechless for those who know how to listen. The incombustible residue that survives the flames can, indeed, tell a great deal about the substance from which it was derived. What it tells is the subject matter of this book.

1.1.1 Ash and the early food analysts

Ash of food and other biological materials has attracted the attention of scientists for many hundreds of years. In the late seventeenth century, Robert Boyle, who in the English-speaking world is known as the 'Father of Modern Chemistry', developed a method which used ash to test whether or not medicinal plants and other substances had been adulterated by sellers of *materia medica*. Boyle's technique was adopted by several of his fellow *Chymists* and continued to be used into the eighteenth century[2].

Though the equipment and the facilities have changed, ashing is still today, three hundred years after Boyle's time, part of the repertoire of the food analyst. When, in the mid-nineteenth century, methods for the 'Proximate Analysis of Food' were developed by Professor Hennenberg and his colleagues at the Weende Agricultural Institute in Germany, ashing was included among their protocols[3]. Determination of 'percentage ash' is taught today to novice analysts, and 'Crude Ash' still appears in reports of the 'Proximate Composition' of foods in current scientific literature[4].

The primary purpose of Boyle, and others who measured ash in biological samples until well into the latter half of the nineteenth century, was to detect adulteration. It was only in relatively recent times that scientists came to appreciate that ash could tell a great deal more about food than whether or not it contained a fraudulent addition. There were, it has to be admitted, good reasons why adulteration was the main concern of those who were interested in the composition of food in former times.

Fraud and adulteration of foods were facts of life on a scale which would horrify today's consumers' watchdogs. The halfpenny newspapers, cheap pamphlets and popular books of the time provided their readers with sensational stories about food poisonings – a scenario not too different from our more recent era of BSE and *E. coli* media scares. It was not just popular hacks who helped stir up public alarm. Even responsible scientists played their part. In 1820 the London physician Frederick Accum published his *Treatise on the Adulterations of Food and Culinary Poisons, exhibiting the Fraudulent Sophistication of Bread, Beer, Wine, Spirituous Liquors, Tea, Coffee, &c.* Though the book was written in sober style and in the scientific language of the day, it was widely read by the non-technical public and also provided ammunition for reforming Members of Parliament who were trying to have legislation passed to prevent the sale of adulterated food.

Even with further support provided by reports published in the *Lancet*, the journal of the eminently respectable British Medical Association, many years were to pass before the UK Parliament could be persuaded to look seriously at the problem. More than half a century later, after a number of false starts, the problem finally began to be tackled by the passing of the *Sale of Food and Drugs Act* of 1875.

The Act is the basis of all modern British food laws and was adopted, in whole or in part, by many other countries, especially in the English-speaking world. Among its many important provisions was the requirement that public analysts had to be appointed by all local authorities. This requirement for professionally trained analysts, with statutory powers to seize and test suspect foods wherever they were sold, was one of the keys to the successful implementation of the Act.

However, it may also have had a negative effect. By putting emphasis on the duty of analysts to protect the public against fraud, the Act encouraged them to adopt a narrow focus in their work and concentrate on the detection of adulteration and contamination. This emphasis is clearly seen in many of the practical manuals and laboratory handbooks they used. One of the best known of these, first published in 1870, was *Foods: their Composition and Analysis. A Manual for the Use of Analytical Chemists and Others, with an Introductory Essay on the History of Adulteration*, by Alexander Winter Blyth. The author held qualifications in surgery, chemistry and law, and was Medical Officer of Health and Public Analyst in the County of Devon and in the St Marylbone District in London. He was recognised as one of the leading experts in analytical chemistry of his time, and the methods he taught were adopted as standard practice in public health laboratories in Britain and in several other countries.

1.1.2 A nineteenth-century view on food ash

Blyth, like the majority of his fellow public health analysts, paid a great deal of attention to ash. 'As a general rule', he wrote, 'testing the ash for abnormal metals and alkaline earths is necessary, and more especially if the ash present any unusual character, whether in weight, colour, or solubility'. The analyst should determine not just the total percentage of ash, but also how much of it was water- or acid-soluble, its alkalinity and how much chlorine and phosphoric acid was present. When these operations had been completed, 'for the purposes of the food analyst, the general constitution of the ash will be sufficiently known'[5].

Though this was all that was required to enable public analysts to carry out their official duties, Blyth knew well that there was more to ash and its constituents than could be revealed by the relatively simple techniques he taught. In fact, he looked forward to a time when, with the help of new analytical equipment, such as the spectroscope,

> *ash constituents of food have been thoroughly and scientifically worked out. A very careful search after the rarer metals and elements in the ash-constituents of plants would, in all probability, be rewarded with the discovery of – if not new elements – yet of a wide dispersion of the elements that are presumed not to be widely disseminated*[6].

1.1.3 Ash in the modern food laboratory

What Blyth predicted has occurred. Application of the spectroscope, in its various manifestations, from the flame photometer to its inductively coupled plasma successors, has revealed the extraordinary range of elements that are the constituents of food ash. These instrumental advances have allowed analysts to go far beyond merely the detection of fraud and contamination, to an understanding of the role in human metabolism of the inorganic components of food. They have allowed them to undertake what Blyth called 'physiologico-chemical enquiry [which] cannot be decided by chemical analysis, or, at all events, by ordinary analysis in which a few constituents are alone estimated'[7].

1.2 The metals in food

About 80 of the 103 elements listed in the periodic table of the elements are metals. The uncertainty as to the exact number arises because the borderline between what are considered to be metals and non-metals is ill-defined. While all the elements in the *s*, *d* and *f* blocks of the table are considered to be metals, there is some doubt about the elements of *p*-block. Usually seven of these (aluminium, gallium, indium, thallium, tin, lead and bismuth) are included among the metals, but the dividing line between them and the other 23 *p*-block elements is uncertain.

It is in most cases easy to distinguish between metallic and non-metallic elements by their physical characteristics. Metals are typically solids that are lustrous, malleable, ductile and electrically conducting, while non-metals may be gases, liquids or non-conducting solids. However, there is also an intermediate group of elements, which are known as metalloids or half-metals, which are neither clearly metals nor non-metals, but share characteristics of both. The metallic characteristics of the elements decrease and non-metallic characteristics increase with increased numbers of valence electrons. In contrast, metallic characteristics increase with the number of electron shells. Thus the properties of succeeding elements change gradually from metallic to non-metallic from left to right across the periods of the periodic table as the number of valence electrons increases, and from non-metallic to metallic down the groups with increasing numbers of electron shells. There is, as a consequence, no clear dividing line between elements with full metal characteristics and those that are fully non-metallic. It is in this indistinct borderline that the metalloids occur.

In this book no distinction will be made between metals and metalloids since they are normally considered together by those who formulate food regulations and industrial codes of practice, as well as by food analysts. The distribution in the periodic table of the metallic, non-metallic and metalloid elements is shown in Figure 1.1.

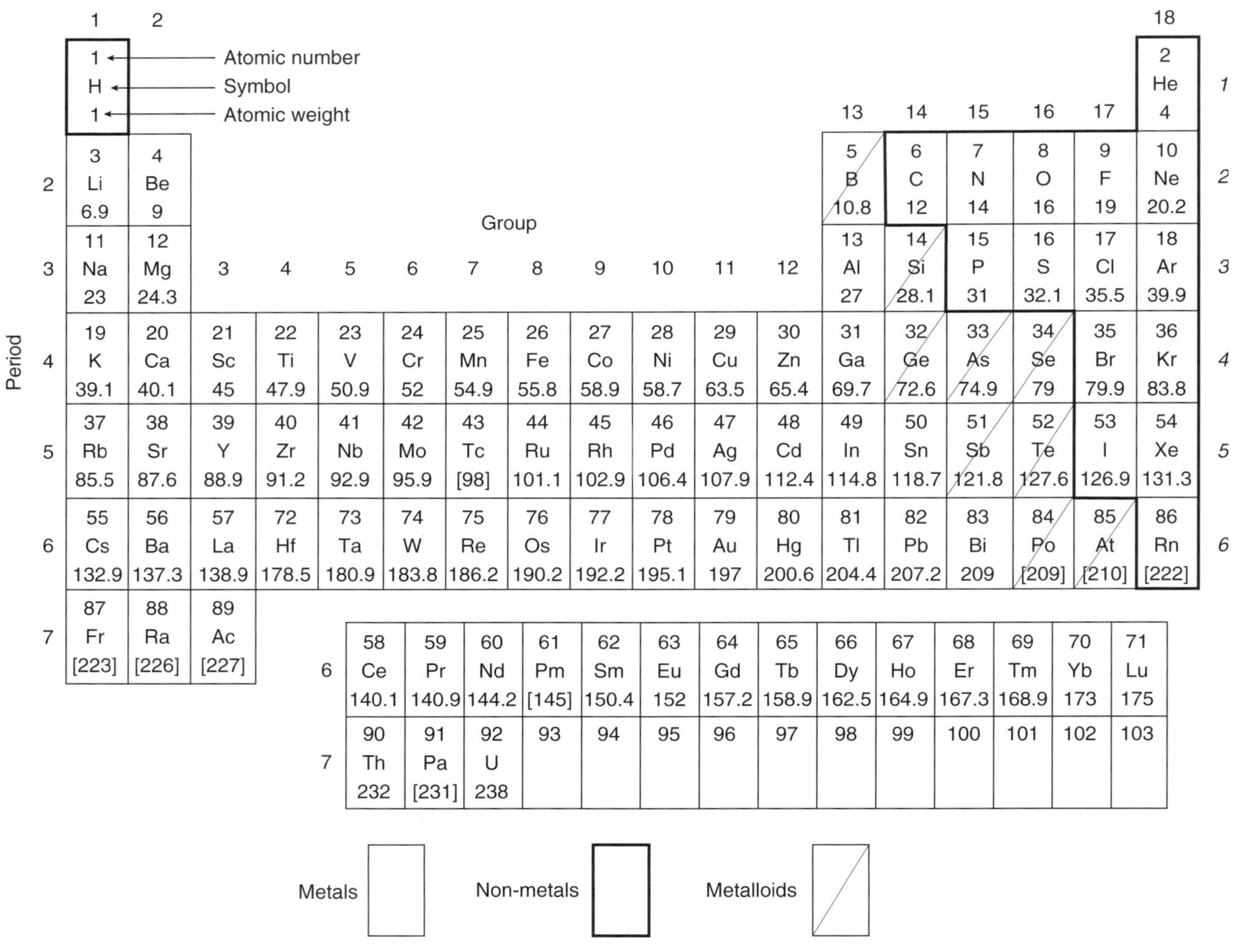

Figure 1.1 Periodic table of the elements.

1.2.1 *Chemical properties of the metals*

In the context of food, it is not the physical characteristics but rather the chemical properties of an element that are of prime importance. These properties differ considerably between metals and non-metals. The differences are mainly related to the fact that atoms of non-metals are generally electronegative and can readily fill their valence electrons by sharing electrons with, or transferring electrons to, other atoms. Thus a typical non-metal, such as chlorine, will combine with another atom by adding one electron to its outer shell of seven, either by taking the electron from another atom or by sharing an electron pair. In contrast, a typical metal, such as sodium, is electropositive, with positive oxidation states, and enters into chemical combination only by the loss of its single valence electron. There are other important chemical differences between metals and non-metals, relating to reduction/oxidation potential, acid/base chemistry and structural or ligand coordination properties, details of which can be found in a modern inorganic chemistry textbook[8].

1.2.2 *Representative and transition metals*

The metals are classified by chemists, in terms of their electronic structure, into two sub-groups which are important in relation to their biological functions. Metals such as sodium and lead which have all their valence electrons in one shell are known as *representative* metals, while those such as chromium and copper which have valence electrons in more than one shell are *transition* metals. Representative metals which have one or two valence electrons have one oxidation state, (e.g. sodium, +1; calcium, +2) while those with three or more valence electrons have two oxidation states (e.g. lead, +2 and +4; bismuth, +3 and +5). In contrast transition metals exhibit a variety of oxidation states. Manganese, for instance, which has two electrons in its outermost shell and five electrons in the next underlying 3*d* shell, exhibits oxidation states of +1, +2, +3, +4, +6 and +7.

There is another important difference between the representative and transition metals which is of significance in relation to their biological functions. Because succeeding elements in the transition series differ in electronic structure by one electron in the first underlying valence shell, the properties of succeeding elements do not differ greatly from left to right across the periodic table. In contrast, since their electronic structures differ by one electron in the outer shell, the properties of succeeding representative metals in a period differ considerably.

The metal zinc is, strictly speaking, in terms of the definition of transition and representative metals, a representative metal. However, it has many characteristics analogous to those of the transition metals and often functions like them in biological systems. Consequently it is often classified along with the transition elements. It has one difference from them that can be of some importance to the analyst. Most of the transition metal ions with incomplete underlying electron shells are coloured, both in solid salts and in solution. The colour depends on the oxidation state of the metal. For example, Fe^{3+} is yellow while Fe^{2+} is green. However, when, as is the case with zinc, and generally with all representative metals, the underlying electron shell is filled, the substance containing the ion is colourless. As we shall see, this absence of coloured compounds probably helped to account for some of the early lack of progress in investigations of the metabolic role of zinc.

1.3 Distribution of the metals in the environment

Unlike the gaseous elements, the metallic elements are not distributed evenly throughout the environment. For a variety of reasons, including the result of human activities, some metals have become concentrated in certain places, while others are almost absent. This uneven distribution has important consequences for the levels of metals that occur in foods and other biological tissues.

1.3.1 *Metals in human tissue*

Analysis of the ash produced when the tissues of a human body are incinerated will show the presence of most of the metallic elements, some in reasonably large quantities, others as almost unquantifiable traces. The ash produced from a total adult body of 70 kg weight could be expected to contain as much as 1 kg of calcium, 250 g of potassium, 150 g of sodium and 50 g of magnesium. Other metals will be present in considerably smaller amounts, ranging from 2–4 g of iron, to 2 mg or less of selenium and other trace elements. There will also be just detectable amounts of what are described as *ultratrace* elements, such as caesium and uranium.

1.3.2 *Metals in soil*

The presence of so many different metals in human tissue is not surprising since the foods we eat also contain a wide range of metals, reflecting the metal composition of the soil on which the foods were produced. Because of the intimate connection between soil and the metals in our bodies, it is appropriate to give some attention to the nature of soil and how it contributes to the metal content of food.

Soils are complex assemblies of solids, liquids and gases[9]. The solid portion includes both primary and secondary minerals, ranging in size from $< 2\,\mu m$ in diameter (gravel) to > 2 mm diameter (rocks). The minerals are inorganic compounds of the different metallic elements, with definite physical, chemical and crystalline properties. A primary mineral, such as quartz (SiO_2), is one that has not been altered chemically since it was first deposited; in contrast a secondary mineral, such as the aluminosilicate kaolin $\{Al_2Si_2O_6(OH)_4\}$, has been chemically changed by weathering. Inorganic components can account for more than 90% of the solid matter of the soil. In a typical fertile soil the solid component (made up of about 45% mineral and 5% organic matter) represents about 50% of the total volume, with air and water each making up the remainder in about equal amounts (20–30%).

The metals and metalloids found in the highest concentrations in the soil are aluminium, iron, calcium, potassium, sodium and magnesium. Also present, though at much lower concentrations, are a wide range of other elements as shown in Table 1.1.

1.3.2.1 *Soil as a source of plant trace elements*

Fertile soils supply plants with all the elements they need for growth. These include the trace elements iron, manganese, zinc, boron, copper, molybdenum and possibly nickel. In addition cobalt is required as a micronutrient by legumes to enable them to fix atmospheric nitrogen when synthesising amino acids and protein. If any of these micronutrients are present in the soil in very low concentrations or only in unavailable (insoluble) form, plants may show symptoms of trace element deficiency. However, if high concentrations of some of the trace nutrients, especially copper, nickel and

Table 1.1 Metals in soil.

Metal	Median (mg/kg)	Range (mg/kg)
Aluminium	72 000.0	700–< 10 000
Arsenic	7.2	< 0.1–97
Barium	580.0	10–5 000
Boron	33.0	< 20–300
Beryllium	0.92	< 1–15
Calcium	24 000.0	100–320 000
Cadmium	0.35	0.01–2
Cobalt	9.1	< 3–70
Copper	25.0	< 1–700
Chromium	54.0	1–2 000
Germanium	1.2	< 0.1–2.5
Lithium	24.0	< 5–140
Magnesium	9 000.0	50–> 100 000
Manganese	550.0	< 2–7 000
Molybdenum	0.97	< 3–15
Selenium	0.39	< 0.1–4.3
Sodium	12 000.0	< 500–100 000
Tin	1.3	< 0.1–10
Titanium	2 900.0	70–20 000
Uranium	2.7	0.29–11
Vanadium	80.0	< 7–500
Zinc	60.0	< 5–2 900

Adapted from Sparks, D.L. (1995) *Environmental Soil Chemistry*, p. 24. Academic Press, New York.

cobalt, are present, plants may show signs of metal toxicity. Other non-nutrient metals, which can also accumulate in soil and be taken up by growing plants, may not harm the plants themselves but may be a potential hazard to animals that consume them. Of particular importance in this regard are arsenic, mercury and cadmium.

1.3.2.2 Variations in the metal content of soils

There can be considerable differences in mineral composition of soils in different countries and even in different regions in individual countries. As has been observed by McBride[10], the specific elemental composition of a soil depends on the chemical composition of the parent material from which the soil was formed. However, the original composition may have been modified over time by the effects of weathering, as well as by pollution caused by human activities and by accumulation from natural biogeochemical processes.

1.3.2.3 Soil metal availability

While the total elemental composition of a soil, which can be determined with considerable accuracy by modern analytical techniques, is of primary interest to the agronomist and toxicologist, a more useful measure is 'availability' or 'lability' of the elements the soil contains. This is certainly the case in the context of this book, since it is their availability for uptake and accumulation by crops that determines whether or not certain elements become food contaminants.

The availability to the plants of metals in the soil on which they grow depends on the properties of the metals, such as tendency to complex with organic matter and to become adsorbed onto other soil constituents. Availability will also be affected if the metals are precipitated, for example as insoluble sulphides. The passage of a metal from the soil through the plant roots and into its growing parts occurs in several steps, which have implications for the contribution that a food plant may have for human intake. These steps are described in detail by McBride, but for our purposes it will be sufficient to note that the soil controls the availability of elements to the plant roots largely by limiting their mobility through the soil solution. Absorption by the roots depends on the concentration of the element in the solution that surrounds them. The plant can modify the chemistry of the adjacent soil solution in what is known as the *rhizosphere effect*, by exuding hydrogen ions and organic chelating agents. These increase the solubility of cations and help their uptake into the interior of the root.

1.3.2.4 Metal transport and location within the plant

Once within the roots, some metals, including, fortunately, the toxic metals lead and cadmium, may remain in or near the roots, while others are translocated upwards to the growing parts and into the fruits and other storage organs. Translocation is a complex operation which depends on the plant's own biochemical and physiological operations, but is also affected by other factors, such as the chemical form (or speciation) of elements in the soil. Since the question of chemical speciation, and its effect on availability of soil metals, is of some considerable significance in relation to contamination of food crops, it is worth considering it in more detail here.

1.3.2.5 Soil metal speciation

Metals in soil occur in several oxidation states and in soluble complexes with different organic and inorganic ligands. The soil solution can contain a wide variety of organic and other ligands, such as fulvic acid and the inorganic anions HCO_3^- and OH^-, as well as other anions that can form soluble complexes with metal cations. This complexing facilitates the transport of metals through the soil solution. It also has a significant effect on absorption of metals by the roots. For instance, uptake of copper, which is correlated to the free metal ion's concentration in the soil solution, is greatly increased when organic matter is added to the soil, with soluble organic matter acting, apparently, as a 'cation carrier'.

The same effect is achieved for cadmium, which forms soluble complexes with bicarbonate and carbonate when CO_2 levels and pH are increased in the soil. According to McBride such findings indicate that both free metal activity (or concentration), representing an *intensity* factor, and total sustained soluble metal concentration, representing a *capacity* (or buffering) factor, are important to biological systems.

Most metals are, in fact, relatively immobile in soil because they adsorb strongly onto soil components or form insoluble precipitates. Certain soils are rich in adsorption sites, such as oxides of iron, aluminium and manganese. Organic matter can also bind to cations. High soil pH favours sorption and precipitation of metal cations as oxides, hydroxides and carbonates. In contrast, acid conditions favour sorption and precipitation of a number of metal anions, including molybdate and selenite.

Soil redox potential is another factor of importance in relation to metal mobility. Some elements, such as chromium, manganese and selenium, are much more soluble in one oxidation state than in another. Certain metals which are known as *chalco-*

philes, which include mercury, copper, lead, cadmium, zinc, arsenic and selenium, form insoluble sulphides under reducing conditions which result in the conversion of sulphate to sulphide.

Mobility of metals is also dependent on whether or not there is movement of water through the soil. In arid climates, water movement, when it occurs, is upwards. This results in the transport of soluble metal complexes to the soil surface, where they accumulate as a result of evaporation. In wet climates, water movement is in the opposite direction and mobile metal complexes leach downwards. They can, however, be intercepted by plant roots and deposited in the leaves and other tissues. When eventually the plants die, the metals they have taken in can bioaccumulate in the humus and other debris on the soil surface.

As we shall see when we come to consider in detail the various ways in which metals get into food, soil is not only the primary source of all the metals which naturally accumulate in plants and animals which we consume, but is also a significant source of adventitious or accidental contamination of our diet. In a world in which the environmental spread of potentially harmful elements as a result of human activity is an ever increasing problem, it is important that those who are concerned with food safety consider metals in soil as an integral part of their planning and management processes[11].

References

1. *Concise Oxford Dictionary* (1995) Clarendon Press, Oxford.
2. Boyle, R. (1690) *Medicina Hydrostatica or Hydrostaticks Applied to the Materia Medica, Showing how by the Weight that divers Bodies, used in Physick, have in Water, one may discover whether they be Genuine or Adulterate*. Printed for Samuel Smith at the Sign of the Princes Arms, in St Paul's Churchyard, London.
3. Muller, H.G. & Tobin, G. (1980) *Nutrition and Food Processing*, p. 68. Avi, Westport, Conn.
4. Maia, F.M.M., Oliveira, J.T.A., Matos, M.R.T., Moreira, R.A. & Vasconcelos, I.M. (2000) Proximate composition, amino acid content and haemagglutinating and trypsin-inhibiting activities of some Brazilian *Vigna unguiculata* (L) Walp cultivars. *Journal of the Science of Food and Agriculture*, **80**, 453–8.
5. Blyth, A.W. (1986) *Foods: their Composition and Analysis: A Manual for the Use of Analytical Chemists and Others*, 4th edn. Charles Griffin & Co, London.
6. Blyth, A.W. (1986) *Foods: Their Composition and Analysis: A Manual for the Use of Analytical Chemists and Others*, 4th edn. Charles Griffin & Co, London.
7. Cox, P.A. (1995) *The Elements on Earth: Inorganic Chemistry in the Environment*, Oxford University Press, Oxford.
8. Cox, P.A. (1995) *The Elements on Earth: Inorganic Chemistry in the Environment*, Oxford University Press, Oxford.
9. Sparks, D.L. (1995) *Environmental Soil Chemistry*, pp. 23 ff. Academic Press, San Diego.
10. McBride, M.B. (1994) *Environmental Chemistry of Soils*, pp. 308ff. Oxford University Press, Oxford.
11. Plant, J., Smith, D., Smith, B. & Williams, L. (2000) Environmental geochemistry at the global scale. *Journal of the Geological Society, London*, **157**, 837–49.

Chapter 2
Metals in food

2.1 The metal components of food

As has been noted earlier, the tissues of plant foods, as well as the animals that feed on them, can be expected to contain most if not all the different metals that occur in the soil on which they were grown. Since the quantities of the elements in soil can vary from place to place, the amounts taken up by plants and retained in their tissues can also show considerable variations. This can have implications for the nutritional health of consumers, especially if they normally depend on a particular foodstuff to supply the bulk of their intake of an essential element. This is the case, for example, with regard to selenium in the UK.

Until relatively recently selenium requirements of the population were mainly met by the import of US and Canadian wheat which is normally rich in selenium. Since the mid-1970s American wheat has largely been replaced in the UK market by locally grown and other EU grain. British and Continental soils have relatively low selenium contents and consequently European cereals contain significantly lower levels of the element than do those grown in North America. As a result dietary intake of the element fell in the UK from about 60 μg/d, in the 1960s, to approximately 40 μg/d in the late 1990s[1].

There can be considerable variations in concentrations of metals even within the same class of food, depending on its geographical origins and other factors. Zinc levels in oysters, for instance, can show a thousand-fold range from as little as 1 mg/kg to nearly 1 g/kg[2]. Rice grown in a selenium-rich area of China can contain as much as 1 μg/g of the element, compared to as little as 0.02 μg/g in the New Zealand product[3].

Variations in metal concentrations related to geographical sources are not confined to primary products. They also occur in manufactured foods and beverages. Brazilian cane spirit (Cachaças), for example, from one region of the country has been found to contain 0.060 mg chromium/litre, 0.032 mg manganese/litre and 1.67 mg copper/litre, compared to 0.012 mg chromium/litre, 0.053 mg manganese/litre and 4.40 mg copper/litre in the same product manufactured in another region[4]. Differences in metal concentrations between samples of illicit alcoholic beverages have been used by US excise officers as 'fingerprints' in an attempt to identify their places of origin, with, it seems, only limited success[5].

Even within a single sample of a food, metals may not be distributed evenly, so that there can be differences in their concentrations in different parts. A whole grain of wheat can contain 7.4 μg copper/g, but with only 2.0 μg/g of this in the endosperm. It is for this reason that refining or fractionation of cereals during milling affects the amount of metals in the food as consumed, as illustrated in Table 2.1, which shows the differences which occur in wheat, oats and maize as a result of processing[6].

In spite of such variations, for a variety of reasons, between the metal contents of individual foodstuffs, it is possible, wherever appropriate data from national dietary

Table 2.1 Changes in metal content of cereals following processing.

	Metal content (μg/g, wet weight mean ± SEM)				
Cereal	Al	Mn	Fe	Zn	Cu
Wheat germ	4.13 ± 0.45	194 ± 11	105 ± 4	92 ± 0.5	14.7 ± 0.8
Whole wheat	0.31 ± 0.06	67.2 ± 2	35.4 ± 0.7	21.0 ± 0.4	3.32 ± 0.03
White flour	0.34 ± 0.09	12.9 ± 0.2	12.4 ± 0.2	9.18 ± 0.13	1.63 ± 0.01
Wholemeal oat flour	3.35 ± 0.14	54.8 ± 0.9	53.3 ± 0.7	27.9 ± 0.1	3.70 ± 0.02
Refined oat flour	0.4 ± 0.2	27.6 ± 0.3	12.9 ± 0.1	14.6 ± 0.1	3.18 ± 0.03
Raw maize	1.58 ± 0.25	7.82 ± 0.22	25.4 ± 0.7	24.5 ± 0.7	2.62 ± 0.82
Maize flour	0.33 ± 0.07	1.73 ± 0.09	5.47 ± 0.21	4.65 ± 0.22	0.65 ± 0.02

Data adapted from Booth, C., Reilly, C. & Farmakalidis, E. (1996) Mineral composition of Australian ready-to-eat breakfast cereals. *Journal of Food Composition and Analysis*, 9, 135–47.

and food consumption and composition studies are available, to estimate overall metal intakes in different countries and by different population groups. How this is done will be described in detail later. Thus we know, for instance, that the average intake of dietary lead of 60–65-year-old men in the US in 1991 was 3.46 μg/d[7], while the selenium intake of one-year-old children in Finland in the same year was 16.0 μg/d[8].

2.2 Why are we interested in metals in food?

Among the many different metals that we consume in our diet and accumulate in our bodies, only a small number are believed to be essential for normal life. An inadequate intake of any one of these essential inorganic nutrients may result in specific biochemical lesions within cells of the body and development of characteristic clinical symptoms. The symptoms will normally respond when the deficiency is corrected by supplying an adequate amount of the missing element.

The nutrient metals have been divided traditionally into two classes according to the amounts of each which are required for normal body function. As is shown in Table 2.2, the metals potassium, magnesium, calcium and sodium are classified as macronutrients, and the remainder as micronutrients.

The use of the term *trace element* for this second group is no longer as common as it was in the past. The term originated in the days when the minute amounts of some elements in biological tissues, down to levels of nanograms (10^{-9} g) or picograms (10^{-12} g), were beyond the instrumental capabilities of analysts to quantify exactly. Consequently, they were referred to as occurring in 'traces'[9]. Today the term is usually used for an element occurring at a level of < 0.01% (< 100 μg/g). Other terms sometimes used for a micronutrient are *minor element* or *oligoelement*. The term *ultratrace element* is also met with occasionally. It usually, though not consistently, is used for elements that occur at levels of < 0.01 μg/g (< 10 ng/g).

2.2.1 *Functions of the trace elements*

To some extent the known functions of inorganic macronutrients and micronutrients overlap. Both work in three principal ways: as constituents of bone and teeth; as

Table 2.2 Nutrients and other metals.

Nutrient metals	
Macronutrients	Calcium Magnesium Potassium Sodium
Micronutrients	Chromium Cobalt Copper Molybdenum Nickel Selenium
Possibly essential micronutrients	Arsenic Boron Vanadium
Toxic metals	Beryllium Cadmium Lead Mercury
Non-toxic, non-essential metals	Aluminium Tin

soluble salts which help to control the composition of body fluids and cells; as essential adjuncts to certain enzymes and other functional proteins. The macroelements play major roles in the first two functions, while the trace elements are especially prominent in assisting enzyme function. Very few of the proteins that act as biological catalysts do so entirely on their own. Most need the assistance of a non-protein prosthetic group. If the prosthetic group is detachable it is known as a coenzyme. The group may be an organic molecule, containing one or more atoms of a metal, or may consist solely of a metal. In the latter case, if the metal is detachable from the protein part, it is known as an activator.

Iron and copper occur in the prosthetic group of many enzymes concerned with cellular oxidation–reduction processes. Molybdenum has much the same function in certain other enzymes. Zinc and manganese function as detachable activators on some enzymes and as part of the prosthetic group in others. Most of the other essential trace elements are believed to play similar enzymatic roles and as a consequence these enzymes are known as *metalloenzymes*.

Trace elements are also found in other important functional components of the body, including hormones and vitamins. The production and storage of the hormone insulin in beta cells of the pancreas, for instance, involves the metal zinc. Haemoglobin and myoglobin, oxygen transporters in the blood and muscle tissue, are iron-containing compounds. Cobalamin, or vitamin B_{12}, contains cobalt.

2.2.2 *New trace elements*

As our knowledge of the role of metals in human metabolism has grown, and our ability to determine their concentrations in foods and body tissues has improved, the number of elements that have been proven to come under the heading of *essential trace element* has also grown. Up to the middle of the twentieth century only five of the metals were considered to be in this category: iron, zinc, cobalt, manganese and molybdenum. In the 1970s selenium was added to their number, and since then the case for adding several others, such as boron, has been strengthened[10]. There is no reason to suppose that the list of essential trace elements is now definitive. It would not be surprising if some elements with currently insufficiently known physiological functions were found to have specific biochemical functions in humans. There is growing evidence that arsenic, for instance, is one such element.

There are several reasons for this uncertainty about the role of possibly essential metals. To a large extent it can be traced to analytical problems. In addition there is the difficulty of developing suitable experimental protocols for investigating the essentiality or otherwise of a metal. It is relatively easy to demonstrate that a dietary insufficiency of a macroelement will result in metabolic disorders, but not for a micronutrient. For ethical reasons humans cannot be experimentally deprived of certain elements in the diet in order to see if symptoms of deficiency result. Quite properly, ethical committees frown upon the once widely accepted practice of using volunteers, even from among the researchers themselves, for such investigations. Animal subjects do not always adequately match human requirements or responses to the elements being tested. Computer programs which might replace such *in vivo* trials still seem to lie a long way in the future.

2.3 The toxic metals

As is indicated in Table 2.2, in addition to the essential elements there is another large group of metals that are toxic. It is not always possible to distinguish clearly between metals in this category and others that play essential roles in human metabolism and body function[11]. All metals are probably toxic if ingested in sufficient amounts. Sometimes the margin between toxicity and sufficiency is very small, as is the case, for example, with selenium. It is difficult to consider toxicity of a single metal in isolation from other metals and, indeed, from all the other components of a food. Under normal conditions all metals are capable of interacting with other metals in the body if consumed together. The physiological effects, including its toxicity, of cadmium, for instance, are closely related to the amount of zinc also present. Likewise the function of iron in cells is affected by both copper and cobalt, and to some extent also by molybdenum and zinc. Several other similar interactions between metals in the body are known to occur.

In spite of the above reservations it is possible to differentiate between elements that are known with certainty to be essential and those which cause severe toxic symptoms at low concentrations and have no known beneficial functions. Mercury, cadmium and lead are usually considered to qualify for inclusion among the toxic metals, though there are indications that cadmium and even lead may one day be shown to be essential elements[12].

The term *heavy metal* is often used, especially in official documents, to describe the group of toxic metals. It is a descriptive rather than a scientific classification. Strictly speaking it should only be used for those elements with an atomic weight of 200 or above, such as mercury (201), thallium (204), lead (207) and bismuth (209)[13]. In practice it is usually used to refer to metals of a high specific gravity which have a strong attraction to biological tissues and are slowly eliminated from biological systems[14]. However, a typical listing of heavy metals will often be found to include arsenic, beryllium, boron, selenium and other metals and metalloids, as well as mercury, cadmium and lead. This is probably a convenience and relates more to the metals' toxicity rather than to their specific gravity. A US government document, published in 1972, defined heavy metals as 'common metallic impurities that are colored by hydrogen sulfide (Ag, As, Bi, Cd, Cu, Hg, Pb, Sb, Sn)', a distinction based on chemical properties, which has been adopted by regulatory authorities in several other countries[15].

2.4 Effects of metals on food quality

The concern of food manufacturers and processors is not simply to see that their products are free from toxic metals, or even from essential trace elements, in quantities high enough to cause health problems for consumers, nor is it simply to meet the various legal requirements or codes of practice that apply to metals in food. They must also make sure that foods do not contain metals that might bring about deterioration and quality defects.

Traces of certain metals can cause a variety of undesirable changes in foods during cooking and storage. Even at levels of only a few mg/kg, complexes can be formed between metal ions and organic compounds causing the development of colours, not all of them welcome. There was a time when less respectable cooks and food processors used the complex formed between copper and chlorophyll to give green vegetables a bright and stable colour, by cooking in an unlined copper saucepan or even by dropping a copper coin into the pot. More acceptable is the modern practice of adding a suspension of food-grade titanium dioxide to skim and other low-fat milk products to improve their whiteness, as well as creamy texture[16]. Another colour change caused by metal contamination of food is the development of a black pigment when iron combines with anthocyanins in certain fruits. The blackening of a cherry pie when it is cut with the blade of a steel knife is a well-known example of this reaction. Iron can impart a grey-green colour to cream and have a similar effect on chocolate-containing foods. Aluminium and tin can also affect colour in some foods.

A study by Borocz-Szabo found that iron contamination can adversely affect the sensory qualities of a variety of liquid foods, including fruit juices, milk, beer and wine[17]. Corrosion products of steel, especially iron salts, caused loss of odour as well as development of astringent, metallic or bitter tastes. Milk developed a nauseating taste under the conditions investigated.

Another, and possibly even more serious, effect of metal contamination on certain foods is the production of rancidity in fats. Traces of copper, iron and certain other metals act as catalysts in the oxidation of unsaturated fatty acids. This can result in spoilage of meat products and in the rapid and costly deterioration of cooking oil and fat-containing foods. The role of iron and copper in lipid peroxidation, causing development of off-flavours, especially in pickled meats, has been discussed by Kanner[18]. A similar effect in pre-cooked cereal-based convenience foods has been reported[19].

2.5 How much metal do we consume with our food?

It is a complex task to estimate the dietary intake of any particular metal and there is no one technique for doing so which can be considered a universal method of choice. Just how many different procedures have been employed to assess food intakes by different investigators is indicated by the extensive bibliography on the topic compiled by Krantzler *et al*[20].

A fairly crude estimate of dietary intakes can be obtained if data are available on total food consumption by a particular population and the average level of the metal in foods consumed is known. But more precise estimates are often required by government and other authorities as well as by food manufacturers. As has been observed by Rees and Tennant of the UK Food Safety Directorate:

> *while simple methods can be used for prioritisation, the use of crude and inaccurate intake estimates in food chemical risk assessments can result in suboptimal risk management solutions being adopted. The cost of such errors is borne by the consumer either in an increased health risk from food or in restrictions in the supply and variety of food. It is therefore the duty of regulators and of the food industry to ensure that accurate estimates of intake are available whenever they are required*[21].

2.5.1 *Estimating metal intakes*

In their very useful review of methods for estimating food chemical intakes, on which the following section is partially based[22], Rees and Tennant note that some of the difficulty which can be experienced in estimating the intake of metals and other food components by particular populations can be overcome if a hierarchical approach in line with guidelines recommended by the World Health Organization is adopted[23].

The hierarchical approach is conducted at three levels (or hierarchies), as follows:

Per capita estimation of intake

This is obtained by multiplying per capita food consumption (which in many countries is available from national food utilisation survey data) by the estimated concentration of the metal in food (obtained from industry or other commercial data, or from laboratory analyses of food samples).

This is the simplest and most cost-effective method and provides an estimate of the average intake of everyone in the whole population. The results do not, however, apply to 'non-average' consumers, but they can highlight potential problems, such as high intakes, and can point to the need for further investigation. They also allow comparisons to be made between consumption in different countries, including Finland, Germany and Japan[24].

Total Diet Studies for intake assessments

The Total Diet Survey (TDS) is also known as the Market Basket Survey (MBS) because it is based on a basket of food representing the total diet of consumers, which is known from surveys of food purchases. The foods, purchased in normal retail outlets, are prepared as if for consumption. They are separated into different food groups and the groups are analysed for metals and other components. The method has been used in the UK by the Ministry of Agriculture, Fisheries and Food (now DEFRA) to monitor food constituents since 1966. The procedures, which were modified in 1981, have been described by Peattie and colleagues[25].

Apart from being able to provide useful trend information on intakes and overall background levels of contaminants, the MBS approach can pinpoint food groups which represent major sources of a particular contaminant. However, it does not give information on individual foods and its findings apply to the average, not the extreme, consumer.

When the method was originally adopted by health authorities in several countries it was mainly used to investigate pesticide and radioactive contamination. Its use has since been extended to include other types of contamination, including metals, for example in Finland[26]. It has been used in a similar manner in Spain[27], Italy[28] and Denmark[29].

Hypothetical diets

A 'model diet' approach can be used with reasonable success to estimate dietary intakes when there is inadequate information to allow the TDS method to be adopted. Normally the diet is built around typical serving sizes of foods, but in particular circumstances it may be weighted to allow for certain 'at-risk' groups. It can be a cost-effective way of estimating intakes, especially when the major source of a food component of interest is a single food. Otherwise, the method is open to error, especially when many foods are involved.

The model diet approach can be refined by constructing scenarios that reflect the least to the worst possible cases. Large numbers of variables can be introduced into the modelling to make the estimates more realistic. Such refinements are useful, particularly for predicting trends, or intakes of 'at-risk' groups.

2.5.1.1 *Surveillance methods for assessing intake*

If the cost-effective methods described above provide inconclusive results, more precise procedures which rely on detailed data on food composition and consumption may be required. The procedures are generally expensive since they require considerable input of professional expertise and other resources. In the surveillance approach, reliable food consumption data for the population of interest is combined with validated data on food composition. Information on food consumption may be collected by various methods, such as *food frequency* recordings or *weighed food diaries*. Intake data can be obtained from official maximum permitted levels, from food manufacturers or from direct laboratory analyses. The method is most effective if the dietary survey covers a large representative sample of the population and records all food consumed.

Surveillance of dietary intakes of heavy metals has been carried out in several countries. In the UK consumption data from MAFF's *Dietary and Nutritional Survey of British Adults* (1986–87) has been used[30]. The Survey covered a nationally representative sample of some 2000 adults who completed a seven-day weighed diary of all foods they consumed. Over 4500 foods were coded and a recipe database used to express consumption data on the basis of complete foods or individual food components. In the USA the Nationwide Food Consumer Survey (NFCS) of the Department of Agriculture (USDA) is used to estimate intake of food components by individuals[31]. A similar programme is operated in Germany[32].

2.5.1.2 *Duplicate diet method for intake estimation*

Where a particular risk of high intake of a metal or other food contaminant is indicated, it may be necessary to obtain additional information about individual consumers who are identified as being particularly 'at-risk' by asking them to supply duplicates of the actual meals they consume. In theory this is the ideal means of assessing intakes of food components since it allows direct measurement to be made of 'at-risk' groups. However, its accuracy depends on the reliability of individuals in dividing their meals between themselves and the duplicate plate. It is also an expensive operation to carry out on a large scale. Where resources have permitted and the need for the most accurate data has been acknowledged, the method has been used to determine food contamination risk, especially in particular population groups. A good example of such an application of the duplicate diet method was MAFF's investigation of mercury intake by fishing communities in the UK in the late 1970s[33].

2.5.2 *Comparison of methods of assessment of metal intakes*

Duffield and Thomson have compared several methods of assessing intake of dietary metals in their study of selenium in New Zealand[34]. They noted that both diet record assessment and duplicate diet collections, which have been reported to give the most reliable estimates, demand a high degree of motivation and cooperation from subjects. In contrast, the food frequency questionnaire (FFQ) method makes relatively low demands and has been shown to be highly successful in placing individuals into broad categories of intakes.

In order to evaluate these different methods in relation to assessment of trace element intakes, the authors compared results obtained when they used each of the methods – chemical analysis of duplicate diets, diet records and FFQ, to investigate selenium intakes in New Zealand. They found that the results from FFQ were considerably higher than those from either duplicate diets or diet records. This may have been due to under-recording or a restriction of food intake during the recording period as well as a tendency of FFQ to overestimate intakes of certain food groups. The authors also found that diet record assessment was not adequate for predicting intakes of individuals, though they noted that this may not reflect inadequacy of the method but rather deficiencies in the New Zealand food composition data. Their conclusion was that duplicate diet analysis remains the recommended measure for estimating dietary selenium intakes.

As the authors point out, this study provides a unique contribution to dietary selenium research as the only one reported (up to 1999) in which three different methods of dietary assessment (duplicate diet analysis, diet records and FFQ) have been used. Whether the findings of their evaluation of the three assessment methods can be applied to other elements besides selenium is unclear.

2.6 Assessing risks from metals in food

Government authorities in most countries monitor dietary intakes of metals in foods because of possible health effects. In the UK this activity has been undertaken on a systematic and regular basis for many decades by the Ministry of Agriculture Fisheries and Food (MAFF; now DEFRA) through the Joint Food Safety and Standards Group (JFSSG)[35]. In the USA food surveillance is the responsibility of the Food and Drug Administration (FDA) with monitoring of foods for toxic elements carried out through its Center for Food Safety and Applied Nutrition. In other countries similar organisations perform the same role of monitoring safety of the national food supply.

Information on concentrations of metals and other elements in food and consumption data obtained by these organisations are used to estimate dietary exposure and assess the safety of foods consumed. The risk to health is assessed by comparing these estimates of exposure with recommended safe levels. These recommendations are based on the concept of the Acceptable Daily Intake (ADI) which was introduced by the Joint Expert Committee on Food Additives (JEFCA) of the Food and Agriculture Organization of the United Nations (FAO) and the World Health Organization (WHO) in 1950. The concept has since become the key element in evaluation of the safety of chemical contaminants in food[36]. The ADI is defined as an estimate of the amount of a food additive or contaminant, expressed on a bodyweight basis, that can be ingested over a lifetime without appreciable risk[37]. It is expressed as a range, and is usually established on the basis of the NOAEL (no-observed-adverse-effect level) or LOAEL (lowest-observed-adverse-effect level) in animals. If they are available, relevant human data, from studies with volunteers or as a result of accidental exposure, are also used. A Safety Factor (SF), usually 10 or 100, depending on whether the data source is animal or human, is used to convert the NOAEL or LOAEL into an ADI.

The Reference Dose (RfD) also uses the NOAEL or LOAEL, but refines the SF into a Modifying Factor (MF) and four different Uncertainty Factors (UF), usually ten, three or one[38]. The RfD is a term that considers the systematic toxicity of the substance in question and determines a level unlikely to cause deleterious effects over a lifetime[39].

The term Nutrient Safety Limit (NSL) has been introduced largely in response to problems of toxicity that can result from the growing practice of consuming nutritional supplements in quantities greater than recommended intakes. The NSL is calculated as an intermediate value between the LOAEL and the RNI (recommended nutrient intake). It provides adequate margins of safety below adverse intake levels but avoids identifying safety limits that are below the RNI[40]. In the UK the term Safe Intake (SI) is used to indicate an intake or range of intakes of a nutrient that is enough for almost everyone but not so large as to cause undesirable effects[41].

Another term used to indicate safety limits for food components is Tolerable Upper Intake Level (UL). This was defined by the Food and Nutrition Board (FNB) of the US National Academy of Sciences as the maximum level of total chronic daily intake judged to be unlikely to pose a risk of adverse health effects to the most sensitive members of the healthy population[42]. ULs are set on the basis of identified hazards, dose-response assessment, intake and exposure estimations and uncertainty assessment.

Certain other exposure endpoints, based on public health considerations, have also been developed by JEFCA. The Provisional Tolerable Weekly Intake (PTWI) is used for substances such as heavy metals that are cumulative in their effect and have no technological purpose in food production. The use of the term 'provisional' indicates that there are inadequate toxicological data to allow a more definitive estimate to be made, while 'tolerable' indicates permissibility rather than acceptability. The term 'Provisional Maximum Tolerable Daily Intake' (PMTDI) is used for contaminants that do not accumulate in the body. It is an estimate of permissible human exposure as a result of natural occurrence of the substance in the diet.

References

1. MacPherson, A., Barclay, M.N.I., Scott, R. & Yates, R.W.S. (1997) 'Loss of Canadian wheat imports lowers selenium intake and status of the Scottish population'. In: *Trace Elements in Man and Animals – TEMA 9*, (eds. P.W.F. Fischer, M.R.L'Abbé & R.W.S. Yates), pp. 203–205. NRC Research Press, Ottawa, Canada.

2. Lisk, D.J. (1972) Trace metals in soils, plants and animals. *Advances in Agronomy*, **24**, 267–320.
3. Reilly, C. (1996) *Selenium in Food and Health*, p. 220. Chapman & Hall/Blackie, London.
4. Nascimento, R.F., Beserra, C.W.B., Furuya, S.M.B., Schulz, M.S., Polastro, L.R., Neto, B.S.M. & Franco, D.W. (1999) Mineral profile of Brazilian cachaças and other international spirits. *Journal of Food Composition and Analysis*, **12**, 17–25.
5. Hoffman, G.F., Brunelle, R.L. & Pro, M.J. (1968) Alcoholic beverages: determination of trace component distribution in illicit spirits by neutron activation analysis (NAA), atomic absorption (AA) and gas–liquid chromatography (GLC). *Journal of the Association of Official Analytical Chemists*, **51**, 580–6.
6. Booth, C.K., Reilly, C. & Farmakalidis, E. (1996) Mineral composition of Australian ready-to-eat breakfast cereals. *Journal of Food Composition and Analysis*, **9**, 135–47.
7. Bolger, P.M., Yess, N.A., Gunderson, E.L., Troxell, T.C. & Carrington, C.D. (1996) Identification and reduction of sources of dietary lead in the United States. *Food Additives and Contaminants*, **13**, 53–60.
8. Wang, W-C., Mäkelä. A-L., Näntö, V., Mälelä, P. & Lagström, H. (1998) The serum selenium concentrations in children and young adults: a long-term study during the Finnish selenium fertilisation programme. *European Journal of Clinical Nutrition*, **52**, 529–35.
9. Versieck, J. & Cornelis, R. (1989) *Trace Elements in Human Plasma or Serum*, 2. CRC Press, Boca Raton, Florida.
10. Nielsen, F.H. (2000) 'The dogged path to acceptance of Boron as a nutritionally important mineral element'. In: *Trace Elements in Man and Animals 10*, (eds. A.M. Roussel, R.A. Anderson & A.E. Favier), pp. 1043–7. Kluwer Academic/Plenum Publishers, New York.
11. Ybañez, N. & Montoro, R. (1996) Trace element food toxicology: an old and ever growing discipline. *Critical Reviews in Food Science and Nutrition*, **36**, 299–320.
12. Reichlmayer-Lais, A.M. & Kirchgessner, M. (1991) 'Lead – an essential trace element'. In: *Trace Elements in Man and Animals – 7* (ed. B. Momçiloviç), 35/1–35/2, Institute of Medical Research and Occupational Health. University of Zagreb, Croatia.
13. Baldwin, D.R. & Marshall, W.J. (1999) Heavy metal poisoning and its laboratory investigation. *Annals of Clinical Biochemistry*, **36**, 267–300.
14. Russel, L.H. (1978) 'Heavy metals in foods of animal origin'. In: *Toxicity of Heavy Metals in the Environment* (ed. F.W. Oehme), pp. 3–32. Dekker, New York.
15. National Research Council Committee on Food Protection (1972) *Food Chemicals Codex*, **ix–xii**. National Academy of Sciences, Washington, DC.
16. Phillips, L.G. & Barbano, D.M. (1997) The influence of fat substitutes based on protein and titanium dioxide on the sensory properties of low fat milks. *Journal of Dairy Science*, **80**, 2726–31.
17. Borocz-Szabo, M. (1980) Effects of metals on sensory qualities of food. *Acta Alimentaria*, **9**, 341–56.
18. Kanner, J. (1994) Oxidative processes in meat and products: quality implications. *Meat Science*, **36**, 169–89.
19. Semwal, A.D., Murthy, M.C.N. & Arya, S.S. (1995) Metal contents in some of the processed foods and their effect on the storage stability of pre-cooked dehydrated flaked Bengalgram Dhal. *Journal of Food Science and Technology – Mysore*, **32**, 386–90.
20. Krantzler, N.J., Mullen, B.J., Comstock, E.M., Holden, C.A., Schutz, H.G., Grivetti, L.E. & Meiselman, H.L. (1982) Methods of food intake assessment – an annotated bibliography. *Journal of Nutrition Education*, **14**, 108–19.
21. Rees, N. & Tennant, D. (1994) Estimation of food chemical intake. In: *Nutritional Toxicology* (eds. F.N. Kotenis, M. Mackey & J. Hjelle), pp. 199–221. Raven Press, New York.
22. Rees, N. & Tennant, D. (1994) Estimation of food chemical intake. In: *Nutritional Toxicology* (eds. F.N. Kotonis, M. Mackey & J. Hjelle), pp. 199–221. Raven Press, New York.
23. World Health Organization (1985) *WHO Guidelines for the Study of Dietary Intakes of Chemical Contaminants, Prepared by the Joint UNEP/FAO/WHO Global Environmental Monitoring Programme*. Offset Publication No. 87. World Health Organization, Geneva.

24. Louekari, K. & Salminen, S. (1986) Intake of heavy metals from foods in Finland, West Germany and Japan. *Food Additives and Contaminants*, **3**, 355–62.
25. Peattie, M.E., Buss, D.H., Lindsay, D.G. & Smart, G.A. (1983) Reorganisation of the British Total Diet Study for monitoring food constituents from 1981. *Food and Chemical Toxicology*, **21**, 503–507.
26. Varo, P. & Koivistoinen, P. (1980) Mineral element composition of Finnish food XII. General discussion and nutritional evaluation. *Acta Agricultural Scandanavica*, **22**, 165–71.
27. Urieta, I., Jalon, M., Garcia, J. & de Galdearo, L.G. (1991) Food surveillance in the Basque country (Spain). I. The design of a Total Diet Study. *Food Additives and Contaminants*, **8**, 861–73.
28. Turrini, A., Saba, A. & Lintas, C. (1991) *Nutrition Research*, **11**, 861–73.
29. Solgaard, P., Aarkrog, A., Fenger, J., Flyger, H. & Graabaek, A.M. (1979) Lead in Danish foodstuffs, evidence of decreasing concentrations. *Danish Medical Bulletin*, **26**, 179–82.
30. Gregory, J., Foster, K., Tyler, H. & Wiseman, M. (1990) *The Dietary and Nutritional Survey of British Adults*. HMSO, London.
31. US Department of Agriculture (1994) *CSFII/DHKS, Nationwide Food Consumption Survey, Continuing Survey of Food Intakes by Individuals and Diet Health and Knowledge Survey, 1990*. Machine-readable data set, Accession no. PB94-500063, National Technical Information Service, Springfield, VA.
32. Institut für Ernährungswissenschaft (1992) *Nationale Verzehrsstudie und Verbundstudie Ernährungserhebung und Risikofaktorenanalytik*, Institut für Ernährungswissenschaft, Giessen, Germany.
33. Haxton, J. Lindsay, D.G., Hislop, J. *et al.* (1979) Duplicate diet studies in fishing communities in the United Kingdom: mercury exposure in a critical group. *Environmental Research*, **10**, 351–8.
34. Duffield, A.J. & Thomson, C.D. (1999) A comparison of methods of assessment of dietary selenium intakes in Otago, New Zealand. *British Journal of Nutrition*, **82**, 131–8.
35. This role has now been taken over by the Food Standards Agency established in April 2000.
36. Black, A.L. (1992) Setting acceptance levels of contaminants. *Proceedings of the Nutrition Society of Australia*, **17**, 36–41.
37. World Health Organization (1987) *Principles for the Safety Assessment of Food Additives and Contaminants in Food. Environmental Health Criteria No. 70*, World Health Organization, Geneva.
38. Hatchcock, J.N. (1998) Safety evaluation of vitamins and minerals. In: *Nutrition and Chemical Toxicology* (ed. C. Ionnaides), pp. 257–83. Wiley, Chichester, UK.
39. Barnes, D.G. & Dourson, M. (1988) Reference dose (RfD): description and use in health risk assessments. *Regulatory Toxicology and Pharmacology*, **8**, 471–86.
40. Hathcock, J.N. (1993) Safety limits for nutrient intakes: concepts and data requirements. *Nutrition Reviews*, **51**, 278–85.
41. Department of Health (1996) *Dietary Reference Values for Food Energy and Nutrients for the United Kingdom*. HMSO, London.
42. Food and Nutrition Board (1989) *Recommended Dietary Allowances*, 10th edn. National Academy of Sciences, Washington, DC.

Chapter 3
Metal analysis of food

3.1 The determination of metals in foods and beverages

There have been many improvements in analytical techniques and instrumentation since the first edition of this book appeared twenty years ago. These have permitted analysts to determine concentrations of a very wide variety of inorganic elements in food and drink with an ease and a level of precision previously beyond their capabilities. As a consequence considerable advances have been made in the study of trace element nutrition and toxicology, and industry has been enabled to respond to the increasing demands for food composition data made by regulatory authorities. Two decades ago the equipment and procedures available were sufficient for the determinations required for what used to be the main responsibility of food manufacturers, to ensure product authenticity, quality and safety.

But far more is required today. In the US, for instance, the *Nutrition Labeling and Education Act* of 1990 (NLEA) has placed responsibilities on food manufacturers and processors to provide consumers with a greatly increased range of analytical data on the components of the products they buy. It has been estimated that these new requirements affect up to a quarter of a million different food products, marketed by some 17 000 companies, at a cost of \$2 billion[1]. The NLEA requires mandatory labelling of 14 different nutrients, including three metals (sodium, calcium and iron). In addition another 34, among them seven additional metals (magnesium, zinc, selenium, copper, manganese, chromium and molybdenum) which currently may be labelled voluntarily, are expected to become mandatory in the near future. Similar legislation is being prepared or has already been introduced in other countries. It is only because of advances in instrumentation and analytical techniques that manufacturers can meet these demands in an efficient and cost-effective way.

New instruments and techniques do not, of themselves, lead to improvements in analytical standards and productivity. No matter how good they are, unless analysts follow those fundamental rules for sample preparation and handling which have long been practised by their professional predecessors, the data they produce will be suspect. The need for care in pre-analytical procedures has been stressed again and again by experts in the field[2].

3.1.1 The first step in analysis: obtaining a representative sample

The most fundamental of the preliminary steps required to ensure accuracy in analysis is proper sampling. Foodstuffs are seldom homogeneous in nature. Even liquids can be stratified in layers of different concentrations. It is essential, therefore, that the bulk sample be sufficiently homogenised to ensure that the subsample which is selected is representative of the whole. The same must be done with less bulky samples when, as with animal and plant tissues, they are naturally heterogeneous.

Problems caused by sample matrix variation and methods for their reduction by unit composting were discussed some years ago by Lento in a review that is still pertinent today[3]. As he noted, bulk material is normally selected for analysis by either random or representative sampling. When random sampling is used, portions are taken in such a way as to ensure that every part of the material has an equal chance of appearing in the sample. It is essential that bias due to personal preference or other cause be avoided. This can be done by numbering the containers and then using a table of random numbers to control selection, irrespective of any extraneous factor.

Large bulk containers may have to be subjected to representative or stratified sampling. In this procedure, samples are taken in a systematic way so that each portion selected represents a corresponding portion of the bulk. The container is considered to be divided into different sections or strata and samples are removed from each stratum. The size of the samples removed should correspond to the relative proportions of the imaginary strata of the bulk. Thorough mixing, for example by end-over-end rotation, is preferable to representative sampling in the case of liquids and sometimes even of very fine powders.

3.1.2 Prevention of contamination

Though it seems hardly necessary to stress the need to be scrupulous in preventing contamination of samples during handling and preparation, failure to do so undoubtedly continues to occur in some laboratories[4], in spite of the warnings by experts in the field[5]. It is not unknown for less able analysts to concentrate all their care and efforts on optimising their instrumental techniques and to neglect pre-analytical factors which are crucial to the success of their efforts. Among the sources of possible contamination which should be guarded against are laboratory equipment items such as homogeniser blades, knives and corers, grinders and sample containers. In the case of blenders, for example, even when the blades are made of titanium[6], a high degree of contamination of samples may result unless they have been given a thorough chemical clean-up before use[7].

A particular source of contamination, even in dedicated clean rooms, can be airborne dust and particles coming from staff clothing and even skin, cosmetics and possibly cigarette smoking. The advice given nearly 20 years ago by Sansoni and Iyengar on this point remains pertinent[8]. Atmospheric dust, they pointed out, is a frequent cause of trace metal contamination. Dust in the air of an open laboratory has been found to contain 3 g Al, 1.6 g Zn and 0.2 g Cu per kg[9], while even in a clean room atmospheric contamination can be significant[10].

3.1.3 Drying of samples

Where analytical results are required in terms of 'fresh food' or 'portions as served', samples will only need to be homogenised or ground in a suitable mill. However, if samples have to be stored before analysis, or if 'dry weight' results are required, further preparation is necessary. This can involve simple drying to constant weight in a fan-assisted air oven, at 70–100°C, care being taken to avoid charring. A vacuum oven allows use of safer, lower drying temperatures and speeds up the process. Freeze-drying is ideal for preparing samples for trace element analysis. In addition to reducing the moisture content to a suitable level, it results in a crumbly texture which is convenient for subsequent subsampling and analysis. The freeze-dried samples, sealed in a plastic (preferably polyethylene) bag, can be stored without refrigeration.

3.1.4 *Purity of chemical reagents and water*

It is essential that all chemical reagents used be of the highest quality. Even analytical grade reagents can be a source of contamination, especially when working at very low concentration levels, and it may be necessary to test those that are used for impurities[11]. Veillon's advice to 'check everything' should be followed by all analysts[12]. It may be advisable, when impurities are detected in reagents such as nitric or perchloric acid, to distil them in a quartz still before use.

Just as important as a potential source of contamination is the water used to dilute samples and reagents. This too must be checked adequately. Only deionised water, purified to a very low conductivity, should be used. As has been noted by Dabeka, unless the purity of chemical reagents and water is established, unreliable blanks may result in overlooked errors, even when all other steps have been taken to ensure analytical quality control[13].

3.1.5 *Glassware and other equipment*

It seems hardly necessary to add that, if the highest standards of analytical quality control are to be achieved, all glassware must be thoroughly cleaned. Merely rinsing with distilled or deionised water is not sufficient. A routine should be followed which includes soaking all glassware overnight in an alkaline detergent, then rinsing with deionised water, with a further soaking in 2% hydrochloric acid. The items should then be washed in deionised water, followed by two more rinsings in water. They should then be left overnight for drying in a clean room. Similar care in washing and drying should be used for other containers and vessels made of plastic.

3.2 Preparation of samples for analysis: digestion of organic matter

Most of the more frequently used techniques for the determination of metals in foodstuffs require that samples be prepared destructively before instrumental analysis is carried out. Exceptions are liquids such as beverages, including water, which may only require dilution before analysis. In most other foods, organic matter has to be removed as this would interfere with the analytical process.

Organic matter is usually removed from food samples by some form of oxidation, either by the use of oxidising acids in a wet digestion or by dry ashing in the presence of air or pure oxygen. The method used will depend on the metals to be analysed and the nature of the food. What is aimed at is a procedure that gives reliability and accuracy in the range of concentrations appropriate to the investigation, and can be carried out with reasonable speed and cost effectively. The actual method chosen is often a compromise, for, of the many procedures reported in the literature, it is often found that no single one will have all the desirable qualities at the same time.

3.2.1 *Dry ashing*

This method of sample preparation involves the incineration of food samples in a muffle furnace at a suitable temperature. The resulting ash, free of organic compounds, is dissolved in dilute acid. This is a convenient and easily carried out procedure for preparing samples for instrumental analysis of many, though not all, metals. Ashing is usually performed at temperatures between 400 and 600°C.

However, these temperatures can be too high for volatile metals such as mercury, arsenic, selenium and lead. Other metals, such as tin, may form insoluble refractory compounds during dry ashing. In some cases the addition of ashing aids, including salts of metals as well as acids, may be used to improve the efficiency of ashing and to assist recovery of certain elements.

Dry ashing continues to be employed widely by analysts as a convenient and versatile method for preparing food samples for instrumental analysis. Straightforward dry ashing of blended homogenised food samples has been employed, for instance, in the UNEP/WHO HEAL study in Sweden to assess intake of 17 elements in the diet[14]. With the addition of nitric acid as an ashing aid, dry ashing has also been used by investigators in Slovenia to determine a range of trace elements in composite diet samples[15]. A more complicated application has been in the investigation by Spanish workers of the effects of freezing on the mineral content of frozen asparagus[16]. This involved ashing of the dried and homogenised vegetable material, without the use of any aid, in a series of mineralisation stages at different times and temperatures (90–250°C; 460°C; 460–100°C) in a programmable electric oven. The procedure, according to the report, prevented loss by volatilisation and gave excellent recoveries.

A particular advantage of dry ashing is that it allows use of relatively large sample sizes. It can also minimise contamination from reagents, since normally only dilute acid is used to dissolve the ash. In addition the method requires only minimal attention from the operator. However, there are some disadvantages. It is usually a lengthy and time-consuming operation and, as was pointed out some years ago by Crosby, problems may be caused by incomplete combustion and adsorption on surfaces of the incineration crucibles, in addition to losses through volatilisation[17].

3.2.2 *Wet digestion techniques*

Wet digestion requires the use of strong oxidising acids, such as nitric, sulphuric and perchloric, or, in certain cases, hydrofluoric, phosphoric and hydrochloric acids. The acids are used either alone or in various combinations, depending on the nature of the sample and the metals to be analysed. Compared to dry ashing, acid digestion gives greater flexibility for digestion of a wide range of organic matter with higher recovery rates for many, though not all, trace elements. The disadvantages of the method are that it is only suitable for small sample sizes and relatively large volumes of reagents are required. This can lead to higher blanks and introduce contamination. The use of strong acids also makes it a potentially hazardous method that requires constant attention by the operator.

Until relatively recently, when microwave heating methods became available and closed digestion systems were introduced, wet digestions of biological samples were carried out in open vessels, such as Kjeldahl flasks, preferably with some means of at least partial reflux of the hot acids. Long-necked vessels, such as heating tubes, were also used for the same purpose. Heating was carried out on a sandbath, or in a specially designed device such as an aluminium block digester. Similar methods are used today, especially for one-off or small numbers of analyses, as they have been for more than 40 years since they were described in a report by the impressively entitled 'Metallic Impurities in Organic Matter Subcommittee of the Analytical Methods Committee of the Society for Analytical Chemistry'[18].

3.2.2.1 *Nitric acid digestion*

Digestion using nitric acid alone has been shown to be suitable for digesting many different types of biological samples, with good recoveries, for example, of cadmium, lead and zinc[19]. Some complex biological matrices may require repeated digestions after being reduced to dryness. In the case of some metals, additional steps may be required to bring about complete release of the metal, for example addition of potassium permanganate in the case of chromium[20]. Hydrogen peroxide can also increase the oxidising power of the nitric acid and is especially efficient for the determination of zinc, copper, lead and cadmium in foods[21].

Trace elements, such as selenium, which tend to form stable organic compounds in biological tissues are difficult to digest completely with nitric acid alone. This difficulty can be overcome by using 90% rather than the normal 70% acid[22]. Addition of magnesium nitrate to the acid can also improve recovery[23].

3.2.2.2 *Nitric–sulphuric acids digestion*

The presence of sulphuric acid, which has a boiling point of 330°C, in the acid mixture greatly improves the oxidising power. This digestion mixture has been used successfully to determine copper, iron, zinc and manganese in food[24]. Addition of hydrogen peroxide can improve the determination of arsenic, iron, aluminium, zinc and chromium in plant tissue[25]. Though some workers have recommended the addition of hydrofluoric acid to improve recovery of chromium, this has not been found to be necessary with most food samples[26]. However, hydrofluoric acid has been reported to increase the efficiency of recovery of mercury after nitric–sulphuric acid digestion of seafood[27].

Nitric-sulphuric acid digestion has been less successful for the determination of selenium in biological materials, probably because organic forms of selenium are resistant to breakdown[28]. It was also found to be unsuitable for the determination of lead, because of the formation of insoluble lead sulphate[29].

3.2.2.3 *Use of perchloric acid*

Perchloric acid is a very powerful oxidising agent and its addition to nitric acid or a nitric–sulphuric acid mixture dramatically increases the speed of digestion. However, there are problems with its use because of its explosive nature. It should never be used on its own, and even in mixtures only with caution[30]. Under certain conditions the presence of perchloric acid in a digestion mixture may result in the loss of volatile elements, such as chromium and selenium. The problem can be overcome by using combinations of acid mixtures[31], and by monitoring of the digestion conditions[32].

3.2.2.4 *Hydrofluoric acid*

The addition of hydrofluoric acid to a sulphuric–nitric–perchloric acids digestion mixture has been reported to prevent interference by silicate in recovery of chromium from brewers' yeast by the formation of silicon tetrafluoride[33]. A similar procedure has been found useful for releasing chromium from silicate-rich plant material[34].

3.2.3 *Microwave digestion*

An important development in analysis in recent years has been the widespread introduction of microwave heating as a replacement for traditional methods of digesting food and other samples prior to instrumental analysis. Microwaves are electromagnetic energy that brings about molecular motion, and thus heating, by the migration of ions and rotation of dipoles, without causing changes in molecular structures[35]. The microwave heating system was developed in the 1940s but it was not until the mid 1970s that the method was applied in the laboratory for digestion of biological materials[36]. Initially microwave ovens developed for domestic purposes were adapted for use in the laboratory, but since then dedicated microwave heating systems, with appropriate fume extraction and other attachments have become available.

The technique has many advantages over other digestion systems. There is a significant reduction in digestion time, especially compared to dry ashing. It also requires minimum supervision by a technician. Because the system and equipment readily lend themselves to automation, it is increasingly being adopted by analysts. Since samples in microwave digestion systems are usually contained in closed vessels, such as low- or high-pressure Teflon® tubes fitted with pressure relief valves, losses due to volatilisation are minimised and the likelihood of contamination from external sources is considerably reduced. It is particularly suitable for digestion of samples with a high fat content which are difficult to digest using other wet systems[37].

Microwave digestion has been used successfully for analysis of a very wide range of elements. It is routinely used in the UK Total Diet Study to determine some 30 different inorganic elements[38]. The method is particularly effective for volatile elements such as arsenic and selenium, with good recoveries and a significant reduction in digestion time[39]. It is also suitable for the analysis of mercury in foods[40].

An excellent example of the use of microwave heating in the determination of metals in foodstuffs has been given by Tahán and his colleagues[41]. Two types of high-pressure reaction vessels were used in their sophisticated procedure, a special 45 ml digestion bomb with a Teflon® PFA (tetrafluoroethylene with a fully fluorinated alkoxy side chain) inner cup for mercury and lead, and a 120 ml Teflon PFA digestion vessel with a pressure release valve for the other metals. The microwave oven used was designed for laboratory use, with adjustable power output, a programmable microprocesssor-based digital computer, Teflon®-coated cavity and other refinements.

3.3 End-determination methods for metal analysis

Rather than attempt to comment on as wide a range as possible of laboratory instruments for identifying and determining metals in foods, consideration will be given here to a limited number of end-determination methods that can be considered to be of the greatest practical significance. Readers who wish to learn about the many other less commonly used and often more expensive and sophisticated instruments may consult the specialist literature on the subject. Particularly useful updates on such equipment can be found in the *Technical Report Series* of the International Atomic Energy Agency, Vienna, in the annual *Atomic Spectrometry Update* and in other specialist journals and reports[42]. An excellent text for those who are especially interested in atomic absorption spectrometry is the monumental 1000-page work by Welz and Sperling, with its 6595 references[43].

Particular attention will be given here to three major topics: the continuing role of the atomic absorption spectrophotometer in food analysis, the development of techniques for multielement analyses and the growing interest in species-selective determination of elements in food. A brief mention will also be made of spectrofluorimetry, because of its current use for certain elements, alongside AA (absorption spectrometry), in many laboratories.

3.3.1 *Atomic absorption spectrophotometry (AAS)*

The atomic absorption spectrophotometer has, since its introduction in the mid-1950s, remained one of the most widely used instruments for the determination of major and minor inorganic elements in agriculture, biology, medicine, food studies, mining and environmental studies[44]. A high proportion of the analytical data being produced today has, as a review of the current literature will show, been obtained by AAS. The method is based on the fact that when a metal is introduced into a flame, an atomic vapour is produced and light, of a wavelength characteristic of the metal, is emitted. This emission is used in the analytical procedure of *emission spectrophotometry*.

Not all the metal atoms in the vapour, however, are excited sufficiently to emit light, even in a hot flame. The unexcited atoms can be made to absorb radiation of their own specific resonance wavelength from an external source. Thus, if light of this wavelength is passed through a flame containing atoms of the element, part of the light will be absorbed and the absorption will be proportional to the density of the atoms in the flame. This was the basis of the original atomic absorption spectrophotometer which was developed simultaneously by Walsh in Australia and Alkemade and Milaz in the Netherlands in the mid-1950s[45].

The reason for the continuing popularity of AAS is, primarily, that the instrument is relatively inexpensive and its use is easily learned by investigators who have not been trained formally in analytical chemistry. In its different modes it can be used to determine more than 60 metallic elements, in a wide range of concentrations and in different matrices. *Flame atomic absorption spectrophotometry* (FAAS) is rapid and sufficiently sensitive to permit determination of most of the trace elements in food at the μg/g range. *Graphite furnace AAS* (GFAAS) allows determination of a wider range of elements, down to the μg/kg range. Certain elements are more effectively analysed by *hydride generation AAS* (HGAAS)[46]. This involves introduction of gaseous compounds of a volatile element, such as arsenic or selenium, into a flame or electrothermally heated graphite tube (ETAAS). *Cold vapour AAS* (CVAAS) is the method of choice for the determination of mercury and is based on the generation of elemental mercury vapour at room temperature.

3.3.1.1 *Background correction*

AAS, like all other analytical techniques, has certain inherent shortcomings. Matrix problems may have to be overcome by matrix modifications, and, in the case of GFAAS in particular, spectral interference can also occur, requiring employment of background correction. This is achieved by the use of a continuum source, usually a deuterium lamp in the UV region and tungsten iodide in the visible. Another, highly efficient, form of background correction relies on the *Zeeman effect*. In this method a magnetic field, applied to the source or the atomiser, is used to split the resonance line into its Zeeman components ($\pi \pm \delta$), in effect converting the single beam into a double beam, thus allowing the background to be monitored on the wings while the analyte signal and background together are monitored with the central component.

3.3.1.2 *Use of slurries and flow injection in AAS*

In recent years considerable progress has been made in reducing the time taken to carry out the various stages of AAS, as well as of other analytical techniques, from sample preparation and through the various steps of end-determination[47]. Sample preparation has been minimised by the use of slurries that can be handled in the same way as liquids[48]. *Flow injection* (FI) and *continuous flow FI* (CFI) modes for introducing the sample have been developed and play an important part in automation. They provide an efficient and time-saving way of manipulating samples and reagents for mixing, diluting and transporting to the point of end-analysis. Glassware, such as containers and measuring devices, can be replaced by tubes, pumps and valves.

3.3.1.3 Speeding up AAS

A number of other developments have been introduced in recent years which, like FI, have resulted in reduction in the time taken to carry out analyses using AAS and other instruments. Some of these are shown in Table 3.1.

Table 3.1 Trends in metal analysis in foods and beverages by atomic absorption spectrophotometry.

Reduction in sample preparation times: solid sampling; use of suspensions, slurries
Microwave digestion: open, closed systems
Sample introduction: flow injection
On-line preconcentration
Fast furnaces
Improvements in digestion modifiers

Adapted from Ybañez, N. & Montoro, R. (1996) Trace element food toxicology: an old and ever-growing discipline. *Critical Reviews in Food Science and Nutrition*, 36, 299–320.

3.3.2 Spectrofluorimetry

Though in principle the spectrofluorimetric method can be applied to the analysis of a number of different elements, in practice its use appears to be confined by food analysts mainly to the determination of selenium. In spite of certain difficulties, such as the need for lengthy sample preparation time, it is, to judge from the literature, widely used for that purpose in preference to GFAAS[49]. The reason is that it is a highly sensitive method and can measure concentrations of selenium down to nanogram levels in many different biological matrices. It has the added advantage of requiring only small sample sizes.

The method depends on the measurement of the fluorescence of Se-2,3-diaminonaphthalene (Se-DAN) complex in cyclohexane extract. The reaction is pH-dependent and requires the addition of ethylenediaminotetra-acetic acid (EDTA) and hydroxylamine hydrochloride as masking agents. Prior sample digestion is necessary to destroy organic matter and reduce Se to inorganic Se(IV). Consequently the method is time consuming. It is also subject to chemical interference, for example by sulphate. However, steps can be taken to overcome such interference[50], and preparation time can be reduced by semi-automation[51].

3.3.3 Inductively coupled plasma spectrometry (ICP-S)

One of the most significant advances made in trace metal analysis in recent decades has been the use of a plasma source for atomisation–excitation of samples. Plasma flames are electrical discharges classified according to whether the discharges occur between two electrodes (*direct current discharges* or *DCP*), or a radiofrequency discharge is used to transfer energy to a gas from an electrical power source (*inductively coupled plasma* or *ICP*).

3.3.3.1 Inductively coupled plasma atomic emission spectrometry (ICP-AES)

The use of ICP in combination with atomic emission (ICP-AES, or ICP-OES, *optical emission spectrometry*), though its characteristics are similar to those of flame AAS, has the great advantage of multielement possibilities. It also significantly reduces matrix interference because of the high temperature of the ICP. The method has been shown to have high precision, give excellent recoveries from a wide range of food matrices and to be applicable in the analysis of a range of elements.

Though ICP-AES equipment is more expensive to install and operate than flame or electrothermal AAS instruments, it has been introduced into many analytical laboratories because of its considerably improved detection limits, accuracy and precision, as well as for its multielement capabilities. A good example of its effectiveness as a mainstream instrument in the laboratory is its use for the simultaneous determination of a range of elements in different food matrices to meet the requirements of the 1990 US *Nutrition Labeling and Education Act*[52].

3.3.3.2 Inductively coupled plasma mass spectrometry (ICP-MS)

The coupling of ICP to mass spectrometry (ICP-MS) has provided one of the most useful and sensitive tools for trace metal analysis. When it was initially developed in the late 1980s it was described as a technique which combined accuracy, precision and long-term stability for even the most difficult samples and was clearly worthy of serious consideration in most routine applications[53]. That prediction has been proved to be correct; however because of the costs involved and the need for expert technical input, ICP-MS is to be found mainly in university and other research laboratories, major government establishments and other well-endowed facilities[54].

In the twenty years since ICP-MS was recognised as being a potentially powerful analytical tool, it has undergone many modifications. Adaptations have involved the marriage of the technique with separation procedures, including *chromatography* and *electrophoresis*. Other developments, for example in methods of sample introduction such as *laser ablation*, *thermal vaporisation* and *flow injection*, have extended its capabilities considerably. Details of these developments can be found in current updates in the literature. An excellent technical review is given in Gill's text on analytical geochemistry[55]. Recent applications relating to analysis of clinical and biological materials as well as food and beverages have been discussed by Taylor *et al.*[56]. Future developments in the technique can be expected to overcome technical problems that occur with present systems and to expand even further the analytical advantages of ICP-MS[57].

As has been commented in a recent review, ICP-MS is one of the most powerful techniques currently available for trace element analysis[58]. Detection limits in the ng/litre range or below are normal, and detection limits for all elements are better, in some cases up to three orders of magnitude, than those obtained by AAS. However, availability of the technique in a laboratory, as well as the type of ICP-MS instrument, will depend to a large extent on the particular analytical tasks that have to be carried out.

From the economic point of view, the investment and the operational cost are not justified for the investigation of limited numbers of samples and elements. In such cases, AAS, in one of its various forms, will continue to be used. If means permit, ICP-MS should be the method of choice for the analysis of large numbers of samples, in which several elements are to be determined simultaneously. It is for this reason that ICP-MS is the instrument of choice in major government analytical centres in many countries, such as the UK's Central Science Laboratory, where the determination of a large number of elements in a variety of food matrices is routinely carried out[59].

3.3.4 *Other analytical techniques for trace elements*

There are many other specialist techniques and types of instrument which in certain circumstances may be as suitable or more effective than those discussed above. However, since they are unlikely to be available in the majority of analytical laboratories and are probably of more interest to the specialist than to most of the readers of this book, they will be mentioned here in passing only. For those who require additional information, this may be obtained in, for example, Gill's text noted above and in the various specialist periodicals devoted to analysis. The most widely used of these other techniques are listed in Table 3.2.

Table 3.2 Analytical techniques for inorganic elements.

Neutron Activation Analysis (NAA)
X-Ray Fluorescence Spectrometry (XRF)
Atomic Absorption Spectrometry (FAAS, GFAAS, HGAAS, CVAAS)
Atomic Emission Spectrometry (AES, ICP-AES, ICP-OES)
Atomic Fluorescence Spectrometry (AFS)
Mass Spectrometry (ICP-MS)
Voltammetry:
Differential Pulse Anodic/Cathodic Stripping Voltammetry (DPASV/DPCSV)
Potentiometric Stripping Analysis (PSA)
Square Wave Anodic Stripping Voltammetry (SWASV)
Fluorimetry

After Schramel, P. (2000) New sensitive methods in the determination of trace elements. In: *Trace Elements in Man and Animals – TEMA 10* (eds. A.M. Roussel, R.A. Anderson & A.E. Favier), pp. 1091–7. Kluwer/ Plenum, New York.

3.4 Determination of elemental species

A highly significant development in recent years in the field of trace element analysis has been the increase in interest in the determination of the chemical species of an element rather than simply total concentrations. As has been noted earlier, the total amount of an element in a food does not necessarily indicate how well it will be absorbed or how it will behave metabolically when consumed. There is, for example, a considerable difference between the absorption of iron in its inorganic and in its organic forms. It is also known that inorganic arsenic is much more toxic than its organic species. Many other examples can be given of such differences between the biological properties of different forms of elements which are customarily reported in food composition tables as total amounts and not as percentages of the different chemical species present.

Until relatively recently it was beyond the capabilities and the financial resources of most analytical laboratories to provide information on more than a very few of the chemical species of metals that occur in foods, but that situation is beginning to change. Techniques and instruments are being developed that, before very long, should make the routine determination of at least the more important of the different metal species possible. This will have implications, not just for research scientists interested in the more abstruse questions of human biology, but also for ordinary consumers, as well as for the food industry and regulatory authorities[60]. Many food scientists and others believe strongly that there is an urgent need for species-specific analytical and toxicological data[61].

3.4.1 *Methodology for the determination of metal species*

Though in recent years a considerable amount of attention has been given by experts in the field to the development of analytical techniques for the determination of metal species, progress is slow. Efforts have been concentrated principally on a small number of elements that are of particular interest, from the point of view of human metabolism, and appear to offer the best chance of success. Most publications on the subject relate to two elements, selenium and cadmium, though some other elements are now beginning to receive attention.

3.4.1.1 *Chemical methods of speciation*

Species determination is not, in fact, an entirely new activity for food analysts. In certain cases it has been undertaken for many years. Magos described a method using CVAAS to determine both inorganic and organic mercury in biological samples[62]. Mercury vapour was released from the samples, after reduction with $SnCl_2$ for inorganic mercury and with $SnCl_2/CdCl_2$ for total mercury analysis, allowing the amount of methylmercury to be estimated easily. A gas chromatography procedure has also been used for the same purpose[63].

Though official tables of food composition, such as the UK's McCance and Widdowson, still give the iron content of foods only in terms of 'total iron', data on the different proportions of the two chemical species, ferrous and ferric, are available for many foods[64]. The analysis is carried out relatively simply by classical wet chemistry methods. Similar classical chemical procedures have been used to determine relative amounts of organic and inorganic arsenic in foods. Such separation methods are, however, time-consuming and are more suitable for one-off studies than for routine and especially large-scale investigations. To meet today's requirements for species analysis, several new instrumental procedures are now available.

3.4.1.2 *Hyphenated techniques for metal speciation*

Recent progress in this rapidly developing area of analysis has resulted from the coupling of a separation process, such as capillary electrophoresis or high-performance liquid chromatography, with an analytical instrument for metal detection, especially ICP-MS, in what is known as a hyphenated technique[65]. Though in theory the technique could be applied to the determination of species of many different metals, most of the work reported to date has been carried out on only a few elements, in particular selenium and cadmium. Moreover, as a report by the Analytical Chemistry Division of the International Union of Pure and Applied Chemistry has pointed out, even with selenium, in spite of considerable advances that have been made, a great deal remains to be done[66].

A technique in which ion exchange chromatography (IEX) is coupled to ICP-MS has been used to separate four different selenium species in different foods[67]. The determination of species of a number of metals, including selenium, in body fluids has been successfully carried out in a technique in which hydride generation has been coupled with ICP-MS[68]. The method could, potentially, be used for methylated and other organic species of all elements that form volatile hydrides. It has been shown to be particularly useful for the determination of mercury metabolites in saliva.

Other combinations of separation and detection systems for speciation analysis have also been reported. A range of different selenium compounds have been determined in mushrooms, using various chromatographic techniques (size exclusion, anion exchange and cation exchange) combined with different detectors (neutron activation analysis and atomic fluorescence spectrometry)[69]. Different species of arsenic in a variety of biological reference materials have been determined using a hyphenated system comprising HPLC and atomic fluorescence spectrometry (AFS), interfaced via a UV-photoreactor and a hydride generation unit[70].

As this brief review indicates, metal speciation analysis is still very much in the research and development stage. While undoubtedly it is a matter of great importance in relation to food and considerable progress can be expected from it in the future, at present there are still many difficulties to be overcome before it becomes an accepted and normal part of the activities of food analysts. It is, moreover, highly unlikely ever to become an easy or low-cost activity, suitable to all analytical laboratories. According to Crews

> *as the techniques available become more sophisticated, and almost certainly more expensive, further developments will depend not only on collaboration between researchers and institutes, but between countries too*[71].

3.5 Analytical quality control

No matter how advanced technically or how modern may be the equipment and procedures used to analyse metals in foods, the quality of the data generated by them must always be checked by a quality assurance programme. Knowledge of the accuracy of its data is the foundation of a laboratory's quality control[72]. Quality assurance must cover all aspects of the analytical process, including sampling, sample pretreatment, standards preparation and method validation, as well as data handling and evaluation. Every step in a programme must be scrutinised, since unreliable analytical data may be the result of a range of faults, including poor methodology, improper instrument calibration, impure reagents or failure in analytical procedure.

According to Dabeka, the only way to validate an analytical result obtained from an unknown sample using a particular instrument and procedure is to use a completely independent method of analysis[73]. This can be done where a laboratory has the use of another instrument of known accuracy, or by calling on the assistance of an independent laboratory to analyse the same sample with its equipment. Where this cannot be done, alternative quality control measures, such as the use of in-house

SRMs and spike samples, should be taken. Whenever possible, this should be accompanied by the use of certified or standard reference materials (C/SRMs).

The importance of appropriate RMs has been underlined by Roelandts in a report on BERM-7, the seventh triennial international symposium on biological and environmental reference materials[74]. They are powerful tools for instrument calibration, validation of analytical procedures and achieving reliability. They also play a key role in quality control and/or assurance (QC/QA).

If RMs are, however, to fulfil these desirable roles, they must match as precisely as possible the food matrix to be analysed and fall within the concentration ranges of the elements expected in the samples. This is essential if the so-called 'matrix effect', encountered frequently in instrumental analyses, is to be avoided. Unfortunately, reference materials available from such organisations as the National Institute of Standards (NIST), the International Atomic Energy Agency (IAEA) and other international and national bodies, do not always meet these requirements. Those that do, however, though often sold by the agencies at below cost, may still be too expensive for some laboratories to use.

As the papers presented at BERM-7 show, considerable efforts are being made to respond to the growing worldwide demand for appropriate natural matrix RMs. Information, including names of issuing agencies and commercial sources, was provided at the symposium on more than 100 reference materials, many of them for trace elements in a variety of matrices, from baby food to human serum, as well as soils and marine organisms. These can be consulted in the issue of *Fresenius' Journal* which was devoted to the symposium proceedings[75].

In the absence of appropriate SRMs, the technique of recovery studies is often, according to Dabeka, the only validation given to an analytical result[76]. Unfortunately, recovery studies do not evaluate the accuracy of the blank signal or unspiked sample signal. They only evaluate the signal of the added analyte and cannot correct for errors due to an invalid blank, invalid baseline or uncorrected background absorption in AAS.

Dabeka proposes two easily performed and very practical quality control measures which can help to compensate for such deficiencies, namely: the inclusion of up to eight reagent blanks in each analytical batch; and the sample weight test. The latter consists of taking two different sample weights, one 2–3 times greater than the other. If the same results are obtained for each sample, there is a very good chance that analytical errors are absent. In his paper Dabeka presents data to show the effectiveness of these two measures in detecting analytical errors in the routine analysis of some 15 different metals by ICP-MS.

References

1. Barnes, K.W. (1998) A streamlined approach to the determination of trace elements in food. *Atomic Spectroscopy*, **19**, 31–9.
2. Veillon, C. (1986) Trace element analysis of biological samples. Problems and precautions. *Analytical Chemistry*, **58**, 851A–66A.
3. Lento, H.G. (1984) Sample preparation and its role in nutritional analysis. In: *Modern Methods of Food Analysis*, (eds. K.K. Stewart & J.R. Whitaker), pp. 71–9. Avi Publishing Co, Westport, Connecticut.
4. Razagui, I.B. & Barlow, P.J. (1992) A chemical cleanup procedure to reduce trace-metal contamination from laboratory blenders. *Food Chemistry*, **44**, 309–12.
5. Parkany, M. (1993) *Quality Assurance for Analytical Laboratories*. Royal Society of Chemistry, London.

6. Bunker, V.W., Delves, H.T. & Fautley, R.A. (1982) A system to minimise trace metal contamination of biological material during homogenisation. *Annals of Clinical Biochemistry*, **19**, 444–5.
7. Razagui, I.B. and Barlow, P.J. (1992) A chemical cleanup procedure to reduce trace-metal contamination from laboratory blenders. *Food Chemistry*, **44**, 309–12.
8. Sansoni, B. & Iyengar, G.V. (1980) Contamination control in trace element analysis. In: *Elemental Analysis of Biological Materials: Current Problems and Techniques with Special Reference to Trace Elements, Technical Report Series*, **197**, 57–71. International Atomic Energy Agency, Vienna.
9. Katz, S.A. (1985) Metals in atmospheric dust, *ICP Report*, pp. 13–19.
10. Cornelis, R. & Schutyer, P. (1984) Analytical problems related to the determination of trace metals in body fluids. *Nephrology*, **38**, 1–11.
11. Moody, J.R. & Beary, E.S. (1982) Purified reagents for trace metal analysis. *Talenta*, **29**, 1003–10.
12. Veillon, C. (1986) Trace element analysis of biological samples. Problems and precautions. *Analytical Chemistry*, **58**, 851A–66A.
13. Dabeka, R. (1997) Knowledge gaps in analytical quality control. In: *Trace Elements in Man and Animals – 9: Proceedings of the Ninth International Symposium on Trace Elements in Man and Animals* (eds. P.W.P. Fischer, M.R. L'Abbé, K.A. Cockell & R.S. Gibson) pp. 406–9. NRC Research Press, Ottawa, Canada.
14. Jorhem, L., Becker, W. & Slorach, S. (1998) Intake of 17 elements by Swedish women, determined by a 24-h duplicate portion study. *Journal of Food Composition and Analysis*, **11**, 32–46.
15. Pokorn, D., Stibilj, V., Gregorič, B., Dermelj, M. & Štupar, J. (1998) Elemental composition (Ca, Mg, Mn, Cu, Cr, Zn, Se and I) of daily diet samples from some old people's homes in Slovenia. *Journal of Food Composition and Analysis*, **11**, 47–53.
16. Amaro, M.A., Moreno, R. & Zurera, C. (1998) Nutritional estimation of changes in mineral content during frozen storage of white asparagus. *Journal of Food Quality*, **21**, 445–458.
17. Crosby, H.T. (1977) Determination of metals in foods. *Analyst*, **102**, 1213–68.
18. Analytical Methods Committee (1960) Methods of destruction of organic matter. *Analyst*, **85**, 643–56.
19. Tinggi, U. & Maher, W. (1986) Determination of trace elements in biological tissues by aluminium block digestion and spike-height flame atomic absorption spectrometry. *Microchemical Journal*, **33**, 304–308.
20. Farre, R. & Lagarda, M.J. (1986) Chromium content in foods. *Journal of Micronutrient Analysis*, **2**, 201–209.
21. Pokorn, D., Stibilj, V., Gregorič, B., Dermelj, M. & Štupar, J. (1998) Elemental composition (Ca, Mg, Mn, Cu, Cr, Zn, Se and I) of daily diet samples from some old people's homes in Slovenia. *Journal of Food Composition and Analysis*, **11**, 45–53.
22. Capelli, R., Minganti, V., Semino, G. & Bertarini, W. (1986) The presence of mercury (total and organic) and selenium in human placentae. *Science of the Total Environment*, **48**, 69–79.
23. Hansson, L., Pettersson, J. & Olin, A. (1987) A comparison of two digestion procedures for the determination of selenium in biological material. *Talenta*, **34**, 829–33.
24. Wenlock, R.W., Buss, D.H. & Dixon, E.J. (1979) Trace nutrients 2. Manganese in British food. *British Journal of Nutrition*, **41**, 253–61.
25. Arafat, N.M. & Glooschenko, W.A. (1981) Method for the simultaneous determination of arsenic, aluminium, iron, zinc, chromium and copper in plant tissues without the use of perchloric acid. *Analyst*, **106**, 1174–8.
26. Tinggi, U., Reilly, C. & Patterson, C.M. (1997) Determination of manganese and chromium in food by atomic absorption spectrometry after wet digestion. *Food Chemistry*, **60**, 123–8.
27. Louie, W.H., Go, D., Fedczina, M., Judd, K. & Dalins, K. (1985) Digestion of food samples for total mercury determination. *Journal of the Association of Official Analytical Chemists*, **68**(5), 891–3.

28. Tinggi, U., Reilly, C. & Patterson, C.M. (1992) Determination of selenium in foodstuffs using spectrofluorometry and hydride generation atomic absorption spectrometry. *Journal of Food Composition and Analysis*, **5**, 269–80.
29. Tinggi, U., Reilly, C., Hahns, S. & Capra, M. (1992) Comparison of wet digestion procedures for the determination of cadmium and lead in marine biological tissues by Zeeman graphite furnace atomic absorption spectrophotometry. *Science of the Total Environment*, **125**, 15–23.
30. Jorhem, L., Becher, W. and Slorach, S. (1998) Intake of 17 elements by Swedish women, determined by a 24-h duplicate portion study. *Journal of Food Composition and Analysis*, **11**, 32–46.
31. Jones, G.B., Buckley, R.A. & Chandler, C.S. (1975) The volatility of chromium from brewers yeast during assay. *Analytica Chimica Acta*, **80**, 389–92.
32. Arafat, N.M. and Glooschenko, W.A. (1981) Method for the simultaneous determination of arsenic, aluminium, iron, zinc, chromium and copper in plant tissues without the use of perchloric acid. *Analyst*, **106**, 1174–8.
33. Chao, S.S. & Pickett, E.E. (1980) Trace chromium determination by furnace atomic absorption spectrometry. *Analytical Chemistry*, **52**, 335–9.
34. Halls, D.J., Mohl, C. & Stoeppler, M. (1987) Application of rapid furnace programmes in atomic absorption spectrometry to the determination of lead, chromium and copper in digests of plant materials. *Analyst*, **112**, 185–9.
35. Neas, E.D. & Collins, M.J. (1988) Microwave heating: theoretical concepts and equipment design. In: *Introduction to Microwave Sample Preparation: Theory and Practice* (eds. H.M. Kingston & L.B. Jassie). ACS Professional Reference Book, American Chemical Society, Washington, DC.
36. Abu-Samra, A., Morris, J.S. & Koirtyohann, S.R. (1975) Wet ashing of some biological samples in a microwave oven. *Analytical Chemistry*, **47**, 1475–7.
37. Tinggi, U. (2001) Cadmium levels in peanuts. *Food Additives and Contaminants*, **15**, 789–92.
38. Ysart, G., Miller, P., Crews, H. *et al.* (1999) Dietary exposure estimates of 30 elements from the UK Total Diet Study. *Food Additives and Contaminants*, **16**, 391–403.
39. Lachica, M. (1990) Use of microwave oven for the determination of mineral elements in biological materials. *Analusis*, **18**, 331–3.
40. Tinggi, U. & Craven, G. (1996) Determination of total mercury in biological materials by cold vapour atomic absorption spectrometry after microwave digestion. *Microchemical Journal*, **54**, 168–73.
41. Tahan, J.E., Sanchez, J.M., Granadillo, V.A., Cubillan, H.S. & Romero, R.A. (1995) Concentration of total Al, Cr, Cu, Fe, Hg, Na, Pb and Zn in commercial canned seafood determined by atomic spectrometric means after mineralization by microwave heating. *Journal of Agricultural and Food Chemistry*, **43**, 910–15.
42. Bacon, J.R., Crain, J.S., Van Vaeck, L. & Williams, J.G. (1999) Atomic mass spectrometry. *Journal of Analytical Atomic Spectrometry*, **14**, 1633–59.
43. Welz, B. & Sperling, M. (1999) *Atomic Absorption Spectrometry*, 3rd edn. Wiley-VCH, Weinheim, Germany.
44. Varma, A. (1984) *CRC Handbook of Atomic Absorption Spectrophotometry*. CRC Press, Boca Raton, Florida.
45. Cantle, J.E. (1982) *Techniques and Instrumentation in Analytical Chemistry*, Vol. 5, *Atomic Absorption Spectrophotometry*. Elsevier, Amsterdam.
46. Federov, P.N., Ryabchuk, G.N. & Zverev, A.V. (1997) Comparison of hydride generation and graphite furnace atomic absorption spectrometry for the determination of arsenic in food. *Spectrochimica Acta, Part B*, **52**, 1517–23.
47. Ybañez, N. & Montoro, R. (1996) Trace element food toxicology: an old and ever-growing discipline. *Critical Reviews in Food Science and Nutrition*, **36**, 299–320.
48. Bendicho, C. & De Loos-Vollebreght, M.T.C. (1991) Solid sampling in electrothermal atomic absorption spectrometry using commercial atomisers: a review. *Journal of Atomic Absorption Spectrometry*, **6**, 353–61.
49. Tinggi, U., Reilly, C. & Patterson, C.M. (1992) Determination of selenium in foodstuffs

using spectrofluorimetry and hydride generation atomic absorption spectrometry. *Journal of Food Composition and Analysis*, **5**, 269–80.
50. Maher, W.A. (1987) Decomposition of marine biological materials for the determination of selenium by fluorescence spectrometry. *Microchemical Journal*, **35**, 125–9.
51. Watkinson, J.H. (1979) Semi-automated fluorimetric determination of nanogram quantities of selenium in biological materials. *Analytica Chimica Acta*, **105**, 319–25.
52. Barnes, K.W. & Debrah, E. (1997) Determination of Nutrition Labeling and Education Act minerals in foods by inductively coupled plasma optical emission spectrometry. *Atomic Spectrometry*, **18**, 41–54.
53. Fulford, J.E. & Gale, B.C. (1987) Trace element analysis by inductively coupled plasma–mass spectrometry. *Analytical Proceedings of the Royal Society of Chemistry*, **24**, 10–12.
54. Jones, J.W. (1988) Radio frequency inductively coupled plasma. *Journal of Research of the National Bureau of Standards*, **93**, 358–61.
55. Gill, R. (1997) *Modern Analytical Geochemistry*. Addison Wesley Longmans, Harlow, Essex.
56. Taylor, A., Branch, S., Halls, D.J., Owen, L.M.W. & White, M. (2000) Clinical and biological materials, foods and beverages. *Journal of Analytical Atomic Spectrometry*, **15**, 451–87.
57. Hieftje, G.M., Myers, D.P., Li G., Mahoney, P.P., Burgoyne, S.J., Ray, S.J. & Guzowski, J.P. (1997) Toward the next generation of atomic mass spectrometers. *Journal of Analytical Atomic Spectrometry*, **12**, 287–95.
58. Schramel, P. (2000) New sensitive methods in the determination of trace elements. In: *Trace Elements in Man and Animals 10* (eds. A.M. Roussel, R.A. Anderson & A.E. Favier), 1091–8. Kluwer Academic/Plenum, New York.
59. Crews, H.M., Baxter, M.J., Lewis, D.J. *et al.* (1997) Multi-element and isotope ratio determinations in foods and clinical samples using inductively coupled plasma–mass spectrometry. In: *Trace Elements in Man and Animals – 9: Proceedings of the Ninth International Symposium on Trace Elements in Man and Animals* (eds. P.W.P. Fischer, M.R. L'Abbé, K.A. Cockell & R.S. Gibson), NRC Research Press, Ottawa, Canada.
60. Crews, H.M. (1998) Speciation of trace elements in foods, with special reference to cadmium and selenium: is it necessary? *Spectrochimica Acta, Part B*, **53**, 213–19.
61. Berg, T. & Larsen, E.H. (1999) Speciation and legislation – where we are today and what do we need for tomorrow? *Fresenius' Journal of Analytical Chemistry*, **363**, 431–4.
62. Magos, L. (1971) Selective atomic absorption determination of inorganic mercury and methylmercury in undigested biological samples. *Analyst*, **96**, 847–53.
63. Capon, C.J. & Smith, J.C. (1977) Gas-chromatographic determination of inorganic mercury and organomercurials in biological materials. *Analytical Chemistry*, **49**, 365–71.
64. Paul, A.A. & Southgate, D.A.T. (1978) *McCance and Widdowson's: The Composition of Foods*. HMSO/Elsevier, London and Amsterdam.
65. Savory, J. & Herman, M.M. (1999) Advances in instrumental methods for the measurement and speciation of trace metals. *Annals of Clinical and Laboratory Science*, **29**, 118–26.
66. Lobinski, R., Edmonds, J.S., Suzuki, K.T. & Uden, P.C. (2000) Species-selective determinations of selenium compounds in biological materials. *Pure and Applied Chemistry*, **72**, 447–61.
67. Crews, H.M., Baxter, M.J., Lewis, D.J., *et al.* (1997) Multi-element and isotope ratio determinations in foods and clinical samples using inductively coupled plasma–mass spectrometry. In: *Trace Elements in Man and Animals – 9* (eds. P.W.P. Fischer, M.R. L'Abbé, K.A. Cockell & R.S. Gibson), NRC Research Press: Ottawa, Canada.
68. Feldman, J., Wickenheiser, E.B. & Cullen, W.R. (1997) Metal/metalloid speciation analysis of human body fluids by using hydride generation ICP-MS methodology. In: *Trace Elements in Man and Animals – 9*, (eds. P.W.P. Fischer, M.R. L'Abbé, K.A. Cockell and R.S. Gibson). NRC Research Press: Ottawa, Canada.
69. Slejkovec, Z., Van Elteren, J.T., Woroniecka, U.D., Kroon, K.J., Falnoga, I. & Byrne, A.R. (2000) Preliminary study on the determination of selenium compounds in some selenium-accumulating mushrooms. *Biological Trace Element Research*, **75**, 139–55.

70. Slejkovec, Z., Van Elteren, J.T. & Byrne, A.R. (1999) Determination of arsenic compounds in reference materials by HPLC-(UV)-HG-AFS. *Talanta*, **49**, 619–27.
71. Crews, H.M. (1998) Speciation of trace elements in foods, with special reference to cadmium and selenium: is it necessary? *Spectrochimica Acta, Part B*, **53**, 213–19.
72. Alvarez, R. (1984) NBS Standard Reference Materials for food analysis. In: *Modern Methods of Food Analysis* (eds. K.K. Steward & J.R. Whitaker) Avi, Westport, Conn.
73. Dabeka, R. (1997) Knowledge gaps in analytical quality control. In: *Trace Elements in Man and Animals – 9: Proceedings of the Ninth International Symposium on Trace Elements in Man and Animals* (eds. P.W.P. Fischer, M.R. L'Abbé, K.A. Cockell and R.S. Gibson) pp. 406–9. NRC Research Press, Ottawa, Canada.
74. Roelandts, I. (1998) Seventh International Symposium on biological and environmental reference materials (BERM-7), Antwerp, Belgium, 21–25 April 1997. *Spectrochimica Acta, Part B*, **53**, 1365–8.
75. *Fresenius' Journal of Analytical Chemistry*, **360** (1998).
76. Dabeka, R. (1997) Knowledge gaps in analytical quality control. In: *Trace Elements in Man and Animals – 9: Proceedings of the Ninth International Symposium on Trace Elements in Man and Animals* (eds. P.W.P. Fischer, M.R. L'Abbé, K.A. Cockell and R.S. Gibson) pp. 406–9. NRC Research Press, Ottawa, Canada.

Chapter 4
How metals get into food

In a recent review Beavington has commented that disregard of the work of the pioneer investigators of contamination of the food chain by trace elements risks re-inventing the wheel[1]. Though some of these pioneers have already been referred to in an earlier section, in a book such as this there is little space to spare for more detailed consideration of forerunners of today's food analysts. If more information is required, this can be obtained in Beavington's very informative paper. Here, while the contributions of the pioneers are acknowledged, attention will be concentrated on present day concerns.

4.1 Metals in the soil

In an earlier section the distribution of metals in soil, and how they are taken up by the roots of plants and accumulated in tissues, have been considered. Though, as we have seen, concentrations of metals in soils can vary considerably, and consequently uptake by plants, which normally reflects soil levels, can be expected to show corresponding variations, food composition data from around the world show remarkably uniform levels of metals (Table 4.1). Travellers need not be concerned

Table 4.1 Iron, copper and zinc levels in US and UK vegetables.

Vegetable	Metal (mg/kg fresh wt) iron	copper	zinc
Broccoli			
(US)	1.1	0.03	0.27
(UK)	1.5	0.07	0.6
Brussel sprouts			
(US)	1.5	0.10	0.87
(UK)	0.7	0.06	0.50
Carrot			
(US)	0.7	0.08	0.52
(UK)	0.6	0.08	0.40

US data from American Dietetics Association (1981) *Handbook of Clinical Dietetics*. Yale University Press, New Haven.
UK data from Paul, A.A. & Southgate, D.A.T. (1978) *McCance and Widdowson's: The Composition of Foods*. HMSO/Elsevier, London and Amsterdam.

that the meal they consume in Sydney will be more than marginally different, from the point of view of its metal content, from a meal of similar type served to them in London or New York, or, indeed, in most other countries. Yet there are occasions on which this would not be true and food contaminated with heavy and other metals can appear on the plate. Why this occurs is what we are concerned with here.

4.1.1 *Uptake of metals by plants*

While all plants will, if growing in a nutritionally balanced soil, take up nutrients to the extent they need for growth, some others possess a special ability that enables them to take up and accumulate, sometimes to high levels, certain elements. This need not cause problems for consumers in every case and may indeed sometimes confer a benefit on them. The shrub, *Camellia sinensis*, for example, the leaves of which are dried to make tea, can accumulate manganese, an essential trace element. Tea, which is probably the most widely consumed beverage in the world, is also one of the most important sources of manganese in the UK diet, providing more than half of the estimated 4.6 mg consumed each day by adults[2].

Unfortunately, the tea plant also accumulates another metal, aluminium[3], about which, as we shall see in a later section, there are some health concerns. Generally levels of aluminium in tea leaves have been found to be between 50 and 500 μg/g[4], though there is an early report of as much as 30 000 μg/g in old leaves[5]. It is unlikely, however, that even those who drink large volumes of tea are at risk since the concentration of aluminium in the brew is considerably lower than in the dried leaves[6].

4.1.1.1 *Accumulator plants*

Not all accumulation of metals by plants from soil is beneficial to the animals, including humans, who may eat their leaves and other parts. Sometimes the metal taken up is toxic and is accumulated to dangerously high levels. A particular type of vetch, *Astragalus racemosus*, is a notorious example of such a plant. Its ability to take up and accumulate large amounts of selenium is well known. Selenium, which we will consider in detail later, though an essential nutrient for animals (but not for plants), can also be highly toxic, to plants as well as to animals. It is widely distributed in soils, usually at concentrations of less than 0.1 mg/kg, low enough to cause no problems to most grasses and fodder plants.

However, in some parts of the world there are areas where considerably higher concentrations occur in certain soils. This prevents the growth of many ordinary pasture plants, but not *Astragalus racemosus*, which not only tolerates high levels of the metal but also accumulates it to dangerous levels in its tissues. Domestic animals, such as horses and sheep, that feed on the plant are poisoned. Selenosis is a major problem in grazing stock in South Dakota and some other parts of the US Midwest[7]. It has also been reported elsewhere where soil selenium levels are high and other accumulator plants grow.

4.1.1.2 *Geobotanical indicators*

Plants such as *Astragalus* that can tolerate high levels of metals in the soil are known in the world of geology as 'indicator plants'. In the past prospectors considered geobotany, the study of such plants and their association with mineral deposits, a valuable addition to their expertise. Even today, though geophysical equipment and high-speed aerial surveying have taken a lot of the foot-slogging out of field work, the presence of indicator plants can provide useful hints to prospectors. Many such plants are known. A species of basil, *Becium homblei*, known to prospectors in parts of Africa as the 'copper flower', has been used to find deposits of copper ore[8]. Other indicator plants have been reported for a variety of metal ores, including uranium, cobalt, gold and silver[9].

Indicator plants are of little more than marginal interest, however, with regard to the problem of metal contamination of food. The occasional animal that eats, for example, an *Astragalus* plant and dies as a result of selenium poisoning, is unlikely to be used for human food. Of far greater interest are those species of ordinary pasture plants and food crops that develop the ability to grow on metal-enriched soil[10]. These plants are of special significance when regenerated metal-working and other industrial 'brown sites' are brought into use for housing or agricultural purposes, and when sewage and other waste is used for soil amendment.

4.1.2 *Effects of agricultural practices on soil metal content*

Amendment of agricultural soil, by the application of various types of top dressing and agrochemicals, can result in significant changes in its metal profile. Some of these changes are deliberate, intended to increase levels of desirable nutrients where these are inadequate. This is what was done in Finland in the early 1980s, when, because of their concern that the national dietary intake of selenium was inadequate, the government decreed that fertilisers were to be enriched with the element. The result was that, within a few years, soil and crop levels of selenium increased significantly and, as a consequence, dietary intakes of the Finnish population more than doubled[11]. Other changes in soil metal concentrations as a consequence of fertiliser use can be less welcome, for example, a build up of levels of cadmium.

4.1.2.1 Metals in agricultural fertilisers

Without the wide scale use of commercial fertilisers, modern agriculture would not survive and food production would be inadequate for the world's needs. However, even such useful products need to be used with care, since their effects may not always be advantageous in the long term. There is increasing evidence that repeated intensive application of certain types of fertilisers can be a significant source of metal contamination of agricultural soils[12].

The long term use of fertilisers on cereal crops in the prairies of south-western Saskatchewan, Canada, has been shown to result in significant enrichment of the soil with a variety of metals, including selenium, antimony, cadmium, lead and uranium. Percentage increases in zinc (25%) and cadmium (66%) were particularly high. Corresponding to the soil changes, zinc and cadmium levels in cereals were 50% higher than in wild plants from unbroken land in the same district[13].

In the Canadian study the fertilisers commonly used were found to have high concentrations of some metals relative to the natural environment. Molybdenum, for example, was six times and cadmium 970 times higher compared to the average crustal rock. A Swedish study found that some widely used commercial fertilisers were particularly rich in cadmium, with levels ranging from 18 to 30 mg/kg[14].

High levels of cadmium in fertilisers have been recognised as a source of contamination in foods in Australia[15]. Phosphate from the Pacific island of Nauru which has been used by Australian farmers for many years, contains between 70 and 90 mg cadmium/kg. This has resulted in a build-up of cadmium in soils in some regions of the country and, as a result, high levels of the metal have been detected in wheat and potatoes, as well as in liver and kidney of cattle and sheep[16]. In recent years low-cadmium fertilisers have generally replaced material from contaminated sources. Unfortunately, once soil has become enriched in cadmium it is very difficult to remove the metal.

The persistence of cadmium in the environment and its ability to cause long-term health hazards is shown by a study of mussels and other crustaceans in fresh water in an area of Western Australia where superphosphate fertilisers had been in continuous use for some 16 years. During that time it was calculated that 273 tonnes of cadmium had been applied to the soil and much of this has been leached into streams and other waterbodies. There it enters the food web through algae and benthic animals, some of which, such as freshwater mussels and crayfish, are used as human food. Analyses of edible tissues from these crustaceans were found to exceed the Australian/New Zealand guidelines for Maximum Permissible Concentrations for human consumption[17].

4.1.2.2 Metals in sewage sludge

Sewage sludge, the organic-rich material that is left behind after human or animal waste has been processed in treatment plants, is produced in enormous quantities throughout the world. The estimated production in England and Wales 20 years ago, for instance, was over a million tonnes, and has been increasing ever since. Even greater amounts are produced in other larger industrialised countries. As a result there is a major problem of disposing of the waste, with increasing restrictions on the use of landfills and sea dumping and less than enthusiastic community support for construction of large-scale incinerators.

A major and generally acceptable use of sludge is as top dressing on agricultural land. The practice has considerable economic and ecological attractions. The sludge is rich in organic matter, on average about 40% of its dry weight. It can contain as much as 2.4% nitrogen and 1.3% phosphate, two essential plant nutrients which, in the absence of sludge, would have to be obtained from expensive commercial fertilisers. Sewage sludge has been found to be a cheap, easily applied treatment that is ideal for improving soil structure and increasing productivity in agricultural land.

Unfortunately, in addition to its desirable qualities, sewage sludge has a negative side. It can contain considerable quantities of a variety of different metals, many of which are toxic to plant and animal life. These metals can come from several different sources, especially industrial, but even domestic waste can make a substantial contribution. Mercury, zinc, lead and cadmium are among metals that occur in household rubbish and other waste. Boron, a constituent of many domestic washing powders, is a major source of the element in waste water that will eventually reach the sewage treatment plant. Street run-off, which in many municipal waste systems is sent to the main waste treatment plants, can contain, in addition to lead and other heavy metals, several of the platinum group of metals which are used in catalytic converters in vehicle engines.

The wide range of metals and their concentrations that have been reported to occur in sewage sludge are shown in Table 4.2. The concentrations are considerably higher than might be expected in normal agricultural land, with more than 300 times as much zinc and about 100 times as much copper and boron[18].

When such material is added even in moderate amounts to agricultural land, it can significantly increase levels of toxic metals and other metals in the soil. The added metals are generally difficult to remove from the soil and can continue to have an effect on crops for many years. The persistence of cadmium, in particular, is a special

Table 4.2 Metals in sewage sludge.

Metal	Content (mg/kg)
Boron	15–1 000
Cadmium	5–1 500
Chromium	6–8 800
Cobalt	2–260
Copper	53–8 000
Iron	6 000–62 000
Lead	120–3 000
Manganese	150–2 500
Mercury	3–77
Molybdenum	2–30
Nickel	2–5 300
Scandium	2–15
Silver	5–150
Titanium	1 000–4 500
Vanadium	20–400
Zinc	700–49 000

Data adapted from Pike, E.R., Graham, L.C. & Fogden, M.W. (1975) Metals in crops grown on sewage-enriched soil. *Journal of the American Pediology Association*, **13**, 19–33, and Capon, C.J. (1981) Mercury and selenium content and chemical form in vegetable crops grown on sludge-amended soil. *Archives of Environmental Contamination and Toxicology*, **10**, 673–89.

problem and there is evidence that as a result of inputs from sludge and other soil amendments, concentrations in many agricultural lands are slowly increasing[19].

Cadmium is by no means the only, or indeed most important, of the metal contaminants added to soils and crops by the use of sewage sludge. Several other potentially toxic metals also cause problems. One of these, molybdenum, is easily taken up from sludge by growing forage crops, especially legumes, and can adversely affect the health of farm animals.

Several other potentially toxic elements (PTEs), including copper, mercury, nickel and zinc, can be present in significant amounts in sewage sludges[20]. Because of the dangers to health that PTEs in sludge can pose to the health of consumers, authorities in many countries have introduced legislation to regulate their use in agriculture. In the UK, this is done by *The Sludge (Use in Agriculture) Regulations 1989* (as amended)[21] which implements a European Community Directive on the subject[22]. In the USA levels of PTEs permitted in sludges used for soil amendments are controlled by the Environmental Protection Agency, under EPA Regulations 503[23]. A Directive of the Commission of the European Communities limits the inputs of potentially toxic metals to soil from sewage sludge[24].

While enrichment of agricultural soils with PTEs through the use of sewage sludge can result in an increased uptake of metals by crops, it should be noted that not all metals are taken up to the same extent. In addition, there can be considerable differences between levels of uptake even of one type of metal by different species of plants[25]. There is some evidence also, that use of sewage sludge may have a less serious outcome with regard to uptake of toxic metals by crops than the application of metal-contaminated commercial fertilisers. Metals in sludge are generally organically bound and less available for plant uptake than are the more mobile metal salt impurities found in fertilisers[26]. A review by Dudka and Miller, which looks at, among other matters, differences in uptake of metals from soils between species and cultivars of plants, should be consulted for further information on the relationship between sludge-amended soil and metal transfer to the human food chain[27].

4.1.2.3 Metal uptake from agrochemicals

A number of different metallic compounds are used as agricultural chemicals and can, as a result, contaminate crops. One of the earliest to be used was copper, as copper sulphate mixed with lime and water in Bordeaux mixture. It was originally sprayed on grapes, it is said, to deter thieves, and only later was found to prevent mildew. Potato growers have long used the mixture to prevent blight and other fungal diseases. Copper salts are still widely used, by home gardeners as well as by commercial growers, for that purpose. In the case of potatoes, the practice is unlikely to cause any problems for consumers, since the mixture is sprayed on the overground haulms and the external peel of the tubers is usually discarded before they are eaten. With grapes there is a possibility that adhering copper salts on the fruit skin will end up in the wine, if sufficient time is not allowed between spraying and harvesting for it to be washed off by rain.

The use of organic mercurial compounds has been the cause of serious and well-documented cases of food poisoning. Such mercury-containing compounds were once widely used as antifungal seed dressings. The misuse of treated seed wheat for making flour caused an epidemic of mercury poisoning in Iraq in the late 1960s, the effects of which persisted among some of those who were poisoned, for many decades[28]. As a result of this tragedy, and similar accidents elsewhere, the use of mercurial compounds in agriculture is now banned in many countries.

A solution of lead arsenate was formerly widely used in orchards to control pests such as ermine moth on apples and sawflies on pears. It was very effective for this purpose but, unfortunately, if used too close to harvest time it left a residue on the skin of the fruit. This would account for the unacceptable levels of arsenic and lead which were relatively frequently found in the past on apples and pears, as well as in cider and perry[29]. There has been some concern that prolonged use of lead arsenate insecticides can lead to persistent contamination of orchard soil and could be taken up into the fruits. However a study of uptake of arsenic and lead by apple and apricot trees growing in such soil has found that though uptake occurs, neither metal reaches levels which are a danger to consumer health[30].

Arsenic can also enter the food chain through another agricultural practice which uses arsanilic acid as a growth promoter. Though the amounts used are very small, the compound is very effective in increasing growth rates when added to the rations of chickens and other animals. The elevated arsenic concentrations sometimes found in the past in chicken liver probably resulted from use of such growth promoters[31]. This practice is now banned in the UK and some other countries.

4.1.3 Industrial contamination as a source of metals in food

Metal contamination of food as a result of mining and other industrial activities is widespread. Modern mining is normally subject to stringent environmental protection regulations and is required to use various devices such as holding dams and filtering systems to prevent metal-contaminated waste escaping into adjoining areas where they might contaminate crops. This was not always the case and abandoned mine workings continue to cause contamination problems in some countries.

4.1.3.1 Metal contamination from mining operations

There is a long history of metal poisoning of farm animals in the metalliferous regions of the Harz Mountains of north-eastern Germany. Silver, lead and copper ores have been mined and smelted there for many hundreds of years and there is widespread soil contamination with these metals. High levels of lead in locally produced animal fodder are still reported today and concerns are expressed about possible effects of high intakes of metals by residents[32].

The legacy of the British Industrial Revolution, with its proliferation of mining activities and metal working throughout the country, continues to cause environmental problems. High levels of cadmium in soil and in locally grown vegetables in the village of Shipham, in Somerset, have been traced to contamination from an abandoned mine[33]. Even when the old workings are in isolated areas, far from farmland, problems can still occur. Increasing flooding in recent decades, a possible effect of global warming, has caused mobilisation of contaminated materials from upland deposits in parts of the old lead mining areas of the Yorkshire dales. As a result lead-containing sediment has been washed downstream in rivers and streams and there is concern that crops and fodder in the basin of the River Ouse may become dangerously contaminated[34]. A similar problem has arisen in southern Spain with the collapse of the tailing pond dike of the Aznalcollar zinc mine north of the Guadalquivir marshes. Farm and wetland as far as 40 km downstream were contaminated with cadmium and zinc. Plants were found to contain more than 7 μg cadmium/g and 3384 μg zinc/g, 40 and 100 times greater respectively than levels in normal tissue[35].

Cadmium in waste water discharges from non-ferrous metal mines and smelters in the Jintzu area of Japan was responsible for the notorious outbreak of *itai-itai* disease reported in the 1960s. As a result of consuming rice irrigated with water from the mine, local residents accumulated excessive amounts of cadmium in their bodies, with serious consequences for their health[36]. These included sometimes fatal renal tubular dysfunction[37], as well as osteomalacia which results in bone brittleness and, in extreme cases, collapse of the skeleton[38].

Mercury contamination of the environment, with potential serious consequences for human health, is widespread in the Amazon basin of Brazil as a result of unregulated gold mining. The mercury is used during the processing of the gold-bearing gravels and in many illegal or poorly organised workings, much of it is lost with the waste water discharge. The mercury enters the aquatic ecosystem and causes considerable contamination of river sediments and of fish and other aquatic organisms. Cattle and pigs have been found to have high blood mercury concentrations and excessive levels have also been detected in blood, urine and hair of local residents[39].

4.1.3.2 Metal contamination from metal industries

The refining and industrial use of metals, in a wide range of industries from electroplating to vehicle manufacture, can contribute significantly to environmental contamination with potential for impact on human health. This connection between industry and metal poisoning has been recognised for many hundreds of years. Today, in spite of national and international efforts to control the problem, environmental and health damage resulting from our industrial use of metals continues to occur throughout the world.

Mercury used as a catalyst in the manufacture of plastic in a factory in Japan in the 1950s caused a tragedy that cost the lives of many and brought the name Minamata onto the headlines of the world press. For many people the Minamata outbreak was the equivalent of Rachel Carson's *Silent Spring* and marked a defining moment in the global environmental movement. The mercury, which was discharged in waste water into Minamata Bay on the west coast of Japan, eventually, after a chemical transformation, entered into the human food chain through fish and was consumed by local fishermen and their families. The result was an epidemic of organic mercury poisoning, a debilitating and, in many cases, fatal illness. As we shall see, mercury is a very persistent and tenacious environmental pollutant and fish caught in Minamata Bay even today, many years after discharge from the factory has ceased, carry a high mercury burden in their tissue[40].

Pollution of the environment with mercury as a result of escapes from industry continue to cause health concerns in many countries. Release of the metal by paper-making plants, in which mercury compounds are used in processing wood pulp, has resulted in contamination of lakes in Sweden and Canada. Fish caught in these waters have been found to have accumulated high levels of mercury in their tissues. The danger to human health is considered so great by Canadian health officials that fishing has been banned in some of these contaminated lakes[41].

Large-scale pollution of water by other metals besides mercury and cadmium also occurs as a result of industrial activity. Typical of such incidents were the levels and variety of metals released in the late 1960s by a steel-making plant on the River Tees in north-west England. The river was contaminated with high levels of aluminium, chromium, copper, iron, manganese and zinc, as well as with lesser amounts of antimony, arsenic, cobalt, tin, titanium, tungsten, uranium, vanadium and tungsten. It was estimated that each day 7500 kg of iron, 2300 kg of manganese, 850 kg of zinc, 310 kg of lead and 25 kg of copper were released[42].

Such extreme levels of pollution are now seldom met with in industrially developed countries, except as the result of spills and other accidents. Steps have been taken by many governments in recent decades to control the situation, for example, in the UK, by the *Deposit of Poisonous Waste Act (1972)* and the *Control of Pollution Act (1974)*.

One of the worst offenders with regard to release of toxic metals into the environment has been, traditionally, the metal-finishing industry. Metal-rich sludges produced by plating shops cause particular problems. These can contain as much as 2.5% nickel, 1.5% zinc, 1% chromium and copper, 0.25% cadmium and 0.2% lead[43]. Economic as well as social pressures have in recent years led the industry to clean up its activities by detoxifying its waste and collecting as much as possible of the metals for reuse.

Because of past poor industrial practices and the persistence of many metals in the environment, there are a number of industrialised areas in the world where, in spite of current efforts to prevent further contamination, there is a local legacy of metal contamination. Recent reports from some former Eastern Bloc countries in which heavy industries flourished in the Soviet era, show that high concentrations of metals are still encountered in soil and water. In the industrialised areas of northern Bohemia, as well as of Saxony, in the former East Germany, concentrations of chromium, cadmium, lead, nickel and zinc occur at unacceptably high levels in agricultural soils and crops. A major contributor to this contamination is believed to be have been low quality brown coal used in power stations[44].

This kind of pollution still occurs today. In Plast, an industrial city in the South Urals in Russia, some 15 factories are engaged in various kinds of metal-related activities. It has been reported that one of these plants releases 140 tonnes of arsenic each year into the surrounding area. Emissions from the other factories have also been found to contain arsenic as well as cadmium, copper, lead, nickel, selenium and zinc.

Soil in the town and surrounding farm land is significantly contaminated with these metals, with many samples well above Russian Maximum Allowable Concentrations (MAC), with corresponding elevated levels in crops. Arsenic, for example, at levels of 0.3–3.6 µg/g (MAC 0.2 µg/g) was found in 94% of potatoes and in 18–25% of other locally grown foods[45].

4.1.3.3 Emission of metals from coal

Coal-burning, in large-scale electrical generators as well as in small plants and even domestic appliances, is often a source of contamination of food. As has been seen above, the use of brown coal to produce electricity has been responsible for local contamination with toxic metals in some countries. It has been found that much of the coal used in China contains significant amounts of mercury and that this can be emitted into the atmosphere when the coal is burned. It has been estimated that over a period of five years, from 1990 to 1995, nearly 2500 tonnes of mercury were emitted over the whole country from coal-burning boilers, with the highest emissions in Beijing, Shanghai and Tianjin[46].

Even on a small scale, coal burning has been shown to cause health problems in China. A study in south-western China found that some coal used domestically was rich in selenium, arsenic and fluorine and that there was a high incidence of arsenic and selenium poisoning among residents in certain areas. The use of coal in unvented ovens and poorly ventilated houses, with consequent pollution of the air and of food, was believed to be responsible for the endemic selenosis, arsenism and fluorosis observed in local residents[47].

4.1.3.4 Problems of use of brownfield sites

The policy of many governments to encourage the re-use for housing and other related development of former industrial brownfield sites, rather than build on greenfields outside urban areas, can unwittingly expose residents in the redeveloped areas to the danger of high heavy-metal intakes. As we have seen, environmental contamination with heavy metals can persist for a long time after a particular industry has ceased to operate. Even in an area in which there has never been a polluting industry, urban soil may contain many pollutants as a result of their origins and uses. Run-off from heavily used roads can contain a variety of toxic and other metals emitted in exhaust from motor vehicles.

Lead is a particular problem, and its build-up on roadsides has been well established. High levels of lead, believed to be due to use of leaded petrol by vehicles, have been found, for example, in such places as an urban school playground in Manila, in the Philippines[48]. Other metals, such as copper and arsenic, are also known to be emitted in exhaust[49]. Cadmium, copper, manganese and zinc have also been detected in urban air and dusts in appreciable amounts, probably in part due to wear and tear of tyres and brake pads[50]. Atmospheric dust produced by coal-fired power plants and municipal incinerators is another source of metal contamination of urban sites[51].

In the light of such findings it is clear that urban soil management and, in particular, use of brownfield sites for domestic purposes, have to be handled with care[52]. Many countries have introduced legislation to control such activities and, especially, to require extensive clean-up of sites before they can be used for residential buildings. In the US, for instance, where brownfield initiatives are being developed by federal authorities and many states, USEPA site-specific risk assessment procedures and heavy metal toxicity criteria are being used to ensure that the re-use of industrial sites does not pose a health risk to residents[53].

4.1.4 *Geophagia*

Not all the soil that is consumed by humans is taken in by accident. Deliberate soil eating, or geophagia (so named from the two Greek words, geo- for earth, and phag- for eat), has been long practised traditionally in many parts of the world and is still to be found today[54]. Soils, selected for particular qualities, such as flavour and mouth feel, from selected sites, are consumed, for example, in countries in Africa, for a variety of reasons, from religious to medicinal. They are also used as part of the normal diet, either sprinkled over a dish, rather like pepper or salt, or ground up in water and used as a sauce. The practice has been criticised as a potential source of contamination with toxic substances and parasites, though, in theory at least, soil used in this way could be a source of valuable trace elements in the diet[55].

However, lead and other toxic elements may also be present, if the soil is contaminated. In one study soil consumed traditionally in a region of Uganda was found to contain elevated levels of the rare earth cerium which was suspected to be associated with a locally high level of incidence of endomyocardial fibrosis[56]. In spite of such problems it has been argued that, geophagia, which is unlikely to be ever more than a practice of local communities in a limited number of places, can play a valuable public health role by contributing to the correction of deficiencies of certain minerals in the diet[57].

4.2 Metal contamination of food during processing

Though considerable attention has been given in the foregoing section to the problem of metal contamination of primary foodstuffs before they leave the farm or market garden, it would be misleading to give the impression that a high proportion of products arriving at food processing plants contain more than minimal amounts of toxic metals. In fact, current legislation and the monitoring practices of public health inspectorates in most of the developed countries see to it that contaminants in foods intended for processing are negligible or at least at levels which are known, in the US, as GRAS, 'generally recognised as safe'. Where there is external contamination with particles that may contain unacceptable levels of metals, these will be removed normally during the initial steps of removing inedible and unwanted external parts, especially of vegetable products.

Contamination with metals does occur occasionally due to the presence of what is known as 'tramp' or 'rogue' metal, such as fragments of farm equipment which have found their way into the product. Foods are routinely screened for such unwanted materials by one or a combination of standard cleansing operations carried out as a first step at food processing plants. Screens are used to sift out larger particles, and metal detectors, using X-ray or magnetic scanners, are used to pick up fragments that escape the sifting. In-line controls can be set up which are capable of detecting the ferrous and non-ferrous materials most frequently encountered as contaminants[58]. In spite, however, of the best efforts of primary producers to send contaminant-free produce to the processor, and of the processor to screen incoming materials, it is still possible for the end products to contain unacceptable levels of certain metals.

4.2.1 Contamination of food from plant and equipment

Food processing equipment and containers have long been recognised as sources of metal contamination of food. It is believed by some historians that the ancient Romans suffered from chronic lead poisoning as a result of leaching from poorly glazed pottery vessels in which they stored wine[59]. In more recent times use of glazed vessels to hold pickled olives has caused lead poisoning in the former Yugoslavia[60].

Lead has long been a problem in food processing because the metal readily lends itself to the fabrication and repair of cooking and storage utensils. The use of strips of lead sheet to repair cracks in wine vessels, a practice that is still employed in some traditional vineyards, has been shown to cause contamination of wine. Eighteenth-century cider drinkers in England sometimes suffered from 'Devonshire dry gripes' or colic, caused by metal picked up from the lead-lined troughs used by cider makers[61].

Amateur repair work on cooking and storage vessels, in which lead solders are used, have been relatively common causes of food contamination. Even in industry, stopgap repairs of processing equipment with parts cannibalised from elsewhere, can cause problems. The replacement of a corroded section of a stainless steel pipe in a freezer unit by a recycled copper tube in an Australian ice cream plant caused copper contamination of products and several poisonings in children who consumed them[62].

In modern food-processing plants, designed with full regard for hygiene and safety, there is little if any danger of metal pick-up by the products during manufacture. High-quality stainless steel, plastics and other structural materials approved for use in contact with foods, are used. The actual grades of materials used will depend on the particular products to be handled and the processes to which they are subjected. Dairy products, for instance, are particularly sensitive to oxidation when in contact with copper and some other metals[63]. Other foods can also be adversely affected by metals and alloys used in plant fabrication. The subject is a complex one and specialist texts should be consulted for further information[64].

Whitman has discussed interactions that can take place between structural materials used in food plants, and foodstuffs, cleaning agents and other substances[65]. He noted that certain kinds of stainless steel contain traces of arsenic, lead, mercury, cadmium and zinc, all of which can cause food poisoning. Other metals such as copper, nickel, iron and chromium can act as catalysts in oxidation reactions and cause rancidity in fats. It is important, therefore, to use only steels of the highest quality from which transfer of any of these metals to food during processing will not occur.

Adoption of good operating procedures is also essential if metal contamination of products is to be avoided. For example, contact time between foods and equipment should be kept to a minimum. It has been shown that migration of certain metals, such as manganese and nickel, from stainless steel to foods is relative to contact time[66].

4.2.2 Metal pick-up during canning

Every process to which food is subjected and each type of container in which it is placed can make some contribution to the load of chemical contaminants we accumulate in our bodies. Packaging is only one of the sources of metal pick-up[67]. However, mainly for historical reasons, the possibility of food being contaminated by the container in which it is packed looms large in public awareness of food hygiene.

The use of what we know as the food 'can' dates back to the response of scientists to Napoleon's offer of a prize, in the early nineteenth century, to anyone who could develop a practical method of food preservation that would enable him to feed his vast armies while on campaign. One of his subjects, Nicholas Appert, provided the solution based on the principle that food spoilage bacteria are destroyed by heat. In 1804 he produced preserved meat by heating and sealing it in glass jars.

Appert's discovery was developed further by an Englishman, Peter Durand, who used tin-coated iron containers in place of glass jars. In 1810 he was granted a patent for his technique of preserving foods in hermetically sealed tin cans. Initially canned food was produced to meet the needs of the British Army and Royal Navy. The cans were made of heavy-gauge tinplate at the Dartford Iron Works in Kent, and were laboriously sealed by hand using lead solder. They were heavy, and bulky, but remarkably successful as a means of preserving food. As early as the 1820s canned beef was taken on a Royal Navy expedition to the Arctic. Other canned foods were developed later: sardines in 1834, green peas in 1837, and condensed milk in 1856. The first pineapple cannery was established in Hawaii in 1892 and, in 1897, the first canned soup was produced in the US[68].

There were, however, problems with many of these early cans. On occasion the food they contained was contaminated with lead and other metals. A public perception grew that canning was associated with food poisoning. There was even a move in some countries to ban the process. In 1892 a congress of physicians in Heidelberg, Germany, recommended that 'tin plate should be forbidden for making vessels in which articles of food are to be preserved'[69].

Though today the basic principles of canning are the same as they were in the nineteenth century, techniques and the quality of materials have been greatly improved since those pioneering days. Nevertheless the problem of metal contamination of canned foods and beverages has not entirely gone away.

4.2.2.1 'Tin' cans

Technically speaking a 'can' is any hermetically sealed container in which food is subjected to a canning process, namely heat treatment to extend its shelf-life. The tinplate used in cans is a composite packaging material made up of a low-carbon-mild steel base, a very thin layer of iron–tin alloy, a thin layer of pure tin, a very thin layer of oxide and finally a monomolecular layer of 'physiologically safe' edible oil[70].

Each layer has its special role to play. The tinplate is 'passivated', either chemically or electronically, to build up a 'passive' layer on the surface to prevent oxidation. In both techniques chromium is deposited on the surface of the tinplate. The steel base is strong and rigid, giving tinplate its robust characteristics that allow it to withstand high-speed can-making as well as subsequent handling and storage. The tin coating gives an almost inert inner surface, while the oxide layer protects the tin from oxidation and corrosion. The outer layer of oil helps to prevent scratching of the can during manufacture and handling. The tin used is of high quality and under standards approved by ISO and EC regulations must be 99.75% pure. The steel base must not contain more than 0.05% manganese and chromium, 0.05% molybdenum and 0.04% nickel[71].

Cans are often coated internally with lacquer, generally a thermosetting resin polymerised on the surface of the tin during a baking process. This provides additional protection where there is a danger that the contents will corrode the metal. The protective effects of such a barrier are shown by a study of dietary tin intakes in France, which found that tin levels in the contents of unlacquered cans was 76.6 ± 36.5 mg/kg while in lacquered cans it was only 3.2 ± 2.3 mg/kg[72]. However, in the months or even years in which canned foods may be in storage, even this barrier may not be adequate. A slow corrosion can take place, especially if the food is acid and the lacquer layer is imperfect. Other factors, such as the presence of nitrate, storage temperatures, length of storage time, as well as the presence of various oxidising and reducing agents in the food, can determine the amount of corrosion that takes place.

Until relatively recently lead solders were used to join seams and attach lids to food cans. This practice has largely been abandoned in many, though not all countries, and has been replaced by new technologies such as using lead-free solders and solder-free, welded or folded joins[73]. In the US, as well as in several other countries, the remarkable reduction in dietary lead intake observed in recent years has been attributed, at least in part, to the phasing out of lead soldered food cans. Not many years ago the intake of lead from this source could be considerable, as some of the older data shows. A study in Israel found lead levels of 2 mg/litre in canned fruit juice[74], while in New Zealand investigators reported levels of 10 mg lead/kg of canned blackcurrants[75]. Other metals, such as tin and iron, were also present at unusually high levels in some of the cans examined.

Though the migration of lead and other metals from the can into its contents is related to such factors as the length of storage and the ambient temperature, this is not a simple relationship. An important part in the process can also be related to the electrochemical properties of the metals, such as the cathodic behaviour of lead relative to tin. When both metals are in electrical contact in a fluid, the tin will dissolve in preference to the lead and lead already in solution will be replaced by tin. The implications of this are well illustrated by the results of analysis of the contents of a tin of roast veal which had been taken on an Arctic expedition in 1837 and had been left in permanent deep freeze until it was recovered a century later. Though the can was crudely made and large amounts of lead solder had been used to seal the top and sides, only 3 mg lead/kg was found when the veal was analysed, along with 71 mg iron/kg and a massive 783 mg tin/kg. A tin of carrots recovered from the same cache contained 308 mg iron/kg, 2440 mg tin/kg, with no lead detectable[76].

Changes in the metal content of canned foods occur during storage even in modern cans. A recent study of canned evaporated milk found that while copper levels fell from an initial 2.23 ± 0.18 to 0.44 ± 0.01 mg/kg over two years, tin had increased from 28 ± 2 to 114 ± 4 mg/kg, and lead from 0.093 ± 0.005 to 0.29 ± 0.01 mg/kg[77]. The level of nitrate in canned food can be the cause of high levels of tin. An outbreak of food poisoning in Germany was traced to the consumption of canned peaches imported from Italy. The fruit contained about 400 mg tin/kg compared to the usual levels of 44–87 mg/kg in canned peaches produced elsewhere. The Italian cans were found to contain high levels of nitrate that was traced to contamination of water used in processing. The water came from a well that had as much as 300 mg NO_3/litre. It was concluded that the level of tin in the peaches was not of itself likely to be toxic, but that nitrate had increased the solubility of the metal[78].

4.2.2.2 Aluminium containers

The aluminium can, for soft as well as brewed drinks, is a feature of modern life. We make enormous use of aluminium foil for food wrapping and for a variety of food containers. This is a relatively recent development, for though aluminium has been commercially available for many years, technological difficulties prevented its widespread use in the food industry. Initially, also, there was some opposition on the grounds of health. However, this opposition has largely disappeared and today aluminium is probably the most widely used metal for food packaging.

As with other packaging metals, the pure aluminium metal is not used on its own to make cans. Various other metals, including iron, copper, zinc, chromium and manganese, are alloyed with the aluminium to provide strength, improve formability and increase corrosion resistance. These metals, as well as the aluminium itself, may under certain conditions migrate into the can's contents. Uncoated aluminium cans are readily attacked by food acids, causing hydrogen release and swelling of the can. Aluminium acts as a 'sacrifical anode' when used in conjunction with tinplate or steel. This is the reason why aluminium ends are sometimes used on steel cans. The aluminium provides electrochemical protection by replacing any iron or tin which might otherwise enter the contents[79].

Aluminium has been found to be an excellent metal for storage of many types of beverages, alcoholic as well as non-alcoholic. It has been shown that even when solution of the metal occurs, the flavour of the beverage will not be affected until more than 10 mg aluminium/litre has dissolved. It is claimed that beers packed in aluminium cans are superior in colour, flavour and clarity to those packed in tin-plated steel cans[80].

Coating the inside of an aluminium can with a vinyl epoxy or other resin helps to prolong the shelf-life of canned alcoholic beverages and soft drinks. Strong alcoholic beverages such as whisky and brandy and some wines can undergo undesirable reactions with the aluminium if packed in unlacquered cans. Pitting corrosion can occur as well as discoloration of the beverage. As aluminium is dissolved by the liquid, a flocculent precipitate of aluminium hydroxide can form[81].

Aluminium foil is used extensively for food wrapping and packaging as well as during cooking. If properly handled and used in the right conditions, the foil fulfils its roles excellently. However, in the case of frozen foods, for example, bad handling and, especially, poor storage and resulting thawing can cause both deterioration of the food and corrosion of the foil. In addition, if aluminium foil comes into contact with other metals, the electrochemical 'sacrificial' reaction can come into play. Pitting and corrosion of the foil occurs, with aluminium contamination of the food.

4.2.3 Contamination of food during catering operations

Some long-favoured practices used on both a large and small scale in cooking in the kitchen have been implicated in metal contamination of food. This has been principally as a result of the use of metals to fabricate the pots and pans and other kitchen utensils. Ceramic, as well as enamelled, utensils have also been implicated.

4.2.3.1 Metal cooking utensils

Tinned copper pans and pots have traditionally been favoured by chefs. The combination of metals was found to be ideal for cooking: the copper, with its high conductivity, allows rapid heating while the tin protects the easily dissolved copper from attack by food acids. Today stainless steel, aluminium and a variety of non-stick surfaced metals, as well as ceramics and glass utensils have been adopted in the majority of domestic kitchens. Tinned copper pots and pans are still, however, to be found in many hotel and restaurant kitchens and even in some modern homes.

Unfortunately, while the excellent heat transfer qualities of these tinned copper utensils is rightly valued, they can also cause contamination of food. The tin used to plate the copper is, in fact, not pure tin but a lead–tin mixture. The lead is required for technical reasons as a flux so that the tin can spread evenly over the surface of the copper. Even though under UK Food Regulations the amount of lead in the tinplate may not exceed 0.2%, this is enough in some circumstances to contribute a significant amount to foods cooked in tinned utensils, as has been shown in a study of lead contamination of foods prepared in a commercial restaurant[82].

Several different types of food were cooked in the normal manner using three different utensils, an aluminium saucepan and a new as well as an old and well-used tinned copper saucepan. The levels of copper and lead in the cooked food, compared to the raw foods, are shown Table 4.3. As the results show, both of the metals are leached to an extent dependent on the state of wear of the cooking utensils. In another study tomato soup heated in a previously unused tinned copper pan was found to contain 3.53 mg lead/kg, compared to 0.71 mg/kg when prepared in an old pan[83]. When tested by the standard method (measurement of uptake of lead by a 4% acetic acid solution in contact with the tinplate over a 24-hour period), the tinning on the new saucepan was found to contain 0.28% lead, which was above the legal limit, while in the worn pan it contained only 0.1%.

Table 4.3 Copper and lead content of foods before and after cooking in different types of metal saucepans (mg/kg, wet weight).

	Fish		Chicken		Cabbage		Potato	
Saucepan	Pb	Cu	Pb	Cu	Pb	Cu	Pb	Cu
Uncooked	0.31	0.82	0.14	2.21	0.15	1.36	0.19	3.10
Aluminium	0.36	1.37	0.21	2.52	0.18	1.04	0.16	1.87
Tinned copper (old)	0.42	5.70	0.25	6.36	0.29	2.07	0.22	2.39
Tinned copper (new)	1.09	2.24	0.94	4.05	0.79	1.93	0.26	1.88

Data from Reilly, C. (1978) Copper and lead uptake by food prepared in tinned-copper utensils. *Journal of Food Technology*, **13**, 71–6.

Other metals may also be picked up by food from cooking utensils. There is evidence that the traditional use of cast iron pots in some countries, especially over low fires for slow cooking, can contribute to a high level of iron in the cooked food. This practice is believed to be related to iron overload of the liver observed in some population groups in southern Africa[84]. A similar observation has been made in Papua New Guinea[85]. The use of iron pots for cooking the food of infants has been suggested as a simple and effective way of improving their iron intake, especially in developing countries[86]. A condition known as Indian Childhood Cirrhosis (ICC), in which excess copper becomes deposited in the liver, may be related to the use of copper and brass utensils in the preparation of infant foods[87].

As will be noted later, concern has been expressed by some medical authorities and several consumer groups that aluminium consumed in the diet may be associated with the onset of Alzheimer's disease and other degenerative mental states. Aluminium can be picked up by food from aluminium cooking utensils and from aluminium foil used to wrap foods. It is also contributed to the diet through the use of certain aluminium compounds as food additives[88]. Whether, however, the resulting levels of uptake are a risk to health has not been established[89].

Even in domestic situations the use of inappropriate metals in equipment used for food and beverage preparation can cause problems. Coffee percolators made from aluminium have been found to be a source of dietary aluminium, especially when the percolation is prolonged[90]. High levels of other metals may also come from similar sources. A copper coffee urn, in which the grounds had been allowed to build up for some time and which was kept on continuous heat, caused gastric upset in those who consumed the beverage, apparently due to copper ingestion[91]. A high intake of cadmium, also resulting in gastric upset, was caused by use of a zinc-plated water-heating unit for beverage preparation[92]. The zinc was found to contain a high level of cadmium.

Many other examples can be given of incidents of poisoning that have resulted because of the use of inappropriate metals and alloys in domestic and commercial food preparation. In general, where there is any doubt about its suitability for use in contact with food an item of equipment should be excluded from the kitchen. Zinc- and cadmium-plated items have already been mentioned in this connection. In addition beryllium, which is used as a hardener in copper alloys, should be excluded from all possible contact with foods, and only metals and equipment made from them which have been approved for catering should be used for food preparation.

It must be noted, however, that not all metal pick-up from cooking and other equipment poses a health risk[93]. It is probable that much of our intake of the essential element chromium is obtained from leachate from stainless steel vessels[94]. Stainless steel may also be an important source of nickel in our diet. We have already seen how iron cooking utensils can contribute to iron nutrition[95].

4.2.3.2 Ceramic ware

Pottery containers and cooking utensils have long been recognised as potential sources of metal contamination of food and drink. Ceramic vessels are normally given a glass-like glaze to produce a non-porous, watertight surface. The glaze is produced by coating the surface of the originally porous pottery with what is known as 'frit' and heating it to a high temperature in a kiln. The frit consists of a mixture of salts of various metals, including lead, and in the high temperature of the kiln it is vitrified and forms a glass-like layer on the pottery surface. The glaze on properly made and kilned plates and dishes should normally be unaffected by food and heat and not release its lead or other components during cooking. Unfortunately some glazed pottery, especially of the craft and home-made kind, which has not been kilned at a sufficiently high temperature or has been made with poorly formulated frits, has been known to release toxic amounts of lead, cadmium and other toxic metals into food.

A Canadian study of a range of earthenware items on sale in high-street stores, for example, found that 50% of them were poorly glazed and were unsafe for use with food. A more recent study in Mexico concluded that use of lead-glazed pottery was among the most important factors related to high blood lead levels in children[96]. Lead-glazed ceramic ware continues to be considered one of the major potential sources of dietary lead in the US by the FDA[97].

Standards related to leaching of heavy metals from ceramic ware have been established by several governments as well as international bodies. The Council of the European Economic Community issued a directive laying down limits for lead and cadmium leaching under specified conditions for ceramic articles[98]. The International Organisation for Standardisation (IOS) has also established permissible levels for lead and cadmium release[99]. In the US the FDA has established an 'action level' for lead release from glazed ceramic ware of 0.5 mg/litre, and has established the 'Ceramic Dinner Ware Lead Surveillance Program' to protect consumers from excessive lead ingestion from such sources[100]. In the UK MAFF (now DEFRA) has issued similar guidelines[101].

Both MAFF (DEFRA) and the USFDA give a piece of advice that it would be well for all consumers to follow. When it comes to whether or not to use a particular ceramic item to prepare or store food or drink, it is best to avoid craft and home-made or antique ceramics and to stick to commercially made products. With these, unlike the other items, you can normally be assured that the glazing meets legal standards and is unlikely to contaminate food with lead or other toxic metals.

4.2.3.3 Enamelled ware

Enamelled kitchen ware can also cause metal contamination of food. Enamel is similar to glaze on pottery; a glass-like layer is baked onto the surface, serving both as an anticorrosive lining and as decoration on a metal utensil. As is the case with ceramics, enamelled casseroles, plates, saucepans and other kitchen appliances that come from reputable manufacturers and conform to British or other standards will be safe to use. But others of less reputable origin, whose adherence to good standards cannot be guaranteed, are suspect. This is especially true of those that are bright yellow or red in colour, indicating the possible use of lead or cadmium pigments. Old, worn enamelled vessels with cracked surfaces are also suspect.

4.2.3.4 Other domestic sources of metal contamination of food

Printing and decoration used on food and beverage containers may also be a source of metal contamination. A Japanese study found that the pigment used to decorate some types of drinking glasses could release high levels of lead and cadmium into beverages[102]. Print and colour on plastic vessels and wraps can also be a source of metal contamination of the contents. High levels of lead have been detected on the outside of soft plastic bread wrappers. In a study of sources of lead in the diet of children, the risk of lead poisoning associated with the re-use of plastic food wrappings to store food in the refrigerator has been stressed[103]. Other metals may also be added to food in the same way[104].

Wrapping paper and cardboard can also contribute to the metal content of food. Cadmium, chromium and lead have been detected in several different kinds of paper used for a variety of different food-related purposes. In paper teabags, for instance, cadmium was found at a level of 0.3 mg/kg[105]. A study of coloured wrapping papers used for confectionery found that many of them contained unacceptably high levels of lead[106]. One such wrapper on a sweet contained more than 10 g lead/kg of wrapper.

A high level of metals in the colours and print used on wrappers does not, of course, mean that the sweets and other products that are contained in them will necessarily be contaminated. The printing and decoration are normally on the outside of the wrappers and are not normally in contact with the contents. The danger can arise when, in re-use, the wrappings are turned inside out and used to store other types of food. In the case of children, the teeth are sometimes used to open a difficult wrapping and it is not unknown for sweet papers to be chewed. However, even in such cases there may not be any danger of absorption of toxic metals. Whether this occurs will depend on the solubility of the pigments. In many countries there are regulations which set strict limits for migration of materials from plastic and other wrappings into food[107]. A study carried out by the Canadian Department of Health and Welfare found that while a range of metals, including copper, mercury and lead, could be detected in plastic wrappers and containers used on foods sold to the public, none of these were soluble to any significant extent and would be unlikely to be absorbed by food[108].

4.3 Food fortification

The addition of certain nutrients, including trace elements, to food and drink is widely practised around the world as a public health measure and a cost-effective way of ensuring the nutritional quality of the food supply[109]. In certain countries and for a limited number of food products, it is a legal requirement. Additions are also made by food manufacturers in response to pressure from consumers and, not least, to ensure that their products have a competitive edge in the market place.

Iodine was the first nutrient to be added on a large scale to a food as a public health measure to combat the goitre which was then endemic in many parts of the world. In 1922, a hundred years after the Swiss physician Coindet found that iodine could cure the disease[110], the first public health intervention using the element was undertaken in the State of Michigan in the US[111]. Iodised salt was introduced and resulted in a marked reduction in levels of incidence of goitre, especially among children. About the same time several cantons in Switzerland introduced iodised salt, with similar results. Before long the policy was introduced in many countries, either as a voluntary initiative or backed by government legislation which required that table salt sold to the public should be fortified with the element. The remarkable success achieved in many countries, showed that an intervention of this nature, in which a nutrient, naturally present in the diet in inadequate amounts, was added to an appropriate food delivery vehicle to boost intakes, was a valuable public health resource.

During World War II many countries introduced fortification of bread and flour to combat nutritional deficiencies which were expected to result from wartime food shortages and rationing. The addition of iron, as well as three water-soluble vitamins, to flour was introduced in the US in 1940 and was later extended to bread[112]. Similar requirements were introduced in the UK and elsewhere about the same time. The requirements, with the addition of calcium, are still in place in the UK today[113], though it has been recommended that it should no longer be mandatory[114].

Strictly speaking, the addition of iron, calcium and the water-soluble vitamins to bread and flour should be described as *restoration* rather than *fortification*. The latter term refers to the addition of nutrients to a food, whether or not it is naturally present in that food, for the purpose of correcting a deficiency in the population. The practice can also be described as *enrichment*. When, in contrast, a nutrient is added to a food to replace what has been lost during processing, the term *restoration* is appropriate.

Another term used in relation to the addition of nutrients to food is *substitution*. This refers to the adding of a nutrient to a substitute or simulated food, designed to take the place of another product (such as textured vegetable protein, TVP). Nutrients are added to bring levels in the substitute up to those normally found in the product it replaces. *Supplementation* is a more general term used in relation to addition of a nutrient to a food in which it is not normally found. Another term, *standardisation*, refers to the addition of nutrients to a food to compensate for natural variations.

4.3.1 Regulations and current practice regarding fortification of foods

Additions of nutrients are normally made to four categories of foodstuffs[115]. These are:

(1) foods for special dietary uses
(2) foods which have reduced levels of nutrients as a result of processing
(3) food substitutes

(4) staple foods used as vehicles for nutrients which may be at low levels in a population.

Legislation relating to the addition of nutrients to foods and regulations about the kinds and quantities of additions that may be used, as well as about the claims that may be made concerning the resulting enriched products, differ between countries. Usually they follow the general principles established by the Codex Alimentarius Commission of FAO/WHO[116], and the USFDA[117].

4.3.1.1 *UK regulations*

In the UK, the addition of iron, as well as calcium and two water-soluble vitamins, to bread and flour is compulsory. There are also specifications for infant formulae which *de facto* require the addition of several nutrients[118], including iron and copper, to ensure that they are at specified levels in the product[119]. Apart from these statutory requirements, addition of minerals and other nutrients is permitted to all foodstuffs, with the exception of alcoholic drinks, provided that:

(1) the addition is not injurious to health
(2) the product is not controlled by a compositional standard which expressly forbids addition
(3) the product is correctly labelled.

Specific claims on the label may only be made with respect to certain 'scheduled' nutrients, which include the two metals iron and zinc[120].

4.3.1.2 *Australian and New Zealand regulations*

In contrast to the voluntary approach to food fortification followed in the UK, as well as in the US where a wide variety of additives may be used under the GRAS system, some countries, such as Australia and New Zealand, have adopted a more restrictive policy. Standard 1.3.2 of the joint Food Standards Code of the two Antipodean countries regulates the addition of vitamins and minerals to foods and the claims that can be made about them[121]. The Standard stipulates that 'a vitamin or mineral must not be added to a food unless the:

(1) addition of that vitamin or mineral is specifically permitted in this Code
(2) vitamin or mineral is in a permitted form, unless stated otherwise in the Code.'

The permitted minerals (and vitamins), the foods to which they may be added, the amount permitted and the 'maximum claim per Reference Quantity' are listed in a short series of tables. Iron and zinc may be added to biscuits, flour, bread and breakfast cereals. The same elements may also be added to 'vegetable protein foods intended as a meat analogue'. Zinc, but not iron, may be added to 'analogues derived from legumes' and to 'analogues of yoghurt and dairy desserts'. In each case 'a maximum claim per Reference Quantity' is given as a proportion of the (Australian) RDI. This ranges from 15 to 50% of the recommended amount, depending on the nutrient and the type of food to which it is added. Similar restrictive lists of permitted additives, and the foods to which they may be added, are also enforced in several other countries.

4.3.2 Foods commonly fortified

Voluntary fortification of a variety of foodstuffs, in accordance with such guidelines, is undertaken by food manufacturers in the UK and other countries. Examples can be seen in the various ready-to-eat (RTE) breakfast cereals, of which more will be said later. Among other foods which are commonly fortified with minerals are vegetable-based products intended as a replacement for meat, such as textured vegetable protein (TVP) to which iron and zinc are added. Several different trace elements are also added to what are described as 'foods for special dietary uses', for example, as slimming aids and 'sports' foods and drinks. A specially enriched biscuit has been used to bring about a significant improvement in the micronutrient status of school children in South Africa[122].

4.3.2.1 RTE breakfast cereals

RTE cereals are one of the success stories of modern food technology. Their growth in popularity in many countries from their introduction at the beginning of the twentieth century has been remarkable. They have replaced, for great numbers of consumers, the traditional cooked breakfast of bacon and eggs and even of oatmeal porridge[123]. In Australia, which is believed to have the third highest consumption level of RTE breakfast in the world, after the UK and Ireland[124], 80% of the population consume them at least once a week, with 50% consuming them every day[125]. Americans are also big consumers of these products[126]. It has been shown that RTE breakfast cereals can make a significant contribution to intake of several essential trace elements, such as iron, zinc and copper, in countries where consumption is high, such as Ireland[127]. A similar finding has been reported in Australia[128], the UK[129], and the US[130].

Most of the cereals used to manufacture such RTE produce, such as maize and rice and refined wheat, are naturally low in trace elements. It is the added minerals used to fortify the foods which account for the levels of the nutrients listed on the package labels and which are a major selling point promoted by the manufacturers. How significant nutritionally these increased levels can be may be gathered from the fact that a normal serving of cornflakes produced by one of the major cereals companies is capable of supplying 29% of the iron, 2.5% of the copper, 3% of the manganese, and 1% of the zinc requirement of an (Australian) adult[131].

4.3.2.2 Functional foods

The term *functional food* is used to describe food products that contain components which can affect some physiological function and have a greater nutritional potency than is found in ordinary foods. It is not a scientific definition but rather a marketing term. Other terms are also used to describe such nutritionally modified products which, in recent years, have become a rapidly growing part of the health food market. They are known also as *nutraceuticals* and *designer* or *engineered foods* and are promoted by their manufacturers as 'the next generation of food, providing specific physiologic, health-promoting, and even disease-preventing benefits'[132]. In Japan, where the market for such products is enormous, the official term used is 'foods for specified health use' (FOSHU). They are defined as 'foods that can be expected to have certain health benefits and that have been licensed to bear health claims'[133].

Among the many different kinds of functional foods on sale in Japan and other countries are products containing such potentially health-promoting ingredients as dietary fibre, oligosaccharides, gamma-linoleic acid, lactic acid bacteria, various phytochemicals and a variety of minerals. Just as for any other food additive, the use of any such ingredient in a product is subject, in the UK and other countries, to the general food law. In the US the ingredients must have GRAS classification for the particular use intended[134]. Any claims made on labels about the health-promoting qualities of the product will also have to conform to labelling legislation[135].

The use of trace elements and other inorganic components in the production of functional foods has not, so far, been extensive, but is expected to grow as the marketing value of the increased presence of antioxidants and related substances is recognised by manufacturers. Zinc is one element that has already been used in this way in some products. Selenium has also been used to a limited extent, for example, in a drink commercially available in China[136], and were it not for technical difficulties, would probably by now be a component of other functional foods designed to combat cancer[137].

4.3.3 Natural fortification of foods with metals

In recent years there has been a growing interest in using natural means, such as plant breeding and selection, rather than actual physical addition of vitamins and minerals, to increase the nutrient content of staple foods. The prevalence of micronutrient deficiencies in many of the less-developed nations has encouraged health authorities to investigate methods of food fortification which are simpler, more reliable and more cost-effective than the distribution of supplements or the mechanical fortification of commercial food products. Though these efforts are still at an early stage, progress has already been made and there are good grounds for hoping that the outcome will boost nutrition in many countries. The aims and early achievements of the considerable interdisciplinary efforts that are being committed to the project were discussed at an international symposium organised by the Consultative Group on International Agricultural Research (CGIAR) and entitled 'Improving Human Nutrition through Agriculture' at Los Baños in the Philippines in October 1999[138].

One of the topics discussed at the conference was CGIAR's Micronutrients Project, which is aimed at developing a package of tools that plant breeders will need to produce mineral- and vitamin-rich cultivars. The target crops are rice, wheat, maize, phaseolus beans and cassava, all staple foods of enormous importance in many of the poorer regions of the world. The minerals targeted are iron and zinc. Progress has already been made in breeding plant varieties which have special abilities to take up minerals from the soil and concentrate them in their edible parts. Zinc-efficient wheat varieties have been developed in Australia and are beginning to be grown commercially there[139]. In the US an iron-efficient soya bean has been developed. At less advanced stages of development is iron-accumulating rice produced at the International Rice Research Institute in the Philippines, and of maize cultivars with improved iron and zinc levels in South Africa[140]. Several other mineral-efficient staple food crops are also in the pipeline[141].

Besides the contribution to human nutrition which these new varieties and cultivars can make, there is also a major advantage for growers of these crops. Plants which are rich in trace elements are higher-yielding and more resistant to disease and drought, and require less irrigation than do ordinary, low mineral levels varieties[142]. These attributes should make them more attractive to farmers, not only in poor countries but also in the developed world. There, too, foods that have naturally enriched trace element concentrations could have a marketing edge and a premium value to health- and environment-conscious consumers.

References

1. Beavington, F. (2001) Pioneer studies in trace element contamination of the food chain: disregard of prior work risks reinventing the wheel. *Journal of the Science of Food and Agriculture*, **81**, 155–60.
2. Wenlock, R.W., Buss, D.H. & Dixon, E.J. (1979) Trace nutrients. 2. Manganese in British foods. *British Journal of Nutrition*, **41**, 253–61.
3. Baxter, M.J., Burrell, J.A. & Massey, R.C. (1990) The aluminium content of infant formula and tea. *Food Additives and Contaminants*, **7**, 101–7.
4. Saroja, A. & Sivapalan, P. (1990) Aluminium in black tea. *Sri Lanka Journal of Tea Science*, **59**, 4–8.
5. Chenary, E.M. (1955) A preliminary study of aluminium and the tea bush. *Plant and Soil*, **6**, 174–200.
6. Ravichandran, R. & Parthiban, R. (1998) Aluminium content in South Indian teas and their bioavailability. *Journal of Food Science and Technology*, **35**, 349–51.
7. Oldfield, J.E. (1999) *Selenium World Atlas*. Selenium–Tellurium Development Association, Grimbergen, Belgium.
8. Reilly, C. (1969) The uptake and accumulation of copper by *Becium homblei*. *New Phytologist*, **68**, 1081–7.
9. Duvigneaud, P. & Denaeyer-De Smet, S. (1963) La végétation du Katanga et des sols métallifères. *Bulletin de la Société Royal Botanique Belge*, **96**, 93–231.
10. Reilly, C. & Reilly, A. (1973) Zinc, lead and copper tolerance in the grass *Stereochlaena camereonii*. *New Phytologist*, **72**, 1041–6.
11. Reilly, C. (1996) Selenium supplementation – the Finnish experiment. *British Nutrition Foundation Nutrition Bulletin*, **21**, 167–73.
12. Alloway, B.J. (1995) *Heavy Metals in Soils*, 2nd edn. Chapman & Hall, London.
13. Alloway, B.J. (1995) *Heavy Metals in Soils*, 2nd edn. Chapman & Hall, London.
14. Strenstrom, T. & Vahter, M. (1974) Heavy metals in sewage sludge for use on agricultural soil. *Ambio*, **3**, 91–2.
15. Williams, C.H. & David, D.J. (1973) Heavy metals in Australian soils. *Australian Journal of Soil Research*, **11**, 43–50.
16. Rayment, G.E., Best, E.K. & Hamilton, D.J. (1989) Cadmium in fertilisers and soil amendments. *Chemical International Conference, Brisbane, Australia, 28 August–2 September*. Royal Australian Chemical Institute, Australia.
17. Bennet-Chambers, M., Davies, P. & Knott, B. (1999) Cadmium in aquatic ecosystems in Western Australia. A legacy of nutrient-deficient soils. *Journal of Environmental Management*, **57**, 283–95.
18. Berrow, M.I. & Webber, J. (1972) The use of sewage sludge in agriculture. *Journal of the Science of Food and Agriculture*, **23**, 93–100.
19. McLaughlin, M.J., Parker, D.R. & Clarke, J.M. (1999) Metals and micronutrients – food safety issues. *Field Crops Research*, **60**, 143–63.
20. McGrath, D., Postma, L., McCormack, R.J. & Dowdall, C. (2000) Analysis of Irish sewage sludges: suitability of sludge for use in agriculture. *Irish Journal of Agriculture and Food Research*, **39**, 73–8.
21. *The Sludge (Use in Agriculture) (Amendment) Regulations 1990* (S.I. [1990] No. 880). HMSO, London.
22. European Community (1986) Council Directive 86/278/EEC. *Official Journal of the European Communities*, **L181**, 6–12.
23. USEPA (1993) *Clean Water Act*, Section 503. **58**(2), US Environmental Protection Agency, Washington, DC.
24. Commission of the European Communities (1986) Council Directive (86/278/EEC). *Official Journal of the European Communities*, **L181**, annex 1A, 6–12.
25. Pike, E.R., Graham, L.C. & Fogden, M.W. (1975) Metals in crops grown on sewage-enriched soil. *Journal of the American Pedological Association*, **13**, 19–33.
26. Frost, H.L. & Ketchum, L.H. (2000) Trace metal concentrations in durum wheat from

applications of sewage sludge and commercial fertiliser. *Advances in Environmental Research*, **4**, 347–55.
27. Dudka, S. & Miller, W.P. (1999) Accumulation of potentially toxic elements in plants and their transfer to human food chain, *Journal of Environmental Science and Health*, **B34**, 681–708.
28. Bakir, F., Damluji, S.F. & Amin-Zaki, L. (1973) Methylmercury poisoning in Iraq. *Science*, **181**, 230–41.
29. Ministry of Agriculture, Fisheries and Food (1982) *Survey of Arsenic in Food: the 8th Report of the Steering Group on Food Surveillance; the Working Party on the Monitoring of Foodstuffs for Heavy Metals*. HMSO, London.
30. Creger, T.L. & Peryea, F.J. (1992) Lead and arsenic in two apricot cultivars and in Gala apples grown on lead arsenate-contaminated soils. *Hortsience*, **27**, 1277–8.
31. Ministry of Agriculture, Fisheries and Food (1998) *Lead, Arsenic and Other Metals in Food. The 52nd Report of the Steering Group on Chemical Aspects of Food Surveillance. Food Surveillance Paper No 52*. The Stationery Office, London.
32. Bartels, C. (1996) Mining in the Middle Ages and modern era in the Harz Mountains and its impact on the environment. *Naturwissenschaften*, **83**, 483–91.
33. Morgan, H. (1988) *The Shipham Report. An Investigation into Cadmium Contamination and its Implications for Human Health*. Elsevier, London.
34. Longfield, S.A. & Macklin, M.G. (1999) The influence of recent environmental change on flooding and sediment fluxes in the Yorkshire Ouse basin. *Hydrological Processes*, **13**, 1051–66.
35. Meharg, A.A., Osborn, D., Pain, D.J., Sanchez, A. & Navesco, M.A. (1999) Contamination of Donana food-chains after the Aznalcollar mine disaster. *Environmental Pollution*, **105**, 387–90.
36. Ashami, M.O. (1984) Pollution of soils by cadmium. In: *Changing Metal Cycles and Human Health* (ed. J.O. Nriagu), pp. 9–111. Springer, Berlin.
37. Nakagawa, H., Nishijo, M., Morikawa, Y. *et al.* (1993) Urinary β2-microglobulin concentration and mortality in a cadmium-polluted area. *Archives of Environmental Health*, **48**, 428–35.
38. Friberg, L., Piscator, M., Norberg, G. & Kjellstrom, T. (1974) *Cadmium in the Environment*, 2nd edn. Butterworth, London.
39. Palheta, D. & Taylor, A. (1995) Mercury in environmental and biological samples from a gold mining area in the Amazon region of Brazil. *Science of the Total Environment*, **168**, 63–9.
40. Futatsuka, M., Kitano, T. & Shono, N. (2000) Health surveillance in the population living in a methyl mercury-polluted area over a long period. *Environmental Research*, **83**, 83–92.
41. Phelps, R.W., Clarkson, T.W., Kershaw, T.G. & Wheatley, B. (1980) Methyl mercury contamination of fish in a Canadian lake. *Archives of Environmental Health*, **35**, 161–8.
42. Prater, B.E. (1975) Water pollution in the River Tees. *Water Pollution Control*, **74**, 63–76.
43. Working Party on Materials Conservation and Effluents (1975) Report of the Industrial and Technical Committee of the Institute of Metal Finishing. *Transactions of the Institute of Metal Finishing*, **53**, 197–202.
44. Petrokova, V., Ustjak, S. & Roth, J. (1993) Heavy metal contamination of agricultural crops and soils in Northern Bohemia. In: *Trace Elements in Man and Animals – TEMA 8* (eds. M. Anke, D. Meissner & C.F. Mills), pp. 496–7. Verlag Media Touristik, Gersdorf, Germany.
45. Skalney, A. (1993) Interelementary relationships and oncological morbidity in an As-polluted area. In: *Trace Elements in Man and Animals – TEMA 8* (eds. M. Anke, D. Meissner & C.F. Mills). Verlag Media Touristik, Gersdorf, Germany.
46. Wang, Q., Shen, W.G. & Ma, Z.W. (2000) Estimation of mercury emission from coal combustion in China. *Environmental Science and Technology*, **34**, 2711–13.
47. Zheng, B.S., Ding, Z.H., Huang, R.G. *et al.* (1999) Issues of health and disease relating to coal use in southwestern China. *International Journal of Coal Geology*, **40**, 119–32.

48. Sharma, K. & Reutergardh, L.B. (2000) Exposure of preschoolers to lead in the Makati area of Metro Manila, the Philippines. *Environmental Research*, **83**, 322–32.
49. Ho, Y.B. & Tai, K.M. (1988) Elevated levels of metals and other metals in roadside dusts and soils in Hong Kong. *Environmental Research*, **49**, 37–51.
50. Sadiq, M., Alam, I., El-Mubarek, A., & Al-Mohdar, H.M. (1989) Preliminary evaluation of metal pollution from wear of auto tires. *Bulletin of Environmental Contamination and Toxicology*, **42**, 743–9.
51. Tong, S.T.Y. & Lam, K.C. (2000) Home sweet home? A case study of household dust contamination in Hong Kong. *Science of the Total Environment*, **256**, 115–23.
52. De Kimpe, C.R. & Morel, J.L. (2000) Urban soil management: a growing concern. *Soil Science*, **165**, 31–40.
53. Proctor, D.M., Shay, E.C. & Scott, P.K. (1997) Health-based soil action levels for trivalent and hexavalent chromium: a comparison with state and federal standards. *Journal of Soil Contamination*, **6**, 595–648.
54. Reilly, C. & Henry, J. (2000) Geophagia: why do humans eat soil? *BNF Nutrition Bulletin*, **25**, 141–4.
55. Aufreiter, S., Hancock, R.G.V., Mahaney, W.C., Stambolic-Robb, A. & Sammugadas, K. (1997) Geochemistry and mineralogy of soils eaten by humans. *International Journal of Food Sciences and Technology*, **48**, 292–305.
56. Smith, B., Chenery, S.R.N. & Cook, J.M. (1998) Geochemical and environmental factors controlling exposure to cerium and magnesium in Uganda. *Journal of Geochemical Exploration*, **65**, 1–15.
57. Johns, T. & Duquette, M. (1991) Detoxification and mineral supplementation as a function of geophagy. *American Journal of Clinical Nutrition*, **53**, 448–56.
58. Ranken, M.D., Kill, R.C. & Baker, C., eds. (1997) *Food Industries Manual*, 24th edn. Blackie/Leatherhead Food Research Association, Surrey.
59. Gilfillan, S.C. (1965) Lead poisoning in ancient Rome. *Journal of Occupational Medicine*, **7**, 53–60.
60. Beritic, T. & Stahuljak, D. (1961) Lead poisoning from pottery vessels. *Lancet*, **i**, 669.
61. Beech, F.W. & Carr, K.G. (1977) Cider and perry. In: *Economic Microbiology*, Volume 1, *Alcoholic Beverages* (ed. A.H. Rose), pp. 130–220. Academic Press, London.
62. Reilly, C. (1989) unpublished observation.
63. Harper, W.J. & Hall, C.W. (1988) *Dairy Technology and Engineering*. Avi, Westport, Conn.
64. Brennan, J.G. (1976) *Food Engineering Operations*, 2nd edn. Elsevier Applied Science, London.
65. Whitman, W.E. (1978) Interactions between structural materials in food plant, and foodstuffs and cleaning agents. *Food Progress*, **2**, 1–2.
66. Sampaolo, A. (1971) Migration of metals from stainless steel. *Rassegna Chimica*, **23**, 226–33.
67. Selby, J.W. (1967) Food packaging, the unintentional additive. In: *Chemical Additives in Food* (ed. R.W.L. Goodwin), pp. 83–94. Churchill, London.
68. FAO (1986) *Guidelines for Can Manufacturers and Food Canners. Introduction.* Food and Agricultural Organization, Rome.
69. Van Hamel Roos (1891) On tin in preserved articles of food. *Rev. Intn. Fals*, **4**, 179 (as quoted in *Analyst*, **16**, 195).
70. Canned Food Information Service (1988) *Food Canning: an Introduction.* Food Canning Information Service, Melbourne, Australia.
71. Board, P.W. & Vignanoli, L. (1976) Electrolytic tinplate for canned foods. *Food Technology Australia*, **28**, 486–7.
72. Biego, G.H., Joyeux, M., Hartemann, P. & Debry, G. (1999) Determination of dietary tin intake in an adult French citizen. *Archives of Environmental Contamination and Toxicology*, **36**, 227–32.
73. Brouillard, M.J. & England, R.O. (2001) A metal finisher's evaluation of commercially available lead-free solder coatings. *Plating and Surface Finishing*, **88**, 38–45.

74. Beckham, I., Blanche, W. & Storach, S. (1974) Metals in canned foods in Israel. *Var Foeda*, **26**, 26–32.
75. Page, G.G., Hughes, J.T. & Wilson, P.T. (1974) Lead contamination of food in lacquered and unlacquered cans. *Food Technology New Zealand*, **9**, 32–5.
76. Kowal, W., Beattie, O.B., Baadsgaard, H., & Krahn, P.M. (1991) Source identification of lead found in tissues of sailors from the Franklin Arctic Expedition of 1845. *Journal of Archeological Science*, **18**, 193–203.
77. Ramonaityte, D.T. (2001) Copper, zinc, tin and lead in canned evaporated milk, produced in Lithuania: the initial content and its change in storage. *Food Additives and Contaminants*, **18**, 31–7.
78. Board, P.W. (1973) The chemistry of nitrate-induced corrosion of tinplate. *Food Technology Australia*, **25**, 16–17.
79. Jiminez, M.A. & Kane, E.H. (1974) *Compatibility of Aluminum for Food Packaging.* American Chemical Society, Washington, DC.
80. Jiminez, M.A. & Kane, E.H. (1974) *Compabibility of Aluminum for Food Packaging.* American Chemical Society, Washington, DC.
81. Brunner, B., Arnold, R. & Stolle, A. (1999) Migration of aluminium from foil to fish. *Fleiswirtschaft*, **79**, 110–12.
82. Reilly, C. (1978) Copper and tin uptake by food prepared in tinned-copper utensils. *Journals of Food Technology*, **13**, 71–6.
83. Reilly, C. (1976) Contamination of food during catering operations. *Hotel Catering and Institutional Management Review*, **2**, 34–40.
84. MacPhail, A.P., Simon, M.O., Torrance, J.D., Charlton, R.W., Bothwell, T.H. & Isaacson, C. (1979) Changing patterns of iron overload in black South Africans. *American Journal of Clinical Nutrition*, **32**, 1272–8.
85. Drover, D.P. & Maddocks, I. (1975) Iron content of native foods. *Papua New Guinea Medical Journal*, **18**, 15–17.
86. Borigato, E.V.M. & Martinez, F.E. (1998) Iron nutritional status is improved in Brazilian preterm infants fed food cooked in iron pots. *Journal of Nutrition*, **128**, 855–9.
87. Tanner, M.S., Bhave, S.A., Kantarjian, A.H. & Pandit, A.N. (1983) Early introduction of copper-contaminated animal milk feeds as a possible cause of Indian childhood cirrhosis. *Lancet*, **ii**, 992–5.
88. Pennington, J.A.T. & Schoen, S.A. (1995) Estimates of dietary exposure to aluminium. *Food Additives and Contaminants*, **12**, 119–28.
89. Klein, G.L. (1990) Nutritional aspects of aluminium toxicity. *Nutrition Research Reviews*, **3**, 117–41.
90. Lione, A., Allen, P.V. & Smith, J.C. (1984) Aluminium coffee percolators as a source of dietary aluminium. *Food and Chemical Toxicology*, **22**, 265–8.
91. Reilly, C. (1962) unpublished observation.
92. Rosman, K.J.R., Hosie, D.J. & De Laeter, J.R. (1977) The cadmium content of drinking water in Western Australia. *Search*, **8**, 85–6.
93. Reilly, C. (1985) The dietary significance of adventitious iron, zinc, copper and lead in domestically prepared food. *Food Additives and Contaminants*, **2**, 209–15.
94. Smart, G.A. & Sherlock, J.C. (1985) Chromium in foods and the diet. *Food Additives and Contaminants*, **2**, 139–47.
95. Borigato, E.V. & Martinez, F.E. (1998) Iron nutritional status is improved in Brazilian preterm infants fed food cooked in iron utensils. *Journal of Nutrition*, **128**, 855–9.
96. Ancona-Crus, M.I., Rothenberg, S.J., Schnaas, L., Zamora-Munoz, J.S. & Romero-Placeres, M. (2000) Lead-glazed ceramic ware and blood lead levels of children in the city of Oxaca, Mexico. *Archives of Environmental Health*, **55**, 217–22.
97. Bolger, P.M., Yess, N.J., Gunderson, E.L., Troxell, T.C. & Carrington, C.D. (1996) Identification and reduction of sources of dietary lead in the United States. *Food Additives and Contaminants*, **13**, 53–60.
98. European Economic Community (1984) Council Directive No. 84/500. *Official Journal of the European Economic Community*, **27**, 12–16.

99. International Organisation for Standardisation (1981) *International Standard* ISO 6486/1, 2. Geneva, Switzerland.
100. US Food and Drugs Administration (2000) Dangers of lead still linger, *FDA Consumer*, January–February 2000, 1–7.
http:/vm.cfsan.fda.gov/~dms/fdalead.html
101. Ministry of Agriculture Fisheries and Food (1974) *Report of the Inter-Departmental Working Group on Heavy Metals*. HMSO, London.
102. Watanabe, Y. (1974) Cadmium and lead on decorated drinking glasses. *Annual Report of the Tokyo Metropolitan Laboratory of Public Health*, **25**, 293–6.
103. Weisel, C., Demak, M., Marcus, S. & Goldstein, B.D. (1991) Soft plastic bread packaging: lead content and reuse by families. *American Journal of Public Health*, **81**, 756–8.
104. Alam, M.S., Srivastava, S.P. & Seth, P.K. (1988) Factors influencing the leaching of heavy metals from plastic materials used in the packaging of food and biomedical devices. *Indian Journal of Environmental Health*, **30**, 131–41.
105. Castle, L., Offen, C.P., Baxter, M.J. & Gilbert, J. (1997) Migration studies from paper and board food packaging materials. 1. Compositional analysis. *Food Additives and Contaminants*, **14**, 35–44.
106. Heichel, G.H., Hankin, L. & Botsford, R.A. (1974) Lead in coloured wrapping paper. *Journal of Milk and Food Technology*, **37**, 499–503.
107. Grammiccioni, L. (1984) Migration of metals into food from containers. *Rassegna Chimica*, **36**, 271–3.
108. Meranger, J.C., Cunningham, H.M. & Giroux, A. (1974) Plastics in contact with foods. *Canadian Journal of Public Health*, **65**, 292–6.
109. Brady, M.C. (1996) Addition of nutrients: current practice in the UK. *British Food Journal*, **98**, 12–18.
110. Coindet, J.F. (1820) Découverte d'un nouveau remède contre le goitre. *Annales de Chimie et Physique*, **15**, 49–59.
111. Mertz, W. (1997) Food fortification in the United States. *Nutrition Reviews*, **5**, 44–9.
112. McNamara, S.H. (1995) Food fortification in the United States: a legal and regulatory perspective. *Nutrition Reviews*, **53**, 103–7.
113. Ministry of Agriculture, Fisheries and Food (1984) Bread and Flour Regulations 1995. *Statutory Instruments, No. 3202*. HMSO, London.
114. Department of Health and Social Security (1981) *Report on Health and Social Subjects No. 23. Nutritional Aspects of Bread and Flour*. Committee on Medical Aspects of Food Policy. HMSO, London.
115. Richardson, D.P. (1990) Food fortification. *Proceedings of the Nutrition Society*, **49**, 39–50.
116. Codex Alimentarius Commission (1987) *General Principles for the Addition of Essential Nutrients to Food, Alinorm 87/26*, Appendix 5. Food and Agriculture Organization, Rome.
117. United States Food and Drug Administration (1987) *Code of Federal Regulations. Title 21. Nutritional Quality Guidelines for Foods*, Part 104.5. Food and Drug Administration, Washington, DC.
118. BNF (1994) *Food Fortification*, 6. British Nutrition Foundation, London.
119. European Commission (1991) Infant formula and follow-on formula, Commission Directive 91/321/EEC. *Official Journal* No. L 175/35. European Economic Community, Brussels.
120. Ministry of Agriculture Fisheries and Food (1996) *Food Labelling Regulations*. HMSO, London.
121. Australian New Zealand Food Authority (2000) *Draft Food Standards Code: Standard 1.3.2 Vitamins and Minerals*.
http://www.anzfa.gov.au/Draft Food Standards/Chapter 1/Part1.3/1.3.2 htm
122. Van Stuijvenberg, M.E., Kvalsvig, J.D., Faber, J., Kruger, M., Kenoyer, D.G. & Benade, A.J.S. (1999) Effect of iron-, iodine-, and β-carotene-fortified biscuits on the micronutrient status of primary school children: a randomised controlled trial. *American Journal of Clinical Nutrition*, **69**, 497–503.

123. Collins, E.J.T. (1976) Changing patterns of bread and cereal-eating in Britain in the twentieth century. In: *The Making of the Modern British Diet* (eds. D.T. Oddy & D.S. Miller), pp. 26–43. Croom Helm, London.
124. Kellogg Market Research (1995) Kellogg (Australia), Sydney.
125. Commonwealth Department of Health (1986) *National Survey of Adults: 1983: No. 1, Foods Consumed*. Australian Government Publications Service, Canberra.
126. Pennington, J. & Young, B. (1990) Iron, zinc, copper, manganese, selenium and iodine in foods from the United States Total Diet Study. *Journal of Food Composition and Analysis*, **3**, 166–84.
127. Sommerville, J. & O'Reagan, M. (1993) The contribution of breakfast to micronutrient adequacy of the Irish diet. *Journal of Human Nutrition and Dietetics*, **6**, 223–8.
128. Booth, C.K., Reilly, C. & Farmakalidis, E. (1996) Mineral composition of Australian Ready-to-Eat breakfast cereals. *Journal of Food Composition and Analysis*, **9**, 135–47.
129. Crawley, H. (1993) The role of breakfast cereals in the diets of 16–17 year old teenagers in Britain. *Journal of Human Nutrition and Dietetics*, **6**, 205–16.
130. Australian New Zealand Food Authority (2000) *Draft Food Standards Code: Standard 1.3.2 Vitamins and Minerals* http://www.anzfa.gov.au/Draft Food Standards/Chapter 1/Part 1.3/1.3.2 htm
131. Collins, E.J.T. (1976) Changing patterns of bread and cereal eating in Britain in the twentieth century. In: *The Making of the Modern British Diet* (eds. D.T. Oddy & D.S. Muller), pp. 26–43. Croom Helm, London.
132. Mackey, M., Hill, B. & Gund, C.K. (1994) Health claims regulations: impact on food development. In: *Nutritional Toxicology* (eds. F.N. Kotsonis, M. Mackey and J. Hjelle), pp. 251–71. Raven Press, New York.
133. Furukawa, T. (1993) The nutraceutical rules: health and medical claims: 'Foods for Specified Use' in Japan. *Regulatory Affairs*, **5**, 189–202.
134. Anon (1993) Designer foods FDA concerns noted. *Food Chemical News*, **35**, 48–50.
135. Reilly, C. (1994) Functional foods – a challenge for consumers. *Trends in Food Science and Technology*, **5**, 121–3.
136. Reilly, C. (1996) *Selenium in Food and Health*, pp. 288–9. Blackie, London.
137. Reilly, C. (1998) Selenium: a new entrant into the functional foods arena. *Trends in Food Science and Technology*, **9**, 114–18.
138. Scrimshaw, N.S. (2000) Foreword, *Food and Nutrition Bulletin*, **21**, 351.
139. Graham, R.D., Ascher, J.S. & Hynes, S.C. (1992) Selecting zinc-efficient cereal genotypes for soils of low zinc status. *Plant and Soil*, **146**, 241–50.
140. Banziger, M. & Long, J. (2000) The potential for increasing the iron and zinc density of maize through plant-breeding. *Food and Nutrition Bulletin*, **21**, 397–400.
141. Bouis, H.E., Graham, R.D. & Welch, R.M. (2000) The Consultative Group on International Agricultural Research (CGIAR) Micronutrients Project: justification and objectives. *Food and Nutrition Bulletin*, **21**, 374–81.
142. Bouis, H. (1996) Enrichment of food staples through plant breeding: a new strategy for fighting micronutrient malnutrition. *Nutrition Reviews*, **54**, 131–7.

Chapter 5
Metals in food and the law

As long as food remained a private matter for individuals and their families, the law of the realm had nothing to do with it. But once food became an item of commerce, legislators stepped in with regulations and decrees. Originally many of these were probably related to religious observations, though, as perhaps the prescriptions given to the ancient Hebrews to avoid eating the flesh of swine and other 'unclean' animals[1] indicate, some of them may also have had practical implications. Later, as trade increased and people, especially in the new towns and cities, began to rely on commercial growers and traders rather than on their own home production, laws relating to the sale and quality of food were introduced as, inevitably, fraud and sharp practices entered the marketplace.

In medieval England the *Assize of Bread*, a law to ensure that bread sold to the public was of the proper weight, was promulgated. This was followed over the centuries by many other decrees designed to prevent adulteration and fraud in specific foods. In the colonies of New England the laws of the mother country relating to food were generally in force, though, with the growing spirit of independence, some of purely local interest were also enacted. In post-independence USA laws relating to the production and sale of food were passed in the different states. Many of these were similar to regulations and statutes that were by then in place in many other countries. Legislation concerning food has continued to grow in volume throughout the world until today, in addition to a multiplicity of national laws, there are numerous international regulations controlling the quality of foods in the world marketplace.

5.1 Why do we have food legislation?

A sentence in the *Sale of Food and Drug Acts* of 1875 tells us the reason why we have food laws: 'No person shall sell to the prejudice of the purchaser any article of food or any other thing which is not of the nature, substance or quality demanded by such purchaser'. This is the key to Britain's first effective legislation relating to food and is the foundation on which today's food laws are based, both in the UK and in countries that inherited her legal system. It establishes the legal right of purchasers to wholesome and unadulterated food and requires purveyors to ensure that the products they sell meet the requirements of the law. In countries where the legal system owes little, if anything, to Britain, a similar principle is at the core of legislation to protect the rights of purchasers and governs the practices of food manufacturers and traders.

As has been mentioned already, there was very good reason why protection of consumers against fraud and adulteration was in the forefront of the minds of the legislators who pushed through the first food laws in the UK. The evidence is clear that legislation was sorely needed to control levels of contamination in foods in pre-1875 days. Apart from the instances cited in Accum's *Treatise on Adulterations of*

Food and Culinary Poisons, many more examples can be found in the popular as well as the scientific literature of the time.

A major concern of consumers in those days was that the food they purchased could have been deliberately contaminated with non-food materials, including metals and their salts. The adulteration was practised usually for one of two reasons: either inert material was incorporated into a product to increase its bulk and hence the seller's profits, or fraudulent additions were made to improve the appearance of an otherwise low-quality product. Chalk (natural calcium carbonate) or alum (naturally occurring crystalline potassium aluminium sulphate) was mixed with flour as an 'extender'; copper sulphate crystals were added to beer and Prussian Blue (potassium ferriferrocyanide) to tea to improve colour. Other metallic adulterants used were talc and French chalk (both natural forms of magnesium silicate), calcium sulphate, lead chromate and iron oxide.

Early numbers of the *Analyst*, a 'monthly journal devoted to the advancement of analytical chemistry', which began to be published in 1875 by the Society of Public Analysts, make interesting reading. Their reports give many examples of fraudulent or, perhaps, simply ignorant practices of food processors and sellers, and make it clear that the skills of the Public Analysts were sorely needed to ensure that the provisions of the recently enacted 'Food Act' were respected.

The cases reported in the *Analyst* covered a wide range of foodstuffs. Cheese seems to have caused particular problems. Some Canadian samples were, according to one report[2], found to have been coated with 'cheese spice', a mixture which contained 48% zinc sulphate, to prevent 'heaving and cracking' of the product. Other samples contained appreciable amounts of lead. The report also noted that

> *it is well known that the green mould in certain kinds of cheese has been imitated by the insertion of copper or brass skewers ... [and] instances have occurred in which preparations of arsenic have been added to cheese as a preservative.*

The use of copper salts to improve the colour of green vegetables was commented on in another article. There were several prosecutions for the offence of selling green peas treated in this way. The practice was not confined to the UK, as an article reprinted from a German chemical journal on the use of copper salts to preserve the green colour of vegetables showed[3]. Other illegal practices intended to improve the appearance of foods included the use of 'barium-ponceau lake' to give Cayenne pepper a 'brilliant fiery colour' and of several different metallic salts to improve the quality of spices[4].

In addition to such adulterations deliberately made to deceive purchasers, the *Analyst* also made note of different kinds of inadvertent contamination of foods. One was a problem which is still with us today, the transfer of metals from printing inks and colours on wrappings to the product they contain. A translation of an article from a French periodical gave information on the use of cheesecloth and paper coloured with lead chromate to wrap 'chocolate bonbons'[5]. Other articles provided information on the danger of transfer of lead to food from tinplate, pewter and solder, as well as from the rubber rings used to seal the lids of glass jars. Since many of these incidents occurred in the UK well after regulations had been introduced to prevent them, the scale of the problem is easily seen. Clearly, it was not just new laws that were required to protect the public from the sale of adulterated food, but vigorous policing to see that the laws, and their related regulations, were fully observed.

5.1.1 International and national legislation

According to the authoritative *Food Industries Manual*, 'food manufacture worldwide is unusual as an industrial process in that its products are heavily circumscribed by legislation of one kind or another'[6]. These range from national laws, such as the UK's *Food Safety Act* and the US *Nutrition Labeling and Education Act*, to international regulations such as the *EU Directive on the Official Control of Foodstuffs* and the directives and guidelines published by the Codex Alimentarius Commission of FAO/WHO.

It is important for anyone who is concerned with the maintenance of food quality, whether as a health professional, government official or food manufacturer, and who wishes to establish and maintain a secure customer base, to have some knowledge of this legislation. How detailed that knowledge has to be will depend on an individual's particular professional role in relation to food. What will be given here is an overview of the legislation in force in a number of countries, as well as of some of the more important international regulations. For those who need to know more and to follow up certain aspects of food legislation in detail, several excellent textbooks and manuals are available. Two of these that are pertinent to the UK, are Flowerdew's *Guide to the Food Regulations in the United Kingdom*[7], and Jukes' *Food Legislation of the UK*[8].

5.1.1.1 UK legislation on metals in food

The primary purpose of food law in the UK and in most other countries, according to Jukes, is to protect the health of consumers and to prevent them being defrauded[9]. This is achieved by a combination of primary legislation (the Acts of Parliament) and secondary legislative measures (the Regulations or Orders). The primary legislative powers are contained in the *Food Safety Act 1990*. This Act was intended to consolidate and modernise the long series of Food Acts under which food manufacture and sale in Britain had been governed since the 1860 *Act to Prevent Adulteration of Food and Drink*[10].

A good idea of the parliamentary manoeuvrings and developments over a period of nearly 150 years which eventually resulted in the passing of the 1990 Act can be obtained from the chronological list provided in Paulus's study of the sociology of food legislation in Britain[11]. While the Acts usually contain general prohibitions, the more detailed controls required to deal with the scientific and technical aspects of food production are provided by the Regulations. Some of this secondary legislation is issued as 'Orders' rather than 'Regulations'. They are published as Statutory Instruments (SIs) with a reference number indicating year of publication and number. Though it may seem that such a system provides a clear and simple pathway for anyone, such as a food manufacturer or trader, who needs to know whether his or her product meets the requirements of the law, it has been remarked by Flowerdew, 'the very nature and verbosity of the UK food regulations have made this rather difficult to do, and will inevitably mean that ambiguities will arise[12].

Regulations are issued by government ministers, normally after wide consultation with organisations which may be affected by them. The minister will also have consulted with expert advisory committees, such as the Food Advisory Committee (FAC), an amalgamation of the former Food Standards Committee (FSC) and the Food Additives and Contaminants Committee (FACC). Among the Reports which have been of service to ministers in drawing up several regulations related to metals in food are, for example, a series on arsenic, copper and other trace elements published by the FSC in the 1950s, and that on metals in canned foods produced by the FACC in the 1970s and 1980s.

Among Regulations of particular pertinence to the subject matter of this book are those that control the use of additives in food. In the UK this is achieved by the use of a 'positive list' system. Only those additives positively listed in the Regulations which perform a defined function (antioxidant, colour etc.) may be used in food. In some cases the approved use may be restricted to certain foods and quantities.

There are relatively few UK Regulations that refer specifically to metal contaminants in food. Government policy has been to monitor and, when necessary, agree a voluntary policy of eliminating contamination at source rather than establishing numerous regulations for individual metals in different foodstuffs. It is only when a serious and potentially widespread hazard is identified, that legislation is introduced. Under the *Food Safety Act* protection is provided to consumers by the provision that prosecutions can be brought on the basis that a product is injurious to health and that non-statutory limits, including internationally recommended limits, can be used as evidence in court[13].

Three metals, arsenic, lead, and tin, as well as a few radioisotopes, are included among regulations relating to contaminants in food under the *1990 Food Safety Act*. A maximum permitted level of 1 mg/kg is set for arsenic in 'unspecified foods'. This is increased or decreased in a range of 'specified foods', from, for example, 0.1 mg/kg in non-alcoholic beverages and 0.5 mk/kg in alcoholic beverages and fruit juices, to 4.0 mg/kg in chicory and 5.0 mg/kg in dried herbs and spices. In the case of lead the maximum permitted level in 'unspecified foods' is 1.0 mg/kg, with, in 'specified foods', for example, 5 mg/kg in dandelion coffee, and 10 mg/kg in curry powder, mustard, game, game pâté and shellfish, and lower amounts of 0.2 mg/kg in beer, non-alcoholic drinks and infants' food. Regulations for tin are simple, with a clear statement that 'no one may sell or import any food containing a level of tin exceeding 200 mg/kg'.

No statutory limits have been laid down for cadmium in foodstuffs. However, the FACC has recommended that monitoring of levels of the metal in the food supply should continue to be carried out. Where food is imported from countries in which levels of cadmium in foods are controlled, enforcement authorities in the UK are encouraged to have regard to these controls when examining such imported food[14].

Limits for copper in foods have been recommended by the FSC. These are general limits of 2 mg/kg in ready-to-drink beverages, and 20 mg/kg in other foods, with certain exceptions. These include alcoholic and concentrated soft drinks, 7 mg/kg; chicory, 30 mg/kg; cocoa powder, 70 mg/kg; tea, 150 mg/kg; tomato purée, paste and juice, 100 mg/kg; and tomato ketchup and sauce, 20 mg/kg[15]. The use of aluminium metal is permitted as an external coating on sugar confectionery for the decoration of cakes and pastries[16]. Aluminium silicate may also be used in a limited number of foods at certain maximum levels[17].

The Regulations relating to radioactive contamination were originally introduced in response to the Chernobyl nuclear power station accident of 1986 but have been retained to allow rapid implementation in case there is a further incident. They are in line with Regulations established by the European Commission. They set maximum permitted levels for foodstuffs and feeding stuffs (in Bq/kg) for radioactive strontium, iodine, plutonium and transplutonium elements and other nuclides.

Many more metals are included in Regulations related to drinking water, both bottled and natural mineral water. Maximum limits are set for a long list of metals, including aluminium (200 μg/l), iron (200 μg/l), manganese (50 μg/l), copper (3000 μg/l), zinc (5000 μg/l), arsenic (50 μg/l), cadmium (5 μg/l), chromium (50 μg/l), mercury (1 μg/l), nickel (50 μg/l), lead (50 μg/l), antimony (10 μg/l), selenium (10 μg/l), boron (2000 μg/l) and barium (1000 μg/l). In Regulations for natural mineral waters it is stated, under the heading 'Toxic substances', that 'the following maximum are specified and the water must not contain any other substances in amounts which make it unwholesome: mercury 1 g/l; cadmium 5 g/l; antimony, selenium, lead 10 g/l; arsenic, cyanide, chromium, nickel 50 g/l'.

5.1.1.2 US legislation on metals in food

There is a distinct resemblance between the food laws of the US and of the UK that can be traced back to some extent to common roots[18]. They both had their origins in the late nineteenth century and were intended to ensure that consumers had access to clean and wholesome food in the marketplace and were protected against adulteration and fraud. In 1906 the first comprehensive federal food law, the *Food and Drug Act*, was passed by the US Congress and signed by President Theodore Roosevelt. The Act prohibited interstate commerce in misbranded and adulterated foods, drinks and drugs. Over the past century, the Act has been modified and amended in a variety of ways, to improve its effectiveness, to cover unforeseen loopholes and to allow for developments in technology and trade. Its name was changed and its scope extended with the passing of the *Federal Food, Drugs and Cosmetic (FDC) Act* in 1938[19].

The current *Federal Food, Drug and Cosmetic Act (as amended)* remains the basic food law of the US. Like the UK *Food Safety Act 1990*, the *FDC* Act is a single enabling law which allows for the publication of detailed statutes or implementing Regulations to ensure that the provisions of the law are properly carried out. Regulations are promulgated by the Food and Drug Administration (FDA) and are published in the Code of Federal Regulations. These are given Titles, such as *Title 21: Food Composition Regulations* (21 CFR).

It is impossible to make a neat summary of current US regulations and standards for metals in food. As in the UK, information has to be sought under several different headings in official documents, such as compositional standards for different foodstuffs and animal feed and for drinking water. Information on food additives, including metallic compounds, which have been classified as GRAS ('generally recognised as safe') and have been accepted by the Commissioner of the Food and Drug Administration as 'food grade' can be found in the *Food Chemicals Codex*[20]. These GRAS additives include, for example, aluminium calcium silicate (as an anti-caking agent in table salt, to a maximum of 2%), and cuprous iodide (as a source of iodine, in table salt to a maximum of 0.01%).

The *Codex* sets maximum limits for trace impurities, including metals, 'at levels consistent with safety and good manufacturing practice. The maximum limits for heavy metals shall be 40 parts per million, for lead ten parts per million, and for arsenic three parts per million, except in instances where higher levels cannot be avoided (under conditions of good manufacturing practice)'[21]. These are not, in fact, statutory requirements since the *Codex* does not have the force of law, but has quasi-legal status as establishing food grade quality. As well as being used in this way by US legislators, its specifications have also been adopted, under certain conditions, by the FACC in the UK as well as by authorities in Canada and certain other countries.

An important amendment to certain provisions of the *Federal Food, Drug and Cosmetic Act* is the *Nutrition Labeling and Education Act (NLEA)* of 1990. The NLEA was designed to assist consumers to maintain healthy dietary practices by requiring that food manufacturers provide extensive information of the nutritional composition of products. Information must be provided on 14 different nutrients, which include the metals calcium, iron and sodium. Thirty-four other nutrients, including nine additional metals, may be labelled voluntarily. It is expected that eventually the labelling of the voluntary nutrients will become mandatory[22]. The NLEA does not, of course, set standards for levels of these metallic elements in foods. It has, however, significant implications for food manufacturers and processors, by requiring them to provide analytical data on three metals (or, if they choose to include them on the label, on nine additional metals).

5.1.1.3 Legislation in Australia and elsewhere in the English-speaking world

Australia, a country approximately the size of the US, but with a population of some 18 000 000 people, is a federation of six separate states and two territories. Each state and territory has its own parliament and passes its own laws relating to internal affairs. There is also a federal parliament at Canberra, the national capital, responsible for external affairs as well as for a multiplicity of national and inter-state matters.

Health legislation, including that related to food, is the responsibility of the individual Australian state and territory parliaments. Although many of the laws and regulations relating to food had been broadly modelled on the UK 1875 *Sale of Food and Drugs Act*, when they were first passed, local interests and inter-state rivalry resulted in considerable diversity of detail between states. These differences proved to be an impediment to free trade in foodstuffs between states and resulted in increased costs to food producers and customers. It was widely recognised that this was an unsatisfactory situation and a number of attempts were made to rationalise food laws on a national basis. It was only, however, in 1975 that a conference of health ministers from each state and territory and the Commonwealth agreed that a uniform Food Act, suitable for adoption by each of the different governments, should be developed.

Several years of discussion and planning followed. Food laws of each of the Australian states and territories were considered, as well as legislation from the UK, the US, Canada, New Zealand and other countries. Particular attention was paid to the *Model Food Act* that had been proposed jointly by the WHO and FAO. By 1980 planning and development were completed and a proposed *Model Food Act* was adopted by the Health Ministers' Conference.

The Act, like the enabling legislation of the UK and the US, is relatively brief and contains very little in its text about food. What it does is to empower authorised officers and departments to take certain actions related to the purpose of the Act (to ensure that consumers have access to pure and wholesome food), such as making regulations prescribing standards for food. These Model Food Standard Regulations were subsequently prepared by the National Health and Medical Research Council (NHMRC) on behalf of the Health Ministers' Conference and, after some delays, were adopted by the individual states and territories. Some 20 years after the first steps were taken towards a uniform food law and standards, a further dramatic advance was made when a decision was made by Health Ministers of Australia as well as of New Zealand to adopt a new joint Food Standards Code for both countries. The Ministers have established the Australia New Zealand Food Authority (ANZFA), an independent expert statutory body, the role of which is to protect the health and safety of people in the two countries through the maintenance of a safe food supply[23].

ANZFA has issued a number of standards which are at present in draft form but, after consideration by appropriate bodies and, no doubt, modifications, will be adopted by the two countries. Two of these are of particular pertinence to the subject matter of this book. *Standard 1.4.1: Contaminants and Restricted Substances* sets maximum levels (MLs) of specified contaminants, including metals, in nominated food groups[24]. Only four metals, arsenic, cadmium, lead and tin, are included. The number of foods is also limited and includes only those about which there is particular concern. In the case of tin, for instance, an ML of 250 mg/kg is given only for 'all canned goods'. In the case of arsenic the MLs are 1 mg total arsenic/kg cereals and, for inorganic arsenic, 2 mg/kg of crustacea and fish and 1 mg/kg of molluscs and seaweed (edible kelp). Mercury MLs are restricted to seafoods, with a mean level of 1 mg/kg for certain specified fish (including marlin, tuna and all species of shark), and 0.5 mg/kg for other fish and fish products, as well as crustacea and molluscs. The method to be used for sampling marine foods for mercury and calculation of the mean level in a certain number of sample units is also prescribed. The proposed MLs for the other two metals, cadmium and lead, are summarised in Tables 5.1 and 5.2.

Table 5.1 Maximum permitted levels of cadmium in food (Australia and New Zealand).

Food	Cadmium levels (mg/kg)
Chocolate and cocoa products	0.5
Kidney (cattle, sheep, pig)	2.5
Liver (cattle, sheep, pig)	1.25
Meat (cattle, sheep, pig)	0.05
Molluscs	2.0
Peanuts	0.1
Rice	0.1
Vegetables (leafy)	0.1
Vegetables (root and tuber)	0.1

Australia New Zealand Food Authority (2000) *Draft Food Standards Code: Standard 1.4.1. Contaminants and Restricted Substances.*
http://www.anzfa.gov.au/draftfoodstandards/Chapter1/Part1.4/1.4.1.htm

Table 5.2 Maximum permitted levels of lead in food (Autralia and New Zealand).

Food	Lead levels (mg/kg)
Brassicas	0.3
Cereals, pulses, legumes	0.2
Offal (cattle, sheep, pig, poultry	0.5
Fish	0.5
Fruit	0.1
Infant formula	0.02
Meat (cattle, sheep, pig, poultry)	0.1
Molluscs	2.0
Vegetables (other)	0.1

Australia New Zealand Food Authority (2000) *Draft Standards Code: Standard 1.4.1. Contaminants and Restricted Substances.*
http://www.anzfa.gov.au/draftfoodstandards/Chapter1/Part1.4/1.4.1.htm

The second standard pertinent to metals in food in the ANZFA Draft Food Standards Code is *Standard 1.3.2: Vitamins and Minerals*[25]. This regulates the addition of vitamins and minerals to food and the claims that can be made concerning them. It lists the foods to which certain vitamins or minerals may be added and the maximum permitted amounts that may be used. The foods include a range of cereal products (biscuits, bread, breakfast cereals, flour), dairy products (milks, cheese, yoghurts, ice cream), edible oils and spreads, fruit juices and analogues (such as meat replacers made from vegetable protein, and others). Apart from calcium and magnesium, the only trace metals which are approved for addition are iron and zinc, and these are restricted to the cereal products as well as to certain analogue foods.

5.1.1.4 International standardisation and harmonisation of food laws

The history of the development of food laws in most technologically advanced nations is very similar to that of the UK and the USA. The change from home production of foods to dependence on shops and markets which followed industrial development and urban expansion led to the passing of many specific and, later, comprehensive food laws in Germany, France and other countries. The fundamental principle underlying all such legislation was the protection of the health of the consumer and the prevention of fraud. Though sharing a common foundation, national differences in outlook and diversity of economic interests led to the passing in the different countries of a great variety of laws and regulations governing the composition, labelling, handling and other aspects of food production and trade. A consequence of this was, inevitably, barriers to free trade between nations. Since the end of World War II there has been considerable international pressure for reduction in these barriers and for harmonisation of legislation and the development of internationally acceptable food standards. Slowly these barriers have been coming down, especially as a result of the efforts of the United Nations through two of its agencies in particular, the Food and Agriculture Organization (FAO) and the World Health Organization (WHO).

One of the most significant steps towards international harmonisation of food standards has been the development of the Codex Alimentarius. Its origins can be traced back to 1958 when FAO collaborated with the International Dairy Federation to produce a *Code of Principles on Milk and Milk Products*. In the same year the International Commission on Agricultural Industries developed the Codex Alimentarius Europeus (the European Food Code). In 1961 these separate activities were merged in a joint Food Standards Programme under the auspices of the FAO and WHO, and the Codex Alimentarius Commission was established, as an international body, with representatives from nearly all members of the United Nations[26]

The Codex Commission has made considerable strides in developing international food standards. In collaboration with member governments, international agencies and other bodies, and following a detailed and prolonged procedure involving many separate steps to allow adequate consultation with interested parties, a number of standards for a variety of foodstuffs have been proposed by the Commission. The guiding principle behind all of its proposals is the protection of the health of consumers and the ensuring of fair practices in the food trade[27].

Two types of standards have been developed: the so-called 'vertical' or commodity standards for individual foods or groups of foods, and 'horizontal' or general standards, such as the General Standard for Contaminants and Toxins in Food[28]. Examples relating to metal contamination of food which come under these two types of standards are the Codex maximum levels for total lead in orange juice of 0.3 mg/kg and 0.1 mg/kg in rapeseed oil. The Codex guideline for methylmercury sets a guideline of 0.5 mg/kg for all fish, with the exception of predatory fish for which the level is 1.0 mg/kg.

The views and proposals of the Codex Commission play a very important role in the preparation of food legislation by individual countries as well as by other bodies such as the European Commission. National food standards and other agencies draw on the Commission for information and advice. Its standards and guidelines are, in many cases, incorporated into legislation, as, for instance, in the Australian *Model Food Act* and *Draft Food Standards Code*. In addition many countries follow Codex recommendations in, for example, setting tolerance limits for toxic metals in foods and beverages.

5.1.1.5 European Community food regulations

When the European Economic Community was established under the Treaty of Rome in 1957, one of its principal aims was the facilitation of trade between its member states. In the case of legislation relating to food, in spite of the fundamental similarities between the laws in the different states, the intricacies of legislation in each of them made harmonisation between them difficult. To overcome this problem a programme was set up aimed at the development of legislation which would be generally acceptable within the Community.

The task was not easy, especially because of the number of different bodies that had to be involved: the government of each state, the European Parliament, the Council of Ministers and others. As a result of this and other difficulties, the development of harmonised food legislation is a slow process. Only a limited amount has successfully passed through the different steps involved and been promulgated by the Council. A great deal still remains in draft form. Nevertheless, though time-consuming, it is a worthwhile exercise and its eventual benefits to the food industry, not just in Europe but worldwide, will be considerable.

EU legislation applied to food usually takes two forms – the Regulation and the Directive[30]. Regulations normally apply to primary agricultural products under the Common Agricultural Policy (CAP) and come into force in all member states at the same time. Directives, intended to create a market free of internal barriers, deal with largely technical matters such as food standards. Directives have to be implemented by each member state before they become law.

There are similarities, as might be expected, between EU Regulations relating to metals in food and those of the Codex Alimentarius. A good example is the Commission decision relating to the measurement of mercury in fish, which sets an ML of 0.5 mg/kg in fish in general, with the exception of certain species, such as tuna and shark, for which the limit is 1.0 mg/kg[31].

A Community Directive of 1995 on food additives has established levels of trace elements they may contain. For instance, the specifications for chlorophyll (E140) are: arsenic < 3 mg/kg, cadmium and mercury < 1 mg/kg, and lead < 10 mg/kg[32]. There are also directives that list essential trace elements that may be used in infant formulae[33]. Another directive relating to metals deals with the composition of natural mineral waters[34]. Apart from these few items, and a number of others expected to be promulgated soon, including regulations on lead and cadmium levels in specific foods, there is as yet only a framework regulation on contaminants in food[35]. There are, however, numerous other regulations and directives on other aspects of food composition and quality not directly pertinent to the subject of this book. Details may be found in Appendix 4 of Jukes' monograph[36].

The UK, like other member states of the EU, has, since it joined the Union in 1973, been adopting EU legislation into its own laws. EU Directives related to metals in food, mentioned here, which have been incorporated into UK Food Standards Regulations include those on food additives (95/45), infant formulas (91/321), and natural mineral waters (80/777). A European Directive of a different, though closely related, nature on radioactive contamination of food (3954/87) was adopted in the UK in response to the Chernobyl nuclear accident in 1986. It can be expected that many more Directives will be produced by the European Union that will have significant implications for the food industry.

5.2 Codes of practice

In addition to the formal legislation relating to food standards, there are many codes of practice which do not have strictly legal status but provide a guide and can be considered as evidence by a court. Such codes set voluntary agreed standards and have been found to have certain advantages over definitive regulations. Their flexibility can allow a more rapid response to changing circumstances and permit operation outside the code, where this can be justified by changing circumstances.

There are many codes of practice which relate to the food industry, such as that relating to fortification of food. Codes have been published by various national and international bodies, including the FAO and WHO as well as the Codex Alimentarius Commission. In the UK, codes have been approved by LACOTS, the Local Authorities Coordinating Body on Food and Trading Standards. This body brings together the various government and local authority trading standards departments and environmental health departments to assist in the enforcement of legislation at a local level. Other codes are produced by national and international trading organisations and manufacturing associations. A list of those observed in the UK has been published by the Institute of Food Science and Technology[37].

References

1. *Leviticus* Chap.**12**: vv 7, 8: 'and the swine, because it parts the hoof and is cloven-footed but does not chew the cud, is unclean to you. Of their flesh you shall not eat, and their carcasses you shall not touch'. (*The Bible, Standard Revised Version*. Collins Fontana Books, London).
2. Allen, A.H. & Cox, F.H. (1897) Note on the presence of heavy metals in cheese. *Analyst*, **22**, 187–9.
3. Mayrhofer, N.T. (1891) Copper in preserved foods. *Chemische Zeitung*, **15**, 1054–5.
4. Kaiser, H. (1899) Presence of barium salts in Cayenne pepper. *Chemische Zeitung*, **23**, 496.
5. Wolfe, J. (1898) Rapid method of detecting lead chromate in papers used for wrapping eatables. *Annales de Chimie Analytique*, **2**, 105.
6. Ranken, M.D., Kill, R.C. & Baker, C. (1997) *Food Industries Manual*. Blackie/Chapman and Hall, London.
7. Flowerdew, D.W. (1990) *A Guide to the Food Regulations in the United Kingdom*, 4th edn. British Food Manufacturing Industries Research Association, Leatherhead, Surrey.
8. Jukes, D.J. (1997) *Food Legislation of the UK*. Butterworth/Heinemann, Oxford.
9. Jukes, D.J. (1997) *Food Legislation of the UK*. Butterworth/Heinemann, Oxford.
10. Ranken, M.D., Kill, R.C. & Baker, C. (1997) *Food Industries Manual*. Blackie/Chapman and Hall, London.

11. Paulus, I. (1974) *The Search for Pure Food*. Robinson, London.
12. Flowerdew, D.W. (1990) *A Guide to the Food Regulations in the United Kingdom*, 4th edn. British Food Manufacturing Industries Research Association, Leatherhead, Surrey.
13. Jukes, D.J. (1997) *Food Legislation of the UK*. Butterworth/Heinemann, Oxford.
14. Ministry of Agriculture, Fisheries and Food (1973) *Survey of Cadmium in Food: Working Party on the Monitoring of Foodstuffs for Heavy Metals, Fourth Report*. HMSO, London.
15. Ministry of Agriculture, Fisheries and Food (1958) *Food Standards Committee Report on Copper. Revised Recommendations for Limits for Copper Content of Foods*. HMSO, London.
16. Ministry of Agriculture, Fisheries and Food (1995) *The Colours in Food Regulations 1995*. S.I. (1995) No. 3124. HMSO, London.
17. Ministry of Agriculture, Fisheries and Food (1995) *The Miscellaneous Food Additives Regulations 1995*. S.I. (1995) No. 3187. HMSO, London.
18. Schmidt, A.M. (1976) The development of food legislation in the U.K. In: *Food Quality and Safety: A Century of Progress. Proceedings of the Symposium Celebrating the Centenary of the Sale of Food and Drugs Act 1875*, London, October 1975, p. 4. MAFF/HMSO, London.
19. US Department of Health, Education and Welfare (1974) *Milestones in the U.S. Food and Drugs Law History*. DHEW Publication No. (FDA) 75-1005. Food and Drug Administration, Rockville, Md.
20. National Research Council Committee on Food Safety (1990) *Food Chemicals Codex*. National Academy of Sciences, Washington, DC.
21. Ranken, M.D., Kill, R.C. & Baker, C. (1997) *Food Industries Manual*. Blackie/Chapman and Hall, London.
22. Barnes, K.W. (1998) A streamlined approach to the determination of trace elements in foods. *Atomic Spectroscopy*, **19**, 31–9.
23. *ANZFA News* No. 25, December 2000/January 2001, p. 8. ANZFA, Canberra, Australia.
24. Australia New Zealand Food Authority (2000) *Draft Food Standards Code*. http://www.anzfa.gov.au/draftfoodstandards/Chapter1/Part1.4/1.4.1.htm
25. Australia New Zealand Food Authority (2000) *Draft Food Standards Code*. http://www.anzfa.gov.au/draftfoodstandards/chapter1/part1.3/1.3.2.htm
26. Kermode, G. (1971) Food standards for the world: the Codex Alimentarius. In: *Food and the Law* (ed. R.L. Joseph), pp. 49–65. Institute of Food Science and Technology (Irish Branch), Dublin.
27. Codex Alimentarius Commission (1997) *Procedural Manual*, 10th edn., Section 1, pp. 3–17. FAO/WHO, Rome.
28. Codex Alimentarius (1998) *General Standards for Contaminants and Toxins in Food*. Food Safety Information Centre RIKILT-DLO, Wageningen, The Netherlands.
29. Codex Alementarius (1991) *Guidelines Levels for Methylmercury in Fish*, CAC-GL 7. Food and Agriculture Organization, Rome.
30. Jukes, D.J. (1997) *Food Legislation of the UK*. Butterworth/Heinemann, Oxford.
31. European Commission (1993) *Decision 93/351/EEC of 19 May 1993 on Methods of Analysis Sampling and Acceptable Levels for Mercury in Fish Products*, L144:23–24. European Commission, Brussels.
32. European Commission (1995) *Directive 95/45/EC of 26 July 1995 on Specific Criteria for the Purity of Food Colours*, L226:1–45. European Commission, Brussels.
33. European Council (1991) *Directive on Infant Formulae and Follow-up Formulae*, 91/321/EEC May 14, 1991. European Commission, Brussels.
34. EEC Directive (1980) *Natural Mineral Waters*, 80/777.
35. European Council (1993) *Community Procedures for Contaminants in Food*, EC315/93, L37:1–3. European Commission, Brussels.
36. Jukes, D.J. (1997) *Food Legislation of The UK*. Butterworth/Heinemann, Oxford.
37. Institute of Food Science and Technology (1993) *Listing of Codes of Practice Applicable to Food*. IFST, London.

Part II
The Individual Metals

Chapter 6
The persistent contaminants: lead, mercury, cadmium

The three representative metals, lead, mercury and cadmium, are, along with arsenic, the commonest and most widely distributed of environmental metal poisons[1]. They are also among the most troublesome of the metal contaminants encountered by the food industry. Each of them has been responsible for large-scale incidents of poisoning and, in spite of the many steps that have been taken by government authorities, they continue to be potential threats to the safety of consumers in certain circumstances. They are specifically included in the official food safety regulations in many countries and between them have probably been responsible for more official documentation than any other food contaminant. For many consumers these three are the elements that first come to mind when metal poisoning is mentioned.

6.1 Lead

When Georgius Agricola described lead as a 'pestilential and noxious metal' in the sixteenth century he was stating a fact that had long been recognised[2]. Similar views about lead had been expressed by others for more than a thousand years[3]. As early as 400 BC, the Greek physician Hippocrates described a disease known as *saturnism* (after the Greek name for the planet Saturn, the ancient symbol for lead), with symptoms ranging from colic to delirium and paralysis, which affected men who worked with lead. Four centuries later the Roman scholar Pliny noted that shipbuilders who used white lead as a preservative on the hulls of vessels suffered from an illness known as *plumbism* (after the Latin name for lead, *plumbum*).

Over the following centuries others also commented on problems associated with the metal. It was not, however, until the nineteenth century, when detailed records began to be kept of the hazards connected with certain industrial practices, that the full extent of lead poisoning began to be acknowledged. The reports of, among others, Ramazzini in Italy, Thackrath in England and Tanquil des Planches in France showed that nineteenth-century Europe desperately needed laws to control industrial practices which exposed workers to hazards from lead and other poisonous substances.

It was only reluctantly that governments and industrialists accepted the veracity of the reports of these pioneers of occupational health, and began to take steps to remedy the dreadful conditions that had been exposed. Legislation was introduced in several countries to control the use of lead in industries such as pottery making, printing ink production and metal fabrication. In the UK the *Factories (Prevention of Lead Poisoning) Act* of 1883 was followed by regulations restricting the consumption of food and drink in sections of pottery works where lead glazes were employed. These early pieces of legislation were followed in the next century by long series of regulations devoted to prevention of lead poisoning. Similar steps were taken in other countries with the result that lead poisoning on the scale encountered in the eighteenth and nineteenth centuries, is today 'a disease for the history texts'[4].

6.1.1 *Chemical and physical properties of lead*

Lead is element number 82, in Group IVB of the periodic table, with an atomic mass of 207.19. It is one of the heavier of the elements with a density of 11.4. It is a soft, bendable and easily fused metal which can be beaten flat with a hammer and cut with a knife. When first cut its surface is bright and mirror-like, but on exposure to moist air soon tarnishes to a dull grey colour as a surface layer of basic lead carbonate forms. A similar protective film forms when the metal comes into contact with 'hard' water containing carbonates. Lead melts at a relatively low temperature of 327°C. Its boiling point is 1725°C. It is a poor conductor of heat and electricity. Oxidation states of lead are 0, +2 and +4. In organic compounds it is usually in state +2. Most salts of Pb(II), as well as lead oxides and sulphides, are only slightly soluble in water, with the exception of lead acetate, lead chlorate and, sparingly, lead chloride.

Lead forms a number of compounds of technological importance. Litharge, a yellow powdery monoxide (PbO), is formed when the metal is heated in air. On further heating the monoxide is raised to a higher oxidation level and Pb_3O_4 (red lead) is formed. White lead or basic lead carbonate, $Pb_3(OH)_2(CO_3)_2$, is prepared commercially by the action of air, carbon dioxide and acetic acid vapour on metallic lead. Another lead pigment is the yellow chromate ($PbCrO_4$). These compounds, as well as lead linoleate and salts of other organic acids, were formerly widely used in the paint industry. Today their use is restricted because of their toxic nature.

Some of the organic compounds of lead are of considerable economic importance. Among the best known are tetraethyl- and tetramethyl-lead, $Pb(C_2H_5)_4$ and $Pb(CH_3)_4$. Both are liquids at normal temperatures and are used as 'anti-knock' or antipercussive additives in petroleum motor fuel. Lead has the ability to form alloys with other metals. Solder, an alloy of lead and tin and other metals, is of considerable economic importance. Formerly pewter, another lead–tin alloy, was made with 20% or more lead, but modern pewter contains little, if any, of the metal.

6.1.2 *Production and uses*

Lead is found, at least in small amounts, almost everywhere in the world. Soils normally contain between 2 and 200 mg/kg. It is seldom found in the native, metallic state. Its most common ores are galena (PbS), cerussite ($PbCO_3$) and anglesite ($PbSO_4$). It is usually associated with other metals, including zinc, cadmium, iron and silver. Often in the past it was the presence of silver that accounted for the extraction of lead from otherwise economically non-viable lead ores.

Economically workable ore bodies occur in many parts of the world, with major production in the US, Russia, Australia, Canada, Peru, Mexico, China, the former Yugoslavia and Bulgaria. Lead mining was once a major industry in Britain, but the mines have now closed. Lead, however, is still produced in the country, both from imported ores and by recycling of lead scrap. World production of primary lead is now about 2.5 million tonnes per annum and is expected to remain stable at around this figure. Secondary lead production, from scrap, is estimated to increase to about 3 million tonnes per annum over the next decade[5].

The ease with which lead can be extracted from its ores probably accounted for its early exploitation by humans. When galena is roasted in a charcoal fire it is partly oxidised at a low temperature. The oxide then reacts with the unconverted lead sulphide and proceeds, more or less by itself, to complete the conversion to the metallic state. This is the system on which the original ore hearth furnace was based, a method used 2000 years ago by Romans and Greeks, and others before them. Today this technique has been replaced by the blast furnace in which reduction is achieved by carbon monoxide or producer gas made from coke.

Uses of lead fall into two main groups: as metal and in chemical compounds. Because of the ease with which it can be fabricated, as well as its high resistance to corrosion, the metal is used to make pipes, tanks and other containers for corrosive liquids. Lead water pipes have been used since Roman times though much less today than formerly. Lead sheets are still used to roof major buildings, such as churches, though far less commonly than in former times. A major modern use of metallic lead is in the manufacture of lead/acid batteries for automobiles, electric vehicles, emergency lighting plants and similar equipment. It is estimated that by 2006 the battery industry will account for more than 70% of the world lead supply[6].

An interesting, though considerably smaller but still significant, use of lead is in the manufacture of ammunition for firearms. It is reported that up to 12 000 tonnes of the metal are used annually for this purpose in France alone[7]. In the 1970s it was estimated that a staggering 15 billion individual lead shot were fired each season over the duck and geese hunting grounds of North America[8].

The effect on the environment of such quantities of lead, much of which falls to the ground and is subsequently eaten by game birds, is considerable. A Yugoslav study found that 41% of mallard ducks killed during the 1979–82 hunting season had more than 2 mg/kg of lead in their tissues, some from embedded shot, much from shot which had been consumed and ended up in the gizzards of the birds. A more recent study in Greenland found that shot was a major source of lead intake in the country[9].

Lead solders of a variety of compositions are used in a number of industries, especially in plumbing and electrical work. Solders were once used to seal all food cans, but this application has now ceased in many countries. They are still used extensively in the manufacture of automobile bodies as well as in wheel bearings. A former important user of lead was the printing industry. Lead alloys of various kinds were used to make typeface. However, since the print was used over and over again as typeface was melted down and recast, the actual quantity of lead involved was limited. Today computer technology and other forms of printing have seen the relegation of lead type to a limited number of traditional printing works. The use of lead compounds, both organic and inorganic, is extensive. We have seen something above of the lead-containing pigments used in printing and painting. Because of the well-recognised dangers to health, especially of children, that can be caused by white lead-based paints, their use is now restricted largely to priming work. Red lead paints are still used in large quantities as a rust inhibitor for iron and steel.

A major use of lead salts is in glazing of ceramics. Lead salts are also used in the manufacture of flint and crystal glass, as well as enamels. Other uses are in the vulcanising of rubber, as fungicides and insecticides in agriculture and as antifouling compounds on the hulls of ships. But, though overall use of the metal and its salts continues to increase worldwide, many of its older applications have been abandoned, or at least restricted, because of their danger to the health of consumers.

One of the most significant changes in the industrial uses of lead has been in the manufacture of motor fuel. In the early 1920s it was discovered that the problem of 'knocking', or premature ignition before sparking of the mixture of air and petrol vapour in the cylinder of an internal combustion engine, could be overcome by the addition of tetraethyl-lead (TEL) and other alkyl-lead compounds. Manufacturers of automobile fuel began to add 'anti-knock' to their products. The use of lead additives to increase octane rating[10] continued for the following half-century, in spite of some concerns about the environmental and health dangers resulting from the release of lead in exhaust fumes of petrol engines.

By the early 1970s the amount of lead used to make leaded petrol had increased enormously. In the UK alone approximately 40 000 tonnes of the metal were being converted annually into TEL and related compounds for use as fuel additives. Most of this was eventually released to the environment as exhaust emissions. In Australia approximately 8000 tonnes of lead were used in the same way in the late 1970s. It was estimated that about 5600 tonnes of this lead were emitted in exhaust fumes[11]. After another decade, during which restrictions had been introduced to limit the sale of leaded petrol and its use had been reduced nationally by some 45%, nearly 5000 tonnes of lead were still being added to motor fuels in that country[12].

Today legislation is in place in the UK, Australia and most other countries to reduce, and eventually totally eliminate, the use of leaded motor fuels. Unfortunately, though the expected benefits to the environment and to human health are considerable, the consequences of more than half a century of the use of leaded petrol has meant that many urban environments carry a burden of lead which can still cause problems[13].

6.1.3 Lead in the human body

Lead is found in every organ and tissue of the human body. The total amount varies with age, occupation, and environment. It has been estimated that western 'Reference Man', a 70 kg male who has not been exposed to excessive amounts of environmental lead, contains between 100 and 400 mg of the metal, with an average of 120 mg, or 1.7 µg/g of tissue[14]. We probably begin life with a small body store of lead. Transfer across the placenta from the mother to the fetus has been shown to occur and the body's lead store increases with age, as exposure to environmental lead is prolonged.

After birth we absorb lead by ingestion from food and drink and, to some extent, by absorption through our lungs. In adults about 10% of the lead ingested is absorbed from the gastrointestinal tract. In children percentage absorption may be as much as 50%. Several dietary factors can affect absorption. Uptake is increased by a low calcium status, as well as by iron deficiency. A diet high in carbohydrates and low in protein can have a similar effect. This can have important consequences, for example for those living on a poor diet in polluted areas. Children from low socio-economic backgrounds, whose diet is low in calcium, have been found to have a significantly increased risk of a high absorption of lead, compared to those with an adequate diet[15].

Once absorbed into the blood, the lead is transported around the body by being attached to blood cells and constituents of plasma[16]. About 5% is placed in an exchangeable compartment (soft tissues and blood) with the remaining 95% sequestered in bone, as insoluble phosphates. Bone lead has a half-life of 20–30 years, but, under some conditions of stress, may be released into the blood at any time.

About 90% of the lead ingested by adults is lost in faeces. Of the 10% absorbed from the gastrointestinal tract, about three-quarters will eventually leave the body in urine. Other excretory pathways are gastrointestinal secretions, hair, nails and sweat. Absorbed lead may also appear in human milk. Under conditions of gross environmental contamination, this could be a hazard for breastfed infants. Usually, however, lead levels in the milk are low, between 2 and 5 µg/l, well within acceptable ranges of intake[17]. A report of levels of lead up to 0.12 mg/l in breast milk of some Japanese nursing mothers can most probably be discounted as due to analytical inadequacies[18]. However, a recent study of the consequences of environmental pollution in Russia suggests that high levels of lead in human milk are not uncommon where long-term neglect of industrial hygiene has occurred[19].

There is a similarity between the metabolism of lead and of calcium. Both metals are found in bone crystal structure. It was once believed that, with time, lead became buried deeper and deeper within the bone structure and was held there permanently. However, there is now evidence that such 'burying' takes place only after a very long time and that a potentially dangerous pool of available lead persists for many months after high levels of exposure. It is possible, as noted above, that even 'fixed' bone lead may be mobilised under certain conditions, such as shock and illness.

Levels of lead determined in ancient skeletons uncovered in archaeological digs suggest that modern humans may accumulate more of the metal than did their ancestors. While lead in bones of 258 twentieth-century Americans averaged 100 mg/kg, with a range of 7.5–195 mg/kg, levels of less than 5 mg/kg were found in the skeletons of Peruvian Indians dating from about AD 1200. Levels of between 5 and 10 mg/kg were found in the remains of third-century Poles, in contrast to levels of about 12 mg/kg in the bones of modern residents of Poland[20].

Lead levels in teeth can also be used to study accumulation of the metal with time. In addition, measurement of the ratios of naturally occurring lead isotopes in the teeth can provide information about the source of the metal. A study of Norwegian teeth from medieval times found Pb^{206}/Pb^{204} ratios between 18.8 and 18.2, in comparison with present day ratios of between 18.0 and 17.6, indicating the impact of industrialisation and traffic[21].

Lead in hair has been used as an indicator of exposure to external contamination as well as to assess dietary uptake. In a US study, levels of lead in nineteenth-century children's 'anitique' hair that had been kept in lockets were found to be on average 164 mg/kg compared to 16 mg/kg in modern hair. It was concluded that the difference was due to changes in the American way of life, such as reduced use of poorly glazed earthenware vessels for food and beverages[22]. However, as others have shown, lead levels in hair can be affected by several external and internal factors besides diet, such as sex and age, hair colour, shampoos and other cosmetic treatments, as well as the part of scalp from which hair is clipped, length of sample and method of preparation for analysis[23]. Consequently hair analysis is generally considered to be of little use as a diagnostic aid, except in the case of gross lead poisoning[24].

6.1.4 *Biological effects of lead*

Symptoms of acute inorganic lead toxicity are relatively easily recognised, but not those of chronic poisoning which may occur after accumulation of lead in the body by small increments over a long period of time. Lead primarily affects four organ systems: haemopoietic, nervous, gastrointestinal and renal. Acute lead poisoning usually manifests itself in gastrointestinal effects. Anorexia, dyspepsia and constipation may be followed by an attack of colic with intense paroxysmal abdominal pain. This is the 'dry gripes' of the 'Devonshire colic' once known to cider makers who used lead-lined vessels in their fermentation processes. At times the pain caused can be so severe that it may be taken for acute appendicitis. There is also frequently marked general weakness, fatigue and malaise[25]. Lead encephalopathy is rare in adults, but, at least until relatively recently, was more common in children. It is observed especially in children with pica (eating of soil and other non-food materials) in some urban environments.

Today gross symptoms of lead poisoning are seldom met with except in those exposed to certain occupational hazards. Attention is now centred largely on sub-clinical levels of poisoning. One of the effects of chronic low-level poisoning is interference with the pathways of haem biosynthesis. Mild anaemia is still sometimes seen in occupationally exposed workers, but the characteristic 'leadworker's pallor', first described in 1831 by the French physician Laennec, is seldom seen today. The erythroid bone marrow where haemoglobin synthesis is carried out is the 'critical organ' targeted by lead[26]. The metal inhibits the action of several enzymes involved in haem biosynthesis, among them delta-aminolaevulinic acid dehydrase (ALA-D). This catalyses the formation of porphobilinogen, a precursor of haemoglobin, from *o*-aminolevulinic acid (ALA). Measurement of the activity of ALA-D, as indicated by concentration of ALA in the blood, provides a sensitive clinical test of lead poisoning. The test gives an indication of levels of lead uptake long before any obvious physical symptoms appear and is of value in detecting incipient lead poisoning.

Peripheral neuropathy was formerly frequently observed among workers exposed to lead. Lead palsy resulting in weakness, and in foot or wrist drop (painter's palsy) in which paralysis of the muscles of hands or feet occurs, though now rare, is still seen in severely poisoned patients[27]. High intakes of lead are also responsible for autonomic neuropathy resulting in abdominal colic and pain[28]. Intakes of even small amounts of lead over a long period can result in chronic and irreversible kidney disease. The legacy of long-term exposure to lead, ingested in fragments of white lead paint by children in Queensland, Australia, has been the high level of incidence of nephritis, in many cases fatal, formerly seen in adults in the State[29]. Similar conditions continue to be observed today in some lead workers[30].

The effects of exposure to even low levels of lead on the behaviour and intelligence of children has been causing increasing concern in recent years. There is strong evidence that significant damage can be caused to their neurophysiological development[31]. A US study found that children with a high body burden of lead (as shown by levels in the dentine of their teeth) had IQs about four points lower than children with low dentine lead[32]. Groups in other countries, including Australia[33] and the UK[34], have shown similar relationships between body lead burden and performance in intelligence. It has been estimated that doubling of the lead body burden of children is associated with a deficit of 1–2 points on the IQ scale[35]. Though some doubts have been expressed about the validity of such conclusions[36], growing concern about the risks to children posed by environmental lead has resulted in steps being taken in several countries, including the US[37], the UK[38] and Australia[39], to reduce the levels of intake of the metal by children and others. As a result of these moves there have been remarkable decreases in the level of dietary intake of lead in many countries in recent years.

In contrast to inorganic lead, organic lead primarily affects the central nervous system and has little effect on blood lead concentrations. TEL, for example, is an acutely toxic compound, far more dangerous than metallic lead or its inorganic compounds. It is a volatile liquid which is fat-soluble and is readily absorbed through the lungs as well as by the skin and the gastrointestinal tract. After absorption TEL is distributed to various tissues, particularly the brain, where it decomposes to triethyl-lead and small amounts of inorganic lead. The earliest symptom of alkyl-lead intoxication is insomnia. Poisoning is usually acute, developing into toxic psychosis, and may result in death. These effects can be brought about by 'petrol-sniffing', a serious problem among some socially deprived young people[40].

6.1.5 *Lead in food and beverages*

A remarkable achievement of the food industry and of government authorities in a number of countries has been the marked reduction in levels of lead in certain foods and in the overall diet, especially of children, since the early 1970s. Though the dangers of lead ingestion had long been recognised and legislation was introduced in many countries to protect consumers against them, it was really only since the risk to children of even moderate intakes was recognised that the problem began to be tackled effectively. The success of many of the measures that have been taken to reduce lead levels in food suggests that within the foreseeable future, childhood lead poisoning will, except in exceptional cases, be 'a disease for the history texts'[41].

Lead, however, will always remain a normal ingredient of our diet. It is present in all foods and beverages, primarily as a natural component, but also as an accidental additive picked up during processing. Levels of the metal in foods throughout the world, except in situations where there are industrial or other types of environmental contamination, are surprisingly stable and uniform, ranging from about 0.01 to about 0.25 mg/kg, as is shown in Table 6.1.

Table 6.1 Lead content of foods in different countries (mg/kg).

Food	Canada	Spain	Japan	UK
Cereals	0.012–0.078	0.01–0.065	0.092 (mean)	0.01–0.04
Meat/fish	0.011–0.121	0.005–0.065	0.186 (mean)	<0.01–0.10
Dairy products	0.001–0.082	<0.005–0.015	0.032 (mean)	<0.01–0.02
Vegetables	0.006–0.254	0.005–0.045	0.090–0.257	<0.01–0.02

Dabeka, R.W., McKenzie, A.D. & Lacroix, G.M.A. (1987) Dietary intake of lead, cadmium, arsenic and fluoride by Canadian adults: a 24 hour duplicate diet study. *Food Additives and Contaminants*, **4**, 89–102.

Urieta, I., Jalon, M. & Eguileor, I. (1996) Food surveillance in the Basque Country (Spain). II. Estimation of the dietary intake of organochlorine pesticides, heavy metals, arsenic, aflatoxin M, iron and zinc through the Total Diet Study, 1990/91. *Food Additives and Contaminants*, **13**, 29–52.

Teraoka, H., Morii, F. & Kobayashi, J. (1981) The concentration of 24 elements in foodstuffs and estimation of their daily intake. *Eiyo to Shokuryo*, **34**, 221–39.

Ministry of Agriculture, Fisheries, and Food (1998) *Lead, arsenic and other metals in food. Food Surveillance Paper No. 52*. The Stationery Office, London.

6.1.5.1 *Lead in meat and offal*

Other foods, apart from those listed in the Table 6.1, can contain significantly high levels of lead. In a Swedish study it was noted that while lead levels were 0.008 mg/kg in beef and 0.12 mg/kg in pork, they were 0.30 mg/kg in cattle kidney and 0.52 in pig kidney[42]. In the UK lead in carcass meat was 0.01–0.03 mg/kg and 0.04–0.37 in offal[43]. High levels in cattle offal have also been reported in the Netherlands[44].

6.1.5.2 *Lead in canned foods*

Foods packed in cans with lead-soldered side seams contain higher levels of lead than do fresh or frozen foods. A study in Israel found that some fruit juices on sale in stores contained 2 mg lead/kg[45]. A similar investigation in Italy reported levels of up to 20 mg/kg of the metal in canned vegetables[46]. In Canada, in the late 1980s, lead levels in infant formulas in lead-soldered cans were found to be on average 46.2 ng/g, compared to 1.7 ng/g in lead-free cans[47]. As we have seen earlier, because of dangers posed to health by a high lead intake, many countries, including the US, have restricted, or in some cases totally stopped (especially for infant foods), the use of soldered cans[48]. In many countries old-style food containers have been largely replaced by seamless ('two piece') and electrowelded cans. As a consequence there has been a considerable drop in lead levels in canned foods, with a significant effect on dietary lead exposure. In the case of infant foods, for instance, average lead levels were reported to have fallen in the US from 0.15 to 0.013 mg/kg, and in infant juices from 0.30 to 0.011 mg/kg between the early 1970s and the early 1980s[49].

6.1.5.3 Lead in wines

Another source of potentially high dietary lead intake, though for an obviously limited group of consumers, are wines in bottles with lead seals. A study conducted by the US Bureau of Alcohol, Tobacco and Firearms (BATF) in 1990 found that some wines, both domestic and imported, had elevated levels of lead. This, it was concluded, was at least partly due to the use of tin-coated lead foil caps on the bottles. As a consequence, a prohibition on the use of such caps was proposed by the FDA[50]. Similar steps have been taken in other countries. In Europe the EC has banned the use of lead foils containing more than 1% lead for this purpose[51].

An investigation in the UK[52] found that unpoured wines, both red and white, contained less than 250 μg lead/litre (range 27–240 μg/litre of the 52 samples tested) whether they were lead-capped or not. However, when lead-capped wines were poured the lead content increased, in several cases dramatically. Twenty-five samples tested showed an average increase of 3.6 $\pm$ 2.5 μg lead/litre. Another five samples increased by 35, 50, 410, 555 and 1890 μg/litre (nearly twice the UK statutory limit of 1 mg/litre). The increases were attributed to deposition of corrosion products from the lead cap on the bottle glass. Though use of tinned-lead foil and similar lead-containing capsules has been decreasing worldwide in recent years, the practice continues in some traditional wineries and there are many vintage wines in cellars that still have the traditional cap. MAFF (now DEFRA) has given the advice that whenever wines are poured, the bottle top should be carefully wiped to remove all possible corrosion products[53]. This would seem to be a wise precaution for wine lovers in the light of the finding that blood lead concentrations in middle-aged men in the UK are positively associated with alcohol consumption[54].

6.1.5.4 Lead in home-grown vegetables

A somewhat surprising source of high dietary lead intake is the urban garden and smallholding. Home-grown vegetables, especially in soil that has been industrially contaminated, have been shown to contain levels of the metal which exceed the statutory limit[55]. It is believed that the most important source of contamination of agricultural soil in the UK is atmospheric deposition from lead in the exhaust of petrol engines. In 1986 it was estimated that 1500 tonnes of lead were deposited in this way, as opposed to 100 tones from sewage sludge[56].

A study conducted on behalf of MAFF, to assess whether consumers of home-grown vegetables could be at risk of excessive lead intake[57], found that in certain urban areas in Britain, soil lead concentrations ranged from 27 to 1676 mg/kg dry weight (mean 266 mg/kg dw) compared to a mean concentration of 74 mg/kg dw in agricultural soil in England and Wales. Analytical results indicated a considerable range of lead concentrations in vegetables grown in urban allotments and gardens. These were < 0.02–1.7 mg/kg fresh weight for spinach, 0.06–1.5 mg/kg for cabbage and 0.06–1.0 mg/kg for broccoli. Though only 1% of all the vegetables analysed had levels above the statutory limit of 1 mg/kg, it was concluded that individuals whose consumption of vegetables was solely derived from home-grown produce and lived in areas with the highest soil lead levels (such as the suburbs of Hammersmith and Richmond in London) could have elevated lead intakes.

6.1.5.5 *Lead in water*

Most natural waters contain about 5 μg of lead per litre. WHO recommends a maximum limit of 50 μg/litre, which is the same as the EU maximum allowable concentration. Normally municipal supplies will contain well below this limit, but there can be exceptions. A survey carried out in Liverpool found that almost half the samples taken from public water supplies had levels above the WHO maximum. Similar findings have been reported in other cities where old lead domestic plumbing is still in place[58]. In a study of domestic water in Birmingham, average lead concentrations were found to be 28 μg/litre. Of these, 45% had concentrations greater than 20 μg/litre, with 12% exceeding the WHO limit[59].

Lead contamination in domestic water may be due to pollution by industrial and municipal waste. However, it is most often caused by the use of lead in the plumbing system. The problem is most serious in areas where the water is 'soft', with a low pH. Such water is plumbosolvent and can dissolve large amounts of metal from the system. On the other hand, hard water of high pH and containing dissolved salts of calcium and magnesium forms 'scale' on the inner surface of the pipe and this prevents solution of the metal. In parts of the UK, such as the Scottish Highlands where domestic water originates in acidic peat moorland, plumbosolvency can be a serious problem. In Glasgow, for instance, water from taps in some older houses was found to contain more than 100 μg lead/litre, twice the WHO maximum and ten times the average concentration in most households in England[60].

6.1.6 *Adventitious sources of dietary lead*

Ordinary food is not always the main source of dietary lead. As has been indicated earlier, lead may also be taken up from a number of adventitious, non-food sources. Some of these, such as lead solder in food cans and lead plumbing in domestic water systems, can, as we have seen, be responsible for serious health problems. A number of other non-food sources which continue to be of concern, in spite of recent successes in reducing overall lead intakes, have been discussed in a report published by the Centre for Food Safety and Applied Nutrition of the USFDA[61]. Though some of these have already been briefly mentioned, it will be useful to consider them in more detail here because of their persistence as a source of lead contamination of the diet.

6.1.6.1 *Lead in alcoholic beverages*

A survey of dietary lead levels in the Basque region of Spain in 1991 found that alcoholic beverages, particularly wine, contribute significantly to dietary lead intake[62].This was attributed to the high wine consumption (173.8 g/day) of those surveyed and to the fact that the wines had considerably higher lead levels than other beverages, though, it was pointed out, concentrations in Spanish wines are similar to those found in other countries. Levels of 50–100 µg/litre have been reported in wines sold in Italy, West Germany and the UK[63].

In the USFDA Report[64] it was noted that wine can represent 'sizeable and controllable contributions to the total body burden of lead'. The finding that many wines, both domestic and foreign, had elevated levels of lead prompted the FDA to advise the Bureau of Alcohol, Tobacco and Firearms, the body with primary regulatory authority in the US for wine, that it would be ready to support action to remove wines containing more than 300 µg lead/litre from the marketplace.

Lead levels in wine and other alcoholic beverages can be increased by storage in unsuitable containers. Leaded crystalware decanters have been shown to leach lead into wine, including port and sherry, with time[65]. The USFDA has issued advice on the use of crystal glass vessels for long-term storage of alcoholic beverages[66].

Another source of lead in alcoholic beverages which is sometimes encountered is poorly glazed ceramic vessels. A Spanish study found that several cases of lead poisoning could be traced to the use of vitrified and non-vitrified earthenware vessels to store wines[67]. Similar cases have been reported from Sweden[68], Britain[69] and the former Yugoslavia, when 40 drinkers were poisoned by lead leached from a glazed ceramic wine jar[70].

6.1.6.2 *Lead in dietary supplements*

Unexpectedly high levels of lead have been found in calcium supplements made from dolomite and bonemeal[71] The USFDA has advised children and pregnant and lactating women to avoid consumption of such supplements because of risk of excessive lead intake[72].

In 1995 samples of propolis, a natural product produced by honeybees from pollen and exudate found on buds of certain trees, which is sold as a health food and is used in a wide variety of health products, were found to contain high levels of lead. A MAFF study in the UK found lead ranging from 1.7 to 1570 mg/kg in some propolis-containing health foods[73]. Since consumption of these products would obviously have a considerable effect on lead intake, warnings were issued to local environmental health departments in the UK as well as to European authorities and, where detected, the contaminated products were withdrawn from sale.

6.1.6.3 *Lead in plastic packing*

A practice in some households that could have the undesirable effect of increasing dietary lead intake by members of the family is the re-use of plastic food bags and other wrappers. In an earlier section consideration has been given to the transfer of lead and other heavy metals from the pigments and print used on paper and other wrappings to food. This possibility is highlighted in a US study[74] of plastic bread wrappings and their re-use in the home. The wrappings were obtained on widely available brands of bread sold in supermarkets. Lead, at an average of 26 $\pm$ 6 mg per bag, was detected on the outer, printed, surfaces of the bags, though not on the insides. It was calculated that up to 100 µg of lead could be leached from each exterior surface by a weak acid in ten minutes. Bread, as originally packed, inside the bags would not have taken up any of the metal, but this was not the case when the bags were turned inside-out and re-used. This was found to be the case with 16% of 106 families who responded to a survey about their use of such bags.

6.1.7 *Dietary intake of lead*

The Joint UNEP/FAO/WHO Food Contamination Programme, or GEMS/Food in its shorthand form, was established by the United Nations Environment Programme (UNEP) to collect, assess, and disseminate information on contaminants in food which are considered to pose serious threats to health[75]. Prominent among them is lead. It is one of the most frequently monitored contaminants, with about 30 countries in the GEMS/Food network providing data on concentrations in a wide variety of foods.

It has been possible, on the basis of the GEMS data, to construct what its called a 'global diet' from which it has been estimated that world intake of lead is 153 μg/d or 18 μg/kg body weight. The data on which the estimate is based covered the years 1981–8, before efforts to reduce the use of soldered cans and the phasing out of leaded petrol in many countries began to have a significant effect on levels of lead in foods. Nevertheless the GEMS/Food data indicated that a downward trend in lead intake had become apparent in some countries such as the UK and the USA, but not in others. In Japan, for instance, after an initial drop in intake, dietary lead levels had begun to increase once more. The increases were believed to be due in some countries to increased consumption of wine and, in industrialised countries, to high concentrations of lead in tap water. The considerable differences between lead intakes in some countries compared to others, for example Cuba (63 μg/kg body weight/week) and Denmark (5 μg/kg bw/week), were considered to be due to differences in consumption of foodstuffs which contribute most to the total intake of lead by adults (drinking water, beverages, cereals, vegetables and fruit). It was not believed that canned food, in spite of the high levels of lead that they sometimes contain, could be identified as major contributors to overall intake[76].

More recent data on dietary lead intakes in several countries indicate that the downward trend noted in the late 1980s has generally continued, as is shown in Table 6.2. In the UK, for instance, estimates of dietary exposure to lead covering the years 1976 to 1994 show a fall from 0.11 mg/day to 0.024 mg/day. Though it is possible that the decrease in estimated intakes may in part be due to a lowering of the limit of detection from 0.05 mg/kg in 1988 to 0.01 mg/kg in 1991, a real decrease in exposure is also evident. This reflects the success of measures taken to reduce lead contamination of food, such as replacement of soldered cans, banning of lead seals on wine bottles and the phasing out of leaded petrol[77]. A similar downward trend has been seen in the US. In the case of 14–16-year-old males, for instance, between 1982 and 1991, lead intakes fell from 38 to 3.2 g/day. Similar rates of reduction were seen in other age and sex groups. More recent data from the US Total Diet Study indicates that the reductions have either levelled off or continued[78].

6.1.7.1 *Lead in children's diets*

Perhaps the most significant effect of recent efforts to reduce lead contamination of food and the environment has been seen in the case of children. It was largely concern at what dietary lead was doing to the health of young people that triggered the increased interest in lead shown by governments and other agencies in the late 1970s. As has been mentioned above, there is growing evidence that blood lead levels previously accepted as normal are associated with learning deficits and that the developing brain is especially susceptible to the toxic action of the metal[79]. Moreover, absorption and retention of lead from food are higher in infants and young children than in adults[80]. Such concerns were largely responsible for what can rightly be

Table 6.2 Estimated dietary lead intakes in different countries.

Country	Intake (μg/day)	Year
Canada	24	1986–88
Egypt	240	1996
Netherlands	32	1984–86
Spain	39	1990/91
Sweden	17	1987
UK	24	1994
USA	15	1990

Dabeka, R.W. & McKenzie, A.D. (1995) Survey of lead, cadmium, fluorine, nickel and cobalt in food composites and estimation of dietary intakes of these elements by Canadians in 1986–1988. *Journal of the Association of Official Analytical Chemists International*, **78**, 897–909.
Saleh, Z.A., Brunn, H., Paetzold, R. & Hussein, L. (1998) Nutrients and chemical residues in an Egyptian mixed diet. *Food Chemistry*, **63**, 535–41.
Van Dokkum, W., De Vos R.H., Muyst, H. & Westra, J.A. (1989) Minerals and trace elements in total diets in the Netherlands. *British Journal of Nutrition*, **61**, 7–15.
Urieta, I., Jalón, M. & Eguileor, I. (1996) Food surveillance in the Basque Country (Spain). II. Estimation of the dietary intake of organochlorine pesticides, heavy metals, arsenic, aflatoxin M, iron and zinc through the Total Diet Study, 1990/91. *Food Additives and Contaminants*, **13**, 29–52.
Becker, W. & Kumpulainen, J. (1991) Contents of essential and toxic mineral elements in Swedish market basket diets in 1987. *British Journal of Nutrition*, **66**, 151–60.
Ysart, G., Miller, P., Crews, H. *et al.* (1999) Dietary exposure estimates of 30 elements from the UK Total Diet Study. *Food Additives and Contaminants*, **16**, 391–403.
Mackintosh, D.L., Spengler, J.D., Ozkaynak, H., Tsal, L. & Ryan, B. (1996) Dietary exposures to selected metals and pesticides. *Environmental Health Perspectives*, **104**, 202–9.

described as a triumph for public health care – the remarkable decrease in dietary lead intakes, especially by children, seen in many countries in recent decades.

This is well illustrated in the case of Canada, where average lead intakes of infants in the late 1970s were 50.3 μg/day (6.64 μg/kg/day)[81]. These had been reduced to 15 μg/day by 1988[82]. The reduction was largely attributed to improvements in canning procedures. Previously, locally available ready-to-use infant formulas packed in lead-soldered cans contained on average 46.2 ng lead/g. In contrast, formulas in lead-free cans contain only 1.7 ng/g, and in glass jars 2.5 ng/g of the metal.

In the Canadian study, as in a British investigation of lead intake by infants[83], it was found that the quality of the water used to prepare the infant's food could have a considerable effect on levels of lead consumed. In the city of Glasgow, for instance, where lead plumbing is still extensively used in older houses, infants were found to have, on average, an intake of 3.4 mg/week, which is 0.4 mg/higher than the PTWI for adults. An interesting finding in the Glasgow study was that when water from the hot tap was used to make up infant formulas, levels of lead in the food were considerably higher than when the cold tap was used.

Significant reductions in lead in children's diets have also been observed in other countries (see Table 6.3). In the US intakes by children under one year of age fell from 0.37 μg/kg bw/day between 1986 and 1991 to 0.2 μg/kg bw/day in 1990/91[84]. Less dramatic, though still significant, falls occurred in Germany (21.2 to 5.8 μg/day, between 1988 and 1995)[85], and in the UK (from 26.4 to 20.57 μg/day, between 1988 and 1993)[86].

There is little doubt that such data confirm 'the success of the measures taken to

Table 6.3 Dietary lead intakes by infants and children.

Country	Years	Age (years)	Intake (μg/day)	Intake (μg/kg body wt/day)
Canada	1986–88	1–4	15	—
Finland	1980	3	49	3.2
Germany	1988/89	5–8	21.2	0.91
Germany	1995	1.5–5.3	5.8	0.35
Netherlands	1988/89	1–4	10	—
UK	1988	2	26.4	2.2
UK	1993	0.3–1.2	20.57	—
USA	1986–91	0.5–0.9	—	0.37
USA	1990/91	0.5–0.9	1.82	0.2

Adapted from Schrey, P., Wittsiepe, J., Budde, U., Heinzow, B., Idel, H. & Wilhelm, M. (2000) Dietary intake of lead, cadmium, copper and zinc by children from the German North Sea Island Amrum. *International Journal of Hygiene and Environmental Health*, 203, 1–9.

reduce lead exposure and contamination of food'[87]. Nevertheless, the asserted aim of the US Congress to eliminate childhood lead poisoning totally has not yet been achieved. There are still persistent pockets of high levels of lead intake, especially in urban and socially deprived areas of the world. In the US alone it has been estimated that more than a million children still have blood lead levels above the 10 μg/dl which is the US guideline for intervention[88]. It would be a mistake for health authorities and government agencies concerned with public health to relax their efforts to control lead contamination of the diet, in the mistaken belief that the remarkable decreases in dietary lead intake revealed by Market Basket Studies in several countries is evidence that the end of the road has been reached[89].

6.1.8 *Analysis of foodstuffs for lead*

Provided due allowance is made for the risk of contamination, lead is readily determined in all types of foods using a variety of the analytical procedures, either single or multielement, which have been discussed earlier. Both wet and dry ashing can be used for sample pretreatment. The most suitable acid for digestion is nitric. Use of sulphuric acid can result in low recoveries due to the formation of insoluble sulphate. Perchloric acid causes losses in the ashing stage and is not suitable for ETAAS. A sulphuric–nitric acids mixture has been found to be suitable for the determination of lead in marine as in other foods[90]. Recovery is good with dry ashing, provided the temperature is maintained below 500°C. To avoid losses due to volatilisation, and to minimise the possibility of contamination from external sources, use of a closed circuit microwave digestion system is recommended.

In the absence of more advanced equipment and where sample numbers are low, atomic absorption spectrophotometry is normally the method of choice for the determination of lead in food. Because of low levels in the majority of biological samples, ETAAS is recommended, rather than flame spectrometry, which is not sufficiently sensitive. Matrix effects can be overcome by preliminary separation of the lead from the digest, by extraction into 4-methylpentan-2-one (MIBK). The use of background correction is highly desirable. Problems can arise if high levels of anions, including phosphate and carbonate, which depress absorbance, are present in the digest[91]. There have been many recent advances in analytical procedures for lead which involve the use of ICP-MS, including high-resolution ICP. Several of these employ hyphenated techniques, such as high-performance size-exclusion chromatography (HPSEC) coupled with ICP-MS[92].

6.2 Mercury

Mercury is one of the 'ancient' metals. It was known to the Greeks in classical times as *hydrargyrum* (liquid silver or quicksilver) because of its appearance and liquid state. The Romans gave it the name mercury, after the swift messenger of the gods. It was used by the alchemists in their attempts to transmute base metals into gold. It was also long valued for its medicinal uses – some real, some imagined. Two of its other early uses were for silvering mirrors and in the extraction of gold from its ores. The latter is an application which continues to the present day, sometimes with unwelcome consequences for human health and the environment[93]. Today mercury is widely used in industry, mainly because of its chemical properties. It serves as a catalyst in a variety of industrial and laboratory reactions. Its high electrical conductivity makes the liquid metal valuable also to the electrical industry.

From very early times it was recognised that mercury could have sinister side effects. Georgius Agricola wrote in the sixteenth century of 'quicksilver disease' from which mercury miners in the Harz Mountains of Germany suffered[94]. The nineteenth-century makers of felt hats who used mercury to prepare the material showed peculiar mental symptoms which gave rise to the saying 'mad as a hatter'. The term suffered a strange transportation, from Lewis Carroll's *Alice in Wonderland* to the Bendigo gold fields of colonial Australia where miners, affected by the mercury they used to extract the metal from crushed ore, were known as 'hatters'[95]. Today there is concern at the possible effects on health, of patients as well as of workers in dental clinics, of the use of mercury amalgams in teeth[96]. The medicinal use of mercury has also caused problems. It has been argued, for instance, that the death in 1840 of the great Italian violinist Paganini may have been hastened by chronic mercury poisoning from the treatment he was given for syphilis and the excessive use of laxatives containing calomel (mercurous chloride)[97]. A more recent unwelcome side effect of the use of mercury compounds in health care is 'Pink disease', caused by the use of calomel in ointments used to treat nappy rash and in teething powder for infants[98].

Mercury is still considered as one of the most dangerous of all the metals we are likely to meet in our food. Worldwide attention among food and toxicology experts is focused on the metal and levels of intake in the diet are strictly monitored in most countries. Concerns relating to such topics as the persistence of mercury in the environment, its bioaccumulation and transport in the aquatic food chain, its levels in different foodstuffs and the implications for health, were raised in 1989 in a WHO document[99]. They continue to receive serious worldwide attention this century.

6.2.1 *Chemical and physical properties of mercury*

Mercury , given the symbol Hg from its Greek name *hydrargyrum*, is element number 80 in the periodic table. Its atomic weight is 200.6 and its density 13.6, making it one of the heavier of the metals. It is liquid over a wide range of temperatures, from its melting point of $-38.9°C$ to its boiling point of $356.6°C$. Elemental mercury (Hg(0)) is rather volatile. A saturated atmosphere of the vapour contains approximately $18\,mg/m^3$ at $24°C$. The metal is slightly soluble in water and in lipids. Sulphates, halides and nitrates of mercury are also soluble in water. Mercury has two oxidation states, Hg(I) mercurous, and Hg(II) mercuric, and forms two series of compounds, e.g. HgCl (mercurous chloride, or calomel) and $HgCl_2$ (mercuric chloride, or corrosive sublimate). In aqueous solution two other combinations are formed with chlorine, $HgCl_3^-$ and $HgCl_4^{2-}$. Mercurous mercury is rather unstable and tends to disassociate in the presence of biological materials to give one atom of metallic mercury and an Hg^{2+} ion.

Organic compounds of mercury are of considerable interest, because of their wide natural distribution and their toxicity. Volatile compounds are formed between alkylmercury compounds and the halogens. These are highly toxic. Less volatile are the hydroxide and nitrate of short-chain alkylmercuric compounds. Methyl- and ethylmercury chloride has an affinity for sulphydryl groups, which enables it to bind to proteins in biological tissues.

6.2.2 *Environmental distribution of mercury*

Mercury is not widely distributed in the environment. Its average concentration in the Earth's crust is about 0.5 mg/kg and there are only a limited number of places in which it occurs in concentrated form suitable for extraction. These are found in what is sometimes known as the 'mercuriferous belt', a chain of volcanic areas which stretches along the Mid-Atlantic Ridge, the Mediterranean, South and East Asia and the Pacific Ring. In this belt are found the ancient mines of Europe and the Western world at Idria in Yugoslavia, Almaden in Spain and New Almaden in California. Commercially worked ores are mainly sulphides, most commonly cinnabar, the red sulphide (HgS).

Mercury vapour is a normal component of the atmosphere, in which elemental mercury is the principal form. The vapour arises naturally from emissions from volcanoes as well as by volatilisation from soil. It has been calculated that natural degassing of the Earth's crust and oceans releases up to 150 000 tonnes of mercury vapour into the atmosphere annually. It is also produced by combustion of fossil fuels, and from mines, metal refineries, waste incinerators and crematoria. These human activities are estimated to produce about 10 000 tonnes of mercury vapour each year[100].

Atmospheric mercury is oxidised to soluble divalent mercury and returns slowly to earth in rainwater. In surface waters and sediments the divalent mercury can be reduced again, or methylated by micro-organisms to form methylmercury. Methylation of mercury usually takes place in the upper layers of sediment on sea or lake bottoms. The methylmercury so formed is rapidly taken up by living organisms. It enters the food chain via plankton-feeders and goes on up through predators feeding on these. At the end of the chain are the larger carnivores, fish, such as the marine sharks and freshwater pikes, which, with growth in size and age, can accumulate considerable amounts of organic mercury. After these come human consumers, who are then at risk of methylmercury poisoning.

As has been shown by the Swedish Expert Group on environmental hygiene[101], local loading of methylmercury can be caused by industrial pollution, such as that produced by the paper pulp and chloralkali plants or, as has been found in the coastal waters between Australia and New Zealand, by emissions from naturally occurring mercuriferous deposits[102]. While little can be done, from the point of view of human health, with regard to mercury from natural sources, strict pollution control can be effective where industry is the source of contamination.

6.2.3 Production and uses of mercury

The metal is easily extracted from cinnabar, which can contain as much as 70% of the element. The ore is roasted in a current of air to release mercury vapour, which is then condensed into flasks. Annual world production was about 10 000 tonnes in the 1980s, but production has been falling as a result of environmental concerns and changes in use of the element[103]. The principal uses of mercury are industrial. About a quarter of total production is employed in the manufacture of scientific and electrical instruments and another quarter as electrodes in the chloralkali industry. The remainder is used in the production of paints and pigments, as well as in agro-chemicals and in medicinal and pharmaceutical products and other specialist items. These include catalysts used in the manufacture of plastics, 'slimicides' in the production of wood pulp and paper, germicides and pesticides. Organic mercury was formerly used in quantity as a fungicidal dressing for seeds of wheat and other cereals, but this application has now been banned. Until recently up to 3% of total mercury production was used as amalgam in dentistry, but because of concerns about the effects on health of such tooth fillings, this use is now rapidly decreasing.

6.2.4 Biological effects of mercury

Mercury is a cumulative poison. It is stored mainly in the liver and kidney, with a small amount (about 10%) in the brain and blood. The level of accumulation depends on the chemical form of the mercury ingested. Inorganic mercury is poorly absorbed and what is taken in is quickly excreted from the body, equally in faeces and urine, with a small amount exhaled. Mercury vapour, in contrast, is readily absorbed through the lungs. Up to 80% of the vapour inhaled enters the blood. Prior ingestion of alcohol can reduce this to 50% because the alcohol inhibits oxidation of Hg(0) to Hg^{2+} by the enzyme catalase and hydrogen peroxide in the red blood cells[104]. It has been shown that elemental mercury absorbed into the blood of a pregnant woman can pass readily through the placenta to the fetus.

Compounds of inorganic mercury can be much more toxic than the metal itself[105]. Mercuric chloride ($HgCl_2$, corrosive sublimate), for example, is a notorious and dangerous poison. The greater part of the absorbed Hg^{2+} is concentrated in the kidneys where it causes severe damage to the brush border membranes, by disrupting membranes and inhibiting SH-enzymes and damaging DNA. The result is kidney failure. If, however, selenium is also present, it can protect against damage by combining with the mercury and blocking its toxic actions. After recovery from the acute effects, mercury can still be detected in kidney tissues for many years[106].

The biological effects of organic mercury are much more severe than those of either metallic or mercuric mercury[107]. Of the different organic compounds that occur naturally, methyl- and ethyl-mercury are the most toxic. Methylmercury has been listed as one of the six most dangerous chemicals in the environment by the International Program for Chemical Safety[108]. Up to 90% of most organic mercury compounds is absorbed by the intestine from ingested food. Phenylmercury is also absorbed through the skin. Dimethylmercury, described as a 'supertoxic' chemical which can quickly permeate latex gloves, has been responsible for a number of fatalities, including the death of a research scientist in the US in 1997[109]. Most of the methylmercury absorbed from the intestine is bound initially to blood proteins and is then transported in red cells to other tissues. The central nervous system is the target organ for organic mercury toxicity. More than 95% of mercury in the brain has been shown to be organic.

Methylmercury has long been recognised as a neurotoxin for both adults and the fetus[110]. The compound freely crosses the placenta and has a devastating effect on the fetal brain. Even when the mother shows no sign of toxicity, the brain of the fetus can be damaged. Even after birth the infant can continue to be affected by methylmercury ingested by the mother, since the compound also appears in the milk. Excretion of absorbed organic mercury is mainly in bile in faeces, though some is also lost through the kidneys. Absorbed organic mercury will be retained for a long time, causing functional disturbances and damage. There seems to be a latent period between initial intake and the appearance of symptoms.

Clinical signs of methylmercury poisoning are sensory disturbances in the limbs, the tongue and around the lips. The effects are dose related and are 5–10 times more serious in the fetus than in an adult. With increasing intake, the central nervous system is irreversibly damaged, resulting in ataxia, tremor, slurred speech, tunnel vision, blindness, loss of hearing and, finally, death[111]. Levels of mercury in the blood and in urine can be used to estimate the total mercury burden of the body. Methylmercury and total mercury levels in hair have also been used with some success to measure intakes. Hair mercury measurements, in sections cut at different distances from the scalp, have been used to provide a chronological record of intakes during a prolonged period of mercury poisoning[112].

6.2.5 *Mercury in food*

Mercury and its organic derivatives occur naturally in minute amounts in most foods[113]. It is found in three different forms: elemental mercury, mercuric mercury and alkylmercury. The form influences absorption and distribution in body tissues and biological half-life. As we have seen, it also affects toxicity. Normally the level of mercury of any kind in our food is very low, ranging from a few micrograms up to 50 μg/kg. Levels of total mercury that can be expected to occur in foods, in the absence of gross contamination, are illustrated by data from the UK Total Diet Study of 1991 (Table 6.4). They are similar to levels published for other countries, such as the US[114].

Fish and organ meat may sometimes contain high levels of mercury as a result of industrial pollution of water or of soil by use of sewage sludge as an agricultural top dressing. A range 1.0–136 μg/kg in kidneys and 7.0–14.0 μg/kg in livers of cattle in the Netherlands has been attributed to the latter cause[115]. Levels of mercury up to 1.61 mg/kg in fish caught in coastal waters of Egypt have been attributed to industrial pollution[116]. Even vegetables, if they have been grown on sludge-amended soil, can contain high levels of mercury. Up to 40 μg/kg (dry weight), ten times the normal level, has been detected in lettuce grown on mercury-polluted soil[117].

The dietary intake by UK adults has been estimated to be an average of 5 μg/day, with an upper range of 9 μg/day[118]. This level of intake changed little over an 18-year period up to the 1994 Total Diet Study. It is well within the JEFCA PTWI of 0.3 mg/week (43 μg/day) for a 60 kg adult. The UK intakes are not very different from the 8 μg/day found in US adults, but lower than those reported from the Basque region of Spain (18 μg/day)[119], and from Egypt (78 μg/day)[120]. The higher intake in Spain has been attributed to a high intake of fish in the Basque region and to a higher mercury concentration in the fish consumed there, compared to other countries (68–200 μg/kg).

Fish consumption in the UK, and in most other countries, is the major contributor to mercury in the diet. In 1994 some 25% of total mercury intake came from this group. Because of the importance of fish as a source of dietary mercury and the known effects of industrial and other forms of pollution on mercury levels in marine

Table 6.4 Mercury in UK foods.

Food group	Mean concentration (mg/kg fresh weight)
Bread	0.004
Miscellaneous cereals	0.004
Carcass meat	0.003
Offal	0.006
Meat products	0.003
Poultry	0.004
Fish	0.054
Oils and fats	0.003
Eggs	0.004
Sugar and preserves	0.004
Green vegetables	0.002
Potatoes	0.003
Other vegetables	0.002
Canned vegetables	0.004
Fresh fruit	0.003
Fruit products	0.002
Beverages	0.0006
Milk	0.0007
Dairy products	0.002
Nuts	0.003

Adapted from Ysart, G., Miller, P., Crews, H. *et al.* (1999) Dietary exposure estimates of 30 elements from the UK Total Diet Study. *Food Additives and Contaminants*, **16**, 391–403.

organisms, many countries operate special surveillance programmes to monitor the levels of mercury in commonly consumed fish. In the UK, for instance, regular surveys are carried out on fish caught in coastal and other waters as well as on imported fish and shellfish[121]. The importance of such monitoring, from the public health point of view, is that more than 75% of the mercury in fish may be present as methylmercury, the most toxic form of the element[122]. Though published food composition data for mercury normally provides information only on total mercury, the PTWI proposed by JEFCA draws a distinction between the two forms. While the PTWI for total mercury is 5 μg/kg bw/week (equivalent to 0.3 mg/week for a 60 kg adult), not more than 3.3 μg/kg bw/week (0.2 mg/week) should be organic.

Alkylmercury is formed in marine and freshwater sediments from inorganic mercury which may have originated from industrial discharges. From there it is taken up by filter-feeding organisms and on via their predators into fish and other marine organisms. In the process mercury may be concentrated, by a factor of as much as 3000, in tissue of a large predator fish such as a freshwater pike or a marine shark. Accumulation in fish is related to size. Large tuna over 60 kg in weight may have levels of organic mercury up to 1 mg/kg in muscle. This compares to terrestrial animals with normally about 20 μg/kg in their tissue. In some polluted waters fish may accumulate far higher levels of total mercury, most of it in organic form. In a study of fish caught in UK waters, the average mercury content of those from clean coastal waters was found to be 210 μg/kg, while some specimens from industrially contaminated rivers and estuaries contained between 500 and 600 μg/kg[123]. Considerably higher levels have been found in South American tuna, which contained 6.9 mg mercury/kg of which 6.2% was in organic form[124].

The notorious Minimata Bay tragedy in post-World War II Japan was caused by mercury pollution of fishing grounds by industry[125]. What became known as *Minimata Disease* was officially recognised in 1956, but had actually been affecting the residents of the area for many years before that. The debilitating and, in several cases, fatal illness was found to be due to consumption of fish contaminated with mercury that had been discharged from a chemical factory. Levels of up to 29 mg methylmercury/kg were found in locally caught fish and shellfish. Though the factory causing the pollution was closed in the late 1960s and discharge of methylmercury into the bay ceased at that time, the area remains polluted to this day and there are 36.9 certified Minimata-diseased patients per 1000 population in the area[126].

The Japanese tragedy, which was followed by other incidents of mercury poisoning also caused by industrial pollution elsewhere in Japan, caused considerable concern in other countries where fish is consumed in quantity and industrial pollution occurs. Several well-fished lakes in Sweden and Canada, near mercury-using wood pulp factories, were known to be heavily polluted. Levels of methylmercury of up to 10 mg/kg were found in the fish[127]. As a result, fishing was banned in several of these lakes and efforts were made to clean up the pollution[128].

Though fish is the major source of methylmercury in the diet, and there is evidence of a correlation between fish consumption and total mercury intake[129], intoxication has also occurred on a large scale as a result of ingestion of methylmercury from a non-fish source. Alkylmercury compounds were formerly used as antifungal agents for dressing cereal seeds. In Iraq, in 1960, seed wheat which had been dressed in this way, and was not intended for human consumption, was used to make flour. As a result many thousands of people were poisoned. Some died, others were permanently incapacitated and there were many cases, also, of prenatal poisoning[130]. Similar incidents were reported subsequently from Pakistan and Guatemala[131].

Health authorities and food legislators in most countries responded to these and other widely-reported large-scale incidents of mercury, and especially organic mercury, poisoning by introducing strict controls on levels of the element permitted in foods such as fish. In Australia, for instance, there is a maximum permitted level of 0.5 mg/kg (fresh weight) in most fish and molluscs, with the exception of certain larger predators, including shark and tuna, which are permitted to contain 1.0 mg/kg[132]. In the UK, following a European Commission decision, a limit of 0.5 mg/kg in edible parts of fresh fish, with a higher limit of 1.0 mg/kg for tuna, shark, swordfish and halibut, has been adopted. In the US, though the FDA has set a limit of 1 mg/kg for methylmercury in fresh fish[133], the EPA believe that the limit should be lower. The Agency has proposed that intake of mercury should be reduced to not more than 0.1 μg/kg body weight/day[134]. This is considerably less than the JEFCA PTWI of 5 μg/kg bw/week. The proposal has given rise to considerable debate and is opposed by several investigators who question whether low levels of mercury are a health hazard, especially in fish caught in waters where there is no industrial pollution. The critics cite a large-scale study that has found no evidence of neurological damage in children born to mothers who regularly eat mercury-contaminated fish in the Seychelles Islands in the Indian Ocean. Similar evidence pointing to the absence of dangers to health when ocean fish containing mercury of natural origin, rather than from industrial pollution, are consumed, was put forward some years ago by Margolin in his comprehensive review of the question[135].

6.2.6 *Analysis of foodstuffs for mercury*

Flameless or cold vapour atomic absorption spectrophotometry (CVAAS) is one of the more widely used methods for the determination of mercury in foods, especially in investigations of a limited number of elements in relatively small numbers of samples. In the method the element is determined by measuring the transient absorbance produced when mercury vapour released from the sample is led through a cell which replaces the burner in the light path of the instrument. Magos has described its use for the determination of both total and inorganic mercury in undigested biological samples[136]. The mercury vapour is released from the samples, either by reduction in alkali solution by $SnCl_2$ for inorganic mercury, or by $SnCl_2/CdCl_2$ for total mercury analysis. Loss of volatile mercury and matrix interferences which are encountered when using CVAAS with undigested samples can be overcome by microwave digestion in a nitric/sulphuric/hydrogen peroxide or a nitric/sulphuric/hydrochloric acids mixture[137].

A reliable and sensitive, but more complicated, method which separates out the different mercury species before analysis has been described by Capon and Smith[138]. It involves an initial extraction of samples with benzene, followed by further extraction and treatment with ethanolic sodium sulphite and separation of the organic mercury into benzene, leaving the inorganic mercury in the aqueous layer. The inorganic mercury is then converted into methylmercury by treatment with methanolic tetramethyltin. The end analysis is performed by gas chromatography. In the UK Total Diet survey, food samples were digested with nitric acid in plastic pressure vessels using microwave heating and then analysed for mercury by hydride generation–inductively coupled plasma–mass spectrometry (HG-ICP-MS)[139].

6.3 Cadmium

Cadmium, unlike lead and mercury, is a relatively 'new' metal, which was discovered in 1817 by the German chemist, Friedrich Stromeyer. When first isolated it was a chemical novelty, available only in small quantities and with no obvious use. But that has changed and cadmium is now employed in quantity, especially in the plating and chemical industries. Unfortunately, as well as being technically valuable, the metal is also highly toxic and is now, as a result of its wide industrial use, a commonly encountered food contaminant and a hazard to health.

Cadmium poisoning may, in fact, have occurred in humans well before the existence of the metal was established in the nineteenth century. Cadmium occurs in association with several other metals in ores and often is an unsuspected contaminant of the extracted metals. It is commonly present in zinc, for instance, and may have been responsible for many instances of what was believed to be 'zinc poisoning'. Even the minute amounts of cadmium which may be present along with other metals can cause poisoning. Moreover, since it is easily soluble in organic acids, the cadmium present, for example, in the zinc used to galvanise utensils can easily leach into acid foods from buckets and other containers.

Cadmium has been described as 'one of the most dangerous trace elements in the food and the environment of man'[140]. The extent of this danger was brought to the world's attention by the *itai-itai* disease outbreaks in Japan between 1940 and 1975 in people living in cadmium-polluted areas. The tragedy, which was due to industrial pollution and resulted in considerable human suffering and death, stimulated a flurry of investigations and alerted health authorities to the serious nature of the problem of cadmium pollution[141]. Recently the metal has been the subject of increased attention as evidence has been coming to light which suggests that, as in the case of lead and its effects on children, previously accepted levels of intake believed to be safe are too high[142]. Today, as with lead and mercury, cadmium levels in the food supply are monitored by health authorities and permitted levels in foodstuffs are regulated by legislation in many countries.

6.3.1 *Chemical and physical properties of cadmium*

Cadmium is element number 48 in the periodic table, with an atomic weight of 112.4. It is a fairly dense (specific gravity 8.6), silvery-white, malleable metal which melts at 320.9°C and boils at 765°C. It has an oxidation state of +2 and forms a number of inorganic compounds, including chloride, sulphate and acetate. Most of these are water-soluble, with the exception of the oxide and the only slightly soluble sulphide. A number of organo-cadmium compounds have been synthesised, but they are very unstable and none occurs naturally. It forms complexes with organic compounds such as dithizone and thiocarbamate that are the basis for colorimetric determination methods for cadmium. The element can join to protein molecules by sulphydryl links. An important physical property of cadmium is its ability to absorb neutrons. This makes it useful in the manufacture of control rods in nuclear reactors.

6.3.2 *Production and uses of cadmium*

Cadmium is found in the rare mineral greenockite and in small amounts in some zinc ores. However, these sources are not commercially exploitable and most of the metal used industrially comes as a by-product from zinc smelters and from the sludges obtained from the electrolytic refining of zinc. World production is about 20 000 tonnes per year.

One of the major uses of cadmium, which accounts for 55% of world production, is in nickel–cadmium dry cell batteries[143]. Another 8% is used in electroplating. The metal provides a better surface protection to iron than does zinc, and is widely used in the automobile industry on engine parts and other components. Formerly it was also frequently used for plating food and beverage containers, especially where they were exposed to damp conditions, as in cold rooms and refrigerators. However, a number of poisoning incidents from food and drink stored in cadmium-plated containers alerted authorities to the toxicity of cadmium plating and the metal is no longer used for such purposes. Two other important uses of cadmium, which between them take about a third of production, are as stabilisers in polyvinyl chloride (PVC) plastics and as dyes and pigments. Cadmium sulphide and cadmium sulphoselenide are widely used as pigments in paints and plastics.

A smaller but very important use (about 3% of total production) is in the making of alloys. Cadmium forms fusible alloys with a number of other metals and these are widely used industrially. Cadmium–copper is found in high-conductivity cables, in bearing alloys and in automobile components. Cadmium–silver and cadmium–lead are used in some solders. In 1992 a Directive of the European Community banned the use of cadmium in pigments, stabilisers and plating, except in cases where no suitable alternative is available, or where, as in nuclear reactors, it is used for safety reasons[144]. As a result of these and other related measures, primary production of cadmium, in parallel with lead and mercury, has begun to decline, as less toxic substitutes are introduced[145].

6.3.3 Cadmium in food

While cadmium is found in most foodstuffs, this is normally at very low levels, unless contamination has occurred. An estimate of cadmium intake based on a hypothetical 'global' diet produced by WHO gave a range of the metal in foods as 6–300 µg/kg, with the highest levels in offal and shellfish[146]. Levels of cadmium in the different foods are seen in Table 6.5, which is based on findings of the UK Total Diet Study of 1991[147]. UK levels are somewhat lower than data from Australia[148] and Canada[149] but higher than levels reported for Spain (Basque Country)[150].

Table 6.5 Cadmium in UK foods.

Food group	Mean concentration (mg/kg fresh weight)
Bread	0.03
Miscellaneous cereals	0.02
Carcass meat	0.001
Offal	0.07
Meat products	0.007
Poultry	0.002
Fish	0.02
Oils and fats	0.02
Eggs	0.001
Sugar and preserves	0.009
Green vegetables	0.006
Potatoes	0.03
Other vegetables	0.008
Canned vegetables	0.007
Fresh fruit	0.002
Fruit products	0.001
Beverages	0.001
Milk	0.001
Dairy products	0.002
Nuts	0.05

Adapted from Ysart, G., Miller, P., Crews, H. *et al.* (1999) Dietary exposure estimates of 30 elements from the UK Total Diet Study. *Food Additives and Contaminants*, **16**, 391–403.

In spite of some differences between levels of cadmium in different foodstuffs, overall dietary intakes of the metal in most countries appear to be generally similar, according to data published by the WHO[151]. Results from some 20 different countries, with one exception, show that daily intakes are below the PTWI of 7 µg/kg body weight and that levels of intake have generally been stable over a number of years. Other data indicate that intake in most countries does not exceed 20 µg/day, with, for example a range of 10–14 µg/day in Germany[152]. Many investigations of cadmium levels in foods and diets have indicated that meat offal and seafoods are usually richer in cadmium than are other components of the diet. Fish consumption has been shown to be positively correlated with blood cadmium levels[153]. Elsewhere vegetables, and especially potatoes, have been found to contribute a higher proportion of dietary cadmium than do fish to the diet[154].

The high levels of cadmium sometimes found in vegetables can be attributed to two causes, the use of fertilisers which are contaminated with cadmium, and the application of contaminated sewage sludge to agricultural soils. Cadmium occurs in fertilisers as a natural constituent of the phosphate rock from which they are produced. Depending on the place or origin of the phosphate, levels of cadmium may range from 2 to 100 mg/kg. In Australia phosphate from the Pacific island of Nauru, which was widely used until recently, contains between 70 and 90 mg/kg of the metal[155]. Application of these fertilisers can make a significant contribution to soil cadmium. Moreover, the effects of continuing use are cumulative and long-lasting. In Australia this has resulted in a legacy of soil and crop contamination in some areas[156]. High levels in potatoes, especially, have been reported[157] as well as in wheat, beef and sheep offal[158]. Similar problems have been reported in Sweden[159].

The use of sewage sludge as a fertiliser for agricultural land can also be a significant source of cadmium contamination of soil. Although the sludge is largely composed of domestic waste, it may also include industrial discharges, food processing waste and storm run-off from roadways. The processes used to rend down raw sewage and remove harmful bacteria do not remove heavy metals, but rather tend to concentrate them, and the resulting sludge may therefore contain high levels of cadmium[160]. In spite of restrictions in many countries that are intended to limit the amount of heavy metals applied to agricultural land, sewage sludge remains a particularly important source of accumulated cadmium in crops and farm animals in some areas[161].

Industrial pollution is a particularly serious cause of cadmium contamination of the food chain. This was the cause of outbreaks of *itai-itai* disease in Japan. Rice irrigated with industrially polluted water accumulated up to 1 mg cadmium/kg, compared to levels of 0.05–0.07 mg/kg in non-polluted areas. The contamination was not confined to one region only of the country. It was calculated that as a result of serious neglect of industrial hygiene, cadmium released from non-ferrous mines and smelters resulted in 9.5% of paddy fields and 7.5% of orchard soils throughout Japan being severely contaminated with the metal[162].

The long-lasting environmental effects of industrial cadmium pollution are well known in other countries besides Japan. Concern was caused in the 1970s when drainage from long-abandoned base metal mines in the village of Shipham, in Somerset, England, resulted in high levels of contamination of soil with cadmium and other metals[163]. Though soil and pasture plants as well as groundwater were found to contain higher than normal levels of cadmium[164], no convincing evidence was found of health effects in local residents from possible exposure to cadmium[165]. Industrial contamination in Russia and other areas of the former Soviet bloc has led to higher than average levels of cadmium and other toxic metals in soils and crops, and raised concerns about possible health effects in the community. Increased urine and hair cadmium levels were found in the general population of three Russian industrial cities[166]. A significant difference in cadmium levels and in dietary intakes of the metal by people living in West Germany and the former GDR has been attributed to higher levels of cadmium pollution in the east[167].

Transfer of cadmium from contaminated soil to vegetables, and thus into the human diet, is well recognised[168]. Studies on brownfield urban sites in the UK found that two kinds of green leaf vegetables, lettuce and spinach, were 'cadmium accumulators' and could take up as much as 3.1 mg/kg (dry weight) and 10.0 mg/kg (dry weight) respectively[169]. Vegetables grown on soil contaminated by volcanic activity in Northern Chile were similarly enriched with cadmium, as well as with arsenic and lead[170]. A range of concentrations from 0.2 to 40 mg cadmium/kg was found in various vegetables sold in the local markets.

6.3.4 Cadmium in water and other beverages

Water normally makes a minimal contribution to dietary cadmium intake. Levels in domestic water are generally less than 1 µg/litre. However, contamination can occur from the use of zinc-plated (galvanised) pipes and cisterns. A study of water used in boilers in Scottish hospitals found cadmium concentrations up to 21 µg/litre in some samples while average levels were significantly higher than the WHO standard for drinking water of 10 µg/litre[171]. In Australia galvanised tanks are widely used outside metropolitan areas to hold rainwater for domestic use. A study found that this practice can result in some cadmium contamination of the water, especially if the tanks are old[172]. Zinc used for galvanising iron in Australia is primarily produced by the electrolytic process. The cadmium content of the zinc produced in this way has decreased markedly since the 1920s, from about 100 mg/kg to less than 10 mg/kg today. Cadmium levels in water held in galvanised tanks have been found to correlate significantly with the cadmium content of the zinc plate. No tanks of recent manufacture had cadmium levels above the WHO limit, though in two of the oldest tanks they were 2.3 and 3.6 µg/litre. The study concluded that there was little risk to the Australian rural community in using tank water, even when this had been collected off galvanised roofs, provided the tank was relatively new.

Other beverages may also have elevated levels of cadmium due to contact with cadmium plate. Soft drinks dispensed from vending machines with cadmium-plated parts have been found to contain up to 16 mg of the metal/litre. Coffee brewed in a continuously operating urn, in which cadmium-plated pipes had been used, had significantly elevated cadmium levels[173]. Illicit alcoholic beverages, made in crude apparatus using cadmium-plated parts, had up to 38 mg of the metal/litre[174].

6.3.5 Dietary intake of cadmium

A mean intake of 14 µg of cadmium/day was reported for the UK population in 1994[175]. This was similar to intakes found in previous Total Diet Surveys in Britain. These intakes were lower than the WHO limit of 60 µg/day for a 60 kg adult. The UK intakes were comparable to intakes in several other countries, such as The Netherlands (6–19 µg), the USA (18 µg) Germany (10–14 µg) and Canada (13 µg). A European Commission study of the EU found that intakes ranged from 7 to 57 µg/day across 15 of its member states[176].

Several factors, apart from food contamination, may increase intake of cadmium. Cigarette smoking is particularly important in this regard. Some cigarettes can contain between 0.9 and 2.0 µg/g (dry weight) of the metal[177]. Cadmium is readily absorbed from cigarette smoke through the lungs, with, it is estimated, every 20 cigarettes smoked contributing 0.5 to 2.0 µg of the metal[178]. Another non-food source of dietary cadmium is zinc-containing dietary supplements. Cadmium contamination has been shown to occur in a high proportion of such supplements tested in the US[179]. Though the concentrations of cadmium in the supplements were not excessively high, the facts that lead was also present and that they are often consumed daily for a long time underline once more the need for manufacturers to use only highly purified minerals in such pharmaceutical preparations.

6.3.6 Uptake and accumulation of cadmium by the body

About 6% of the cadmium ingested in food is normally absorbed from the gut. The presence of other components of the diet, including phytate, calcium and protein, can affect the level of absorption. After absorption, the cadmium is transported in the blood, bound to albumin. It is taken up by the liver, where it induces the synthesis of metallothionein (MT), to which it binds. MT is a low molecular weight protein that is involved both in transport and selective storage of a variety of metals. Cadmium and zinc, and to a lesser extent iron, mercury and copper, compete for binding sites on MT. After the initial binding of cadmium to MT, the complex is released into the circulation. It is filtered by the kidneys where it is reabsorbed into the proximal tubules, which are the critical organs. Cadmium in the renal tubular cells has a half-life of 17–30 years and is retained until the cells' synthetic capacity for MT is exceeded[180].

Newborn babies have very little cadmium in their bodies. However, accumulation steadily takes place, even from low levels of intake and renal concentration increases over the years, reaching a maximum at about the age of 50 years[181]. By then as much as 30 mg of cadmium will have been accumulated in the body of an average resident in an industrialised country, with most of this in the kidney. In the case of a smoker, or of someone living in a polluted environment, there may be as much as 100 mg/kg in the kidney cortex[182]. When levels reach a maximum or 'critical' level, kidney damage will begin as the MT–cadmium complex is degraded and the metal is released, either to recombine with MT or to begin toxic processes within the tubular cells. The result is irreversible renal damage, with release of the cadmium and an increase in urinary excretion of the metal.

6.3.7 Effects of cadmium on health

Ingestion of cadmium can rapidly cause feelings of nausea, vomiting, abdominal cramp and headaches. Diarrhoea and shock can also occur. About 15 mg cadmium/litre of liquid is enough to bring about these effects. It is probable that a number of cases of acute 'zinc' poisoning reported in children and others who drank lemonade and other fruit drinks which had been stored in galvanised vessels, with almost immediate nausea and vomiting, were actually due to cadmium.

Long-term ingestion of cadmium causes serious health problems. In Japan, cases of *itai-itai* disease were identified among people living in cadmium-polluted areas. The cadmium, consumed in rice which accumulated the metal from polluted irrigation water, caused proximal tubule damage, anaemia and a severe loss of bone mineral resulting in painful fractures. The name of the disease, which means 'ouch! ouch!', is a macabre joke which refers to the gasps of agony of those who suffered these effects. In other countries where similar cadmium pollution occurred and populations were also exposed to excessive intake, symptoms were less severe than in Japan where, it is believed, other dietary influences such as a generally low intake of proteins, exacerbated the effects of the cadmium poisoning[183].

Evidence in support of this view has come from a study of oyster fishers and their families in New Zealand who consume high amounts of cadmium-rich oysters, ingesting nearly as much of the metal as did the Japanese farmers who ingested cadmium-rich rice. The New Zealanders, who had an otherwise nutritious and plentiful diet, neither accumulated high levels of cadmium nor suffered from tubular proteinuria. However, the oysters and the total diet of the New Zealanders are not deficient in protein, calcium, zinc and iron, unlike the Japanese diet[184].

There is some evidence that, at least in industrially exposed workers, cadmium intake may result in a higher than normal level of cancers of the prostate and lungs[185]. There is, as yet, no convincing evidence that cadmium ingested in food will have the same effect. However teratogenic effects have been attributed to cadmium and chromosome aberrations have been observed in some *itai-itai* victims. Congenital abnormalities occur in rats fed cadmium-contaminated water over a long period[186].

Cadmium toxicity may be counteracted by simultaneous ingestion of certain other metals. In animals, cobalt, selenium and zinc have been shown to have this effect. Cadmium competes with zinc and other metals for binding sites on the protein MT, which indicates that metabolically significant interactions occur between these metals[187]. A low intake of copper may reduce tolerance to cadmium[188].

6.3.8 *Analysis of foodstuffs for cadmium*

Though wet digestion sample preparation has been recommended for the determination of cadmium in foods[189], dry ashing, following pre-digestion using a microwave digestion system, has been found to be very effective for the purpose, especially in foods with a high fat content, such as nuts[190]. Dry ashing, followed by graphite furnace atomic absorption spectrophotometry, has been used by a number of investigators in Spain and elsewhere to determine cadmium in a variety of foods[191]. ICP-MS, following microwave digestion with nitric acid in pressure vessels, is the method used in the UK's Total Diet Study for cadmium[192]. A similar multielement method has been used by the USFDA in a study of cadmium and other heavy metals in seafoods[193].

References

1. Baldwin, D.R. & Marshall, W.J. (1999) Heavy metal poisoning and its laboratory investigation. *Annals of Clinical Biochemistry*, **36**, 267–300.
2,. Agricola, G. (1556) *De Re Metallica*, English translation by H.C. & L.H. Hoover, Dover, New York (1950).
3. Caravanos, J. (2000) *Comprehensive History of Lead.* http://www.hunter.cuny.edu/health/eohs/ph702/pbhistory.htm
4. Needleman, H.L. (1991) Childhood lead poisoning: a disease for the history texts. *American Journal of Public Health*, **81**, 685–7.
5. Winckel, J.W. & Rice, D.M. (1998) Lead market trends – technology and economics. *Journal of Power Sources*, **73**, 3–10.
6. Winckel, J.W. & Rice, D.M. (1998) Lead market trends – technology and economics. *Journal of Power Sources*, **73**, 3–10.
7. Perrin, J. (1974) Le plomb du chasse. *Matériaux et Techniques*, November, 518–20.
8. NRCC (1973) *Lead in the Canadian Environment*. NRCC No. 13682. National Research Council of Canada, Ottawa.
9. Johansen, P., Asmund, G. & Riget, F. (2001) Lead contamination of seabirds harvested with lead shot – implications to human diet in Greenland. *Environmental Pollution*, **112**, 501–4.
10. The percentage by volume of iso-octane, C_8H_{18} (2,2,4-trimethylpentane) in a mixture of iso-octane and normal heptane, C_7H_{16}, which is equal to the fuel in knock characteristics under specified test conditions (Uvarov, E.B., Chapman, D.R. & Isaacs, A. (1973) *A Dictionary of Science*. Penguin, Middlesex, UK).
11. AAS (1981) *Health and Environmental Lead in Australia*. Australian Academy of Science, Canberra.
12. NHMRC (1993) *Reducing Lead Exposure in Australia: a Consultation Resource*. National Health and Medical Research Council, Canberra.
13. Sharma, K. & Reutergarth, L.B. (2000) Exposure of preschoolers to lead in the Makati area of Metro Manila, the Philippines. *Environmental Research*, **83**, 322–32.

14. Baldwin, D.R. & Marshall, W.J. (1999) Heavy metal poisoning and its laboratory investigations. *Annals of Clinical Biochemistry*, **36**, 267–300.
15. Bruening, K., Kemp, F.W., Simone, N., Holding, Y., Louria, D.B. & Bogden, J.D. (1999) Dietary calcium intakes of urban children at risk of lead poisoning. *Environmental Health Perspectives*, **107**, 431–5.
16. Baltrop, D. & Smith, A.M. (1985) Kinetics of lead interactions with human erythrocytes. *Postgraduate Medical Journal*, **51**, 770–3.
17. WHO (1989) *Minor and Trace Elements in Breast Milk*. World Health Organization, Geneva.
18. Hotiuchi, K. (1970) Lead in human tissues. *Osaka City Medical Journal*, **16**, 1–28.
19. Snakin, V.V. & Prisyazhnaya, A.A. (2000) Lead contamination of the environment in Russia. *Science of the Total Environment*, **256**, 95–101.
20. Schroeder, H.A. & Tipton, I.H. (1968) Lead in ancient and modern skeletons. *Archives of Environmental Health*, **17**, 965–78.
21. Aberg, G. & Stray, H. (1998) Man, nutrition and mobility: a comparison of teeth and bone from the Medieval era and the present from Pb and Sr isotopes. *Science of the Total Environment*, **224**, 109–19.
22. Weiss, D., Whitten, B. & Leddy, D. (1972) Lead levels in hair of American children. *Science*, **178**, 69–70.
23. Petering, H.G., Yeager, D.W. & Witherup, S.O. (1973) Hair analysis. *Archives of Environmental Health*, **27**, 327–30.
24. Taylor, A. (1986) Usefulness of measurement of trace elements in hair. *Annals of Clinical Biochemistry*, **23**, 364–78.
25. Pagliuca, A., Baldwin, D., Lestas, A.N., Wallis, R.M., Bellingham, A.J. & Mufti, G.J. (1990) Lead poisoning: clinical biochemical and haematological aspects of a recent outbreak. *Journal of Clinical Pathology*, **43**, 277–81.
26. Rossi, E., Costin, K.A. & Garcia-Webb, P. (1990) Effect of occupational lead exposure on lymphocyte enzymes involved in haem biosynthesis. *Clinical Chemistry*, **36**, 1980–3.
27. Graham, J.A., Maxton, D.G. & Twort, C.H. (1981) Painter's palsy: a difficult case of lead poisoning. *Lancet*, **ii**, 1159–62.
28. Lee, W.R. (1981) What happens in lead poisoning? *Journal of the Royal College of Physicians*, **15**, 48–54.
29. Inglis, P., Henderson, D.A. & Emmerson, B.T. (1978) Lead-related nephritis. *Journal of Pathology*, **124**, 65–73.
30. Cardenas, A., Roels, H., Bernard, *et al.* (1993) Markers of early renal changes induced by industrial pollutants. II Application to workers exposed to lead. *British Journal of Industrial Medicine*, **50**, 28–36.
31. WHO (1995) Inorganic lead. *Environmental Health Criteria*, **165**. World Health Organization, Geneva.
32. Needleman, H.L., Gunnoe, C., Leviton, A. *et al.* (1979) Deficits in physiological and classroom performance in children with elevated dentine lead levels. *New England Journal of Medicine*, **300**, 689–95.
33. Baghurst, P.A., McMichael, A.J., Wigg, N.R. *et al.* (1992) Environmental exposure to lead and children's intelligence at the age of seven years: the Port Pirie cohort study. *New England Journal of Medicine*, **327**, 1279–84.
34. Fulton, M., Raab, G., Thomson, G., Laxen, D., Hunter, R. & Hepburn, W. (1987) Influence of blood lead on the ability and attainment of children in Edinburgh. *Lancet*, **i**, 1221–6.
35. Pocock, S.J., Smith, M. & Baghurst, P. (1994) Environmental lead and children's intelligence: a systematic review of the epidemiological evidence. *British Medical Journal*, **309**, 1189–97.
36. Palca, J. (1992) Lead researcher confronts accusers in public hearing. *Science*, **256**, 437–8.
37. Centres for Disease Control (1991) *Strategic Plan for the Elimination of Childhood Lead Poisoning*. Centres for Disease Control, US Departments of Health and Human Services, Washington, DC.

38. Department of the Environment (1990) *UK Blood Lead Monitoring Programme 1984–1987: Results for 1987. Pollution Report No. 28*. HMSO, London.
39. NHMRC (1993) *Reducing Lead Exposure in Australia, a Consultative Document*. National Health and Medical Research Council, Canberra.
40. Brown, A. (1983) Petrol sniffing lead encephalopathy. *New Zealand Medical Journal*, **96**, 421–2.
41. Needleman, H.L. (1991) Childhood lead poisoning: a disease for the history books. *American Journal of Public Health*, **81**, 685–7.
42. Jorhem, L. & Sundström, B. (1993) Levels of lead, cadmium, zinc, copper, nickel, chromium, manganese, and cobalt in foods on the Swedish market, 1983–1990. *Journal of Food Composition and Analysis*, **6**, 223–41.
43. Ministry of Agriculture, Fisheries and Food (1998) *Lead, Arsenic and Other Metals in Food. Food Surveillance Paper No. 52*. The Stationery Office, London.
44. Vos, G., Hovens, J.P.C. & Delft, W.V. (1987) Arsenic, cadmium, lead and mercury in meat, livers and kidneys of cattle slaughtered in the Netherlands during 1980–1985. *Food Additives and Contaminants*, **4**, 73–88.
45. Beckham, I., Blanche, W. & Storach, S. (1974) Metals in canned foods in Israel. *Var Foeda*, **26**, 26–32.
46. Branca, P. (1982) Uptake of metal by canned food with length of storage. *Bulletin Chimique Laboratoire*, **33**, 495–506.
47. Dabeka, W. & McKenzie, A.D. (1988) Lead and cadmium in commercial infants' food and dietary intake by infants 0–1 year old. *Food Additives and Contaminants*, **5**, 333–42.
48. FDA (1993) Lead-soldered food cans. *Federal Register*, **58**(117), 33860–71 (21 June). Food and Drug Administration, Washington, DC.
49. Jelinek, C.F. (1982) Levels of lead in the United States food supply. *Journal of the Association of Official Analytical Chemists*, **65**, 942–6.
50. FDA (1992) Tin-coated lead foil capsules for wine bottles. *Federal Register*, **58**(117), 33860–71 (21 June). Food and Drug Administration, Washington, DC.
51. EC (1991) Council Regulation EEC 2356/91 amending Council Regulation EEC 2392/89. *Official Journal of the European Communities*, **L261**, 1–2.
52. Sherlock, J.C., Pickford, C.J. & White, G.F. (1986) Lead in alcoholic beverages. *Food Additives and Contaminants*, **3**, 347–54.
53. MAFF (1987) Wipe before you pour says Donald Thompson. *Food Facts Press Release*, **FF 18/87**. Ministry of Agriculture, Fisheries and Food, London.
54. Shaper, A.G., Pocock, S.J., Walker, M. *et al.* (1982) Effects of alcohol and smoking on blood lead in middle-aged British men. *British Medical Journal*, **284**, 289–302.
55. Morgan, H. (1988) Special issue: the Shipham Report. An investigation into cadmium contamination and its implications for human health. *Science of the Total Environment*, **75**, 1–135.
56. MAFF (1993) *Review of the Rules for Sewage Sludge Application to Agricultural Land: Food Safety and Relevant Animal Health Aspects of Potentially Toxic Elements. Report of the Steering Group on Chemical Aspects of Food Surveillance*. Ministry of Agriculture, Fisheries and Food, London.
57. Ministry of Agriculture, Fisheries and Food (1998) *Lead, Arsenic and Other Metals in Food. Food Surveillance Paper No. 52*. The Stationery Office, London.
58. Craun, G.F. & McCabe, L.J. (1970) Lead in domestic water supplies. *Journal of the American Water Workers Association*, **67**, 593–9.
59. Smart, G.A., Sherlock, J.C. & Normal, J.A. (1987) Dietary intakes of lead and other metals: a study of young children from an urban population in the UK. *Food Additives and Contaminants*, **5**, 85–93.
60. Department of the Environment (1977) *Lead in Drinking Water. A Survey of Great Britain 1975–1976*. Pollution Paper No. 12. HMSO, London.
61. FDA (1992) Tin-coated lead foil capsules for wine bottles. *Federal Register*, **58**(117), 3860–71 (21 June). Food and Drug Administration, Washington, DC.
62. Urieta, I., Jalón, M. & Eguileor, I. (1996) Food surveillance in the Basque Country (Spain). II. Estimation of the dietary intake of organochlorine pesticides, heavy metals,

arsenic, aflatoxin M, iron and zinc through the Total Diet Study, 1990/91. *Food Additives and Contaminants*, **13**, 29–52.
63. Jorhem, L., Mattson, P. & Slorach, S. (1988) Lead in table wines on the Swedish market. *Food Additives and Contaminants*, **5**, 645–9.
64. FDA (1992) Tin-coated lead foil capsules for wine bottles. *Federal Register*, **58**(117), 3860–71 (21 June). Food and Drug Administration, Washington, DC.
65. De Leacy, E.A. (1988) Lead-crystal decanters – a health risk? *Australian Medical Journal*, **147**, 162.
66. FDA (1991) *Advice on leaded crystalware.* FDA Talk Paper. Food and Drug Administration, Rockville, Md.
67. Mendez, J.H., De Blas, O.J. and Gonzalez, V. (1989) Correlation between lead content in human biological fluids and the use of vitrified earthenware containers for foods and beverages. *Food Chemistry*, **31**, 205–13.
68. Elinder, C.G., Lind, B., Nilsson, B. & Oskarsson, A. (1988) Wine – an important source of lead exposure. *Food Additives and Contaminants*, **5**, 641–4.
69. Whitehead, T.P. and Prior, A.P. (1960) Lead poisoning from earthenware container. *Lancet*, **i**, 1343–4.
70. Beritic, T. & Strahuljak, D. (1961) Lead poisoning from glazed surfaces. *Lancet*, **i**, 669.
71. Bourgoin, B.P., Evans, D.R., Cornett, J.R., Lingard, S.M. & Quattrone, A.J. (1993) Lead content in 70 brands of dietary supplements. *American Journal of Public Health*, **83**, 1155–60.
72. FDA (1982) Advice on limiting intake of bonemeal. *Food and Drug Administration Drug Bulletin*, pp. 5–6.
73. MAFF (1995) Analysis of bee products for heavy metals. *Food Surveillance Information Sheet*, **53**, Ministry of Agriculture, Fisheries and Food, Food Safety Directorate, London.
74. Weisel, C., Demak, M., Marcus, S. & Goldstein, B.D. (1991) Soft plastic bread packaging: lead content and reuse by families. *American Journal of Public Health*, **81**, 756–8.
75. Galal-Gorchev, H. (1991) Dietary intake of pesticide residues: cadmium, mercury and lead. *Food Additives and Contaminants*, **8**, 793–806.
76. Gems/Food (1991) *Summary of 1986–1988 monitoring data.* World Health Organization, Geneva.
77. Ysart, G., Miller, P., Crews, H., *et al.* (1999) Dietary exposure estimates of 30 elements from the UK Total Diet Study. *Food Additives and Contaminants*, **16**, 391–403.
78. Bolger, P.M., Yess, N.J., Gunderson, E.L., Troxell, T.C. & Carrington, C.D. (1996) Identification and reduction of sources of dietary lead in the United States. *Food Additives and Contaminants*, **13**, 53–60.
79. Needleman, H.L. & Bellinger, D. (1984) The development consequences of childhood exposure to lead. In: *Advances in Clinical Child Psychology* (eds. B. Lahey & A. Kazdin). Plenum, New York.
80. Nutrition Foundation's Expert Advisory Committee (1982) *Assessment of the Safety of Lead and Lead Salts in Food.* Nutrition Foundation, Washington, DC.
81. FDA (1993) Lead-soldered food cans. *Federal Register*, **58**(117), 33060–71 (21 June) Food and Drug Administration, Washington, DC.
82. Dabeka, R.W. & McKenzie, A.D. (1995) Survey of lead, cadmium, fluoride, nickel and cobalt in food composites and estimation of dietary intakes of these elements by Canadians in 1986–1988. *Journal of the Association of Official Analytical Chemists*, **78**, 897–909.
83. Sherlock, J.C. & Quinn, M.J. (1986) Relationship between blood lead concentrations and dietary lead intake by infants 0–1 year old. *Food Additives and Contaminants*, **3**, 167–76.
84. Gunderston, E.L. (1997) FDA Total Diet Study, July 1986–April 1991, dietary intakes of pesticides, selected elements, and other chemicals. *Journal of the Association of Official Analytical Chemists International*, **78**, 1353–63.
85. Arnold, R., Kibler, R. & Brunner, B. (1998) Die alimentäre Aufnahme von ausgewählten Schadstoffen und Nitrat-Ergebnisse einer Duplikatstudie in bayerischen Jugend- und Seniorheimen. *Zeitschrift für Ernährungswissenschaft*, **37**, 328–35.

86. Richmond, J., Strehlow, C.D. & Chalkeley, S.R. (1993) Dietary intake of Al, Ca, Cu, Fe, Pb, and Zn in infants. *British Journal of Biomedical Science*, **50**, 178–86.
87. Ysart, G., Miller, P. Crews, H. *et al.* (1999) Dietary exposure estimates of 30 elements from the UK Total Diet Study. *Food Additives and Contaminants*, **16**, 391–403.
88. Bruening, K., Kemp, F.W., Simone, N., Holding, Y., Louria, D.B. & Bogden, N. (1998) Identification and reduction of sources of dietary lead in the United States. *Food Additives and Contaminants*, **13**, 53–60.
89. Reilly, C. (2002) Lead poisoning in children: is this the end of the story? *Nutrition Bulletin*, **26**, in press.
90. Locatelli, C. & Torsi, G. (2001) Heavy metal determination in aquatic species for food purposes. *Annali di Chimica*, **91**, 65–72.
91. Zong, Y.Y., Parsons, P.J. & Slavin, W. (1998) The determination of lead in bone matrices by Zeeman-effect ETAAS. *Spectrochimica Acta, Part B*, **53**, 1031–5.
92. Taylor, A., Branch, S., Halls, D.J., Owen, L.M.W. & White, M. (2000) Atomic spectroscopy update: clinical and biological materials, foods and beverages. *Journal of Analytical Atomic Absorption Spectrometry*, **15**, 451–87.
93. Palheta, D. & Taylor, A. (1995) Mercury in environmental and biological samples from a gold mining area in the Amazon region of Brazil. *Science of the Total Environment*, **168**, 63–9.
94. Agricola, G. (1956) *De Re Metallica*, English translation by H.C. & L.H. Hoover. Dover, New York (1950).
95. Bycroft, B.M., Coller, S.A.W., Deacon, G.B. & Coleman, D.J. (1982) Mercury contamination of the Lederberg River, Victoria, Australia, from an abandoned gold field. *Environmental Pollution*, **28**, 135–47.
96. Weiner, J.A. & Nylander, M. (1995) An estimation of the uptake of mercury based on urinary extraction of mercury in Swedish subjects. *Science of the Total Environment*, **168**, 255–65.
97. O'Shea, J.G. (1988) The death of Paganini. *Journal of the Royal College of Physicians*, **22**, 104–5.
98. Stephenson, J.B.P. (1966) Pink disease. *British Medical Journal*, **I**, 1110–11.
99. United Nations Environment Programme, International Labour Organisation, World Health Organization (1989) Mercury – environmental aspects. *Environmental Health Criteria*, **86**. World Health Organization, Geneva.
100. WHO (1976) *Environmental Health Criteria: Mercury*. World Health Organization, Geneva.
101. Swedish Expert Group (1971) *Methyl Mercury in Fish. A Toxicological–Environmental Appraisal of Risks. Nord. Hyd. Tdskr*, Supplement 4.
102. Working Group on Mercury in Fish (1980) Discussions on mercury limits in fish. *Australian Fisheries*. October, 2–10.
103. Hylander, L.D. (2001) Global mercury pollution and its expected decrease after a mercury trade ban. *Water Air and Soil Pollution*, **125**, 331–44.
104. Baldwin, D.R. & Marshall, W.J. (1999) Heavy metal poisoning and its laboratory investigation, *Annals of Clinical Biochemistry*, **36**, 267–300.
105. Clarkson, T.W. (1997) The toxicology of mercury. *Critical Reviews of Clinical Laboratory Science*, **34**, 369–403.
106. Fowler, B.A. (1992) Mechanisms of kidney cell injury through metals. *Environmental Health Perspectives*, **100**, 57–63.
107. Langford, N.J. & Ferner, R.E. (1999) Toxicity of mercury. *Journal of Human Hypertension*, **13**, 651–6.
108. Bennet, B.G. (1984) Six most dangerous chemicals named. Monitoring and Assessment Research Centre, London, on behalf of UNEP/ILO/WHO International Program on Chemical Safety. *Sentinel*, **1**, 3.
109. Nirenberg, D.W., Nordgren, R.E., Chang, M.B. *et al.* (1998) Delayed cerebellar disease and death after accidental exposure to dimethylmercury. *New England Journal of Medicine*, **338**, 1672–6.

110. Berlin, M.H., Clarkson, T.W. & Friberg, L.T. (1963) Maximum allowable concentrations of mercury compounds. *Archives of Environmental Health*, **6**, 27–39.
111. Harada, M. (1995) Minimata disease: methyl mercury poisoning in Japan caused by environmental pollution. *Critical Reviews in Toxicology*, **25**, 1–24.
112. Bakir, F., Damluji, S.F. & Amin-Zak, L. (1963) Methyl mercury poisoning in Iraq. *Science*, **181**, 230–42.
113. Margolin, S. (1980) Mercury in marine seafood: the scientific medical margin of safety as a guide to the potential risk to public health. *World Review of Nutrition and Dietetics*, **34**, 182–265.
114. MacIntosh, D.L., Spengler, J.D. , Ozkaynak, H., Tsai, L. & Ryan, B. (1996) Dietary exposure to selected metals and pesticides. *Environmental Health Perspectives*, **104**, 202–9.
115. Vos, G., Hovens, J.P.C. & Delft, W.V. (1987) Arsenic, cadmium, lead and mercury in meat, livers and kidneys of cattle slaughtered in the Netherlands during 1980–1985. *Food Additives and Contaminants*, **4**, 73–88.
116. Moharram, Y.G., Moustafa, E.K., El-Sokkary, A. & Attia, M.A. (1987) Mercury content of some marine fish from the Alexandria coast. *Nahrung*, **31**, 899–904.
117. Capon, C.J. (1981) Mercury and selenium content and chemical form in vegetable crops grown on sludge-amended soil. *Archives of Environmental Contamination and Toxicology*, **10**, 673–89.
118. Ysart, G., Miller, P., Crews, H. *et al.* (1999) Dietary exposure estimates of 30 elements from the UK Total Diet Study. *Food Additives and Contaminants*, **16**, 391–403.
119. Urieta, I., Jalón, M. & Eguileor, I., (1996) Food surveillance in the Basque Country (Spain). II. Estimation of dietary intake of organochlorine pesticides, heavy metals, arsenic, aflatoxin M, iron and zinc through the Total Diet Study, 1990/91. *Food Additives and Contaminants*, **13**, 29–52.
120. Saleh, Z.A., Brunn, H., Paetzold, R. and Hussein, L. (1998) Nutrient and chemical residues in an Egyptian total mixed diet. *Food Chemistry*, **63**, 535–41.
121. Margolin, S. (1980) Mercury in marine seafood: the scientific medical margin of safety as a guide to the potential risk to public health. *World Review of Nutrition and Dietetics*, **34**, 182–265.
122. Capon, C.J. (1990) Speciation of selected trace elements in edible seafoods. In: *Food Contamination from Environmental Sources* (eds. J.O. Nriagu & M.S. Simmons), pp. 145–95. John Wiley, New York.
123. Ministry of Agriculture, Fisheries and Food (1973) *Supplementary Report on Mercury in Food*. HMSO, London.
124. Capon, C.J. & Smith, J.C. (1982) Chemical form and distribution of mercury and selenium in edible seafood. *Journal of Analytical Toxicology*, **6**, 10–21.
125. Harada, M. (1978) Methyl mercury poisoning due to environmental contamination ('Minimata Disease'). In: *Toxicity of Heavy Metals in the Environment* (ed. F.W. Oehme), pp. 261–72. Dekker, New York.
126. Futasuka, M., Kitano, T., Shono, N. *et al.* (2000) Health surveillance in the population living in a methyl mercury-polluted area over a long period. *Environmental Research*, **83**, 83–92.
127. Skerfving, S. (1974) Methylmercury exposure, mercury levels in blood and hair, and health status of Swedes consuming contaminated fish. *Toxicology*, **2**, 3–23.
128. Tomlinson, G.H. (1979) Acid precipitation and mercury in Canadian lakes and fish. *Scientific Papers from the Public Meeting on Acid Rain Precipitation*, May 1978, Lake Placid, New York (New York State Assembly, Albany, NY).
129. Turner, M.D., Marsh, D.A., Smith, J.C., English, J. & Clarkson, T.W. (1980) Methylmercury in populations eating large quantities of marine fish. *Archives of Environmental Health*, **35**, 367–78.
130. Bakir, F. (1973) Methyl mercury poisoning in Iraq. *Science*, **181**, 230–41.
131. Magos, L. (1975) Methylmercury poisoning. *British Medical Bulletin*, **31**, 241–2.
132. Australia New Zealand Food Authority (2000) *Draft Food Standards Code. Standard 1.4.1 Contaminants and Restricted Substances.* http://www.anzfa.gov.au/draftfoodstandards/Chapter1/Part1.4/1.4.1.htm

133. USFDA (2000) Mercury in fish: cause for concern? *Food and Drug Administration Consumer*.
http://vm.cfsan.fda.gov/~dms/mercury.html
134. Kaiser, J. (2000) Toxicology: mercury report backs strict rules. *Science*, **289**, 371–2.
135. Margolin, S. (1980) Mercury in marine seafood: the scientific medical margin of safety as a guide to the potential risk to public health. *World Review of Nutrition and Dietetics*, **34**, 182–265.
136. Magos, L. (1971) Selective atomic absorption determination of inorganic mercury and methyl mercury in undigested biological samples. *Analyst*, **96**, 847–53.
137. Tinggi, U. & Craven, G. (1996) Determination of total mercury in biological materials by cold vapour atomic absorption spectrometry after microwave digestion. *Microchemical Journal*, **54**, 168–73.
138. Capon, C.J. & Smith, J.C. (1982) Chemical form and distribution of mercury and selenium in edible seafood. *Journal of Analytical Toxicology*, **6**, 10–21.
139. Baxter, M.J., Crews, H.M., Robb, P. & Strutt, P. (1997) Quality control in the multi-element analysis of foods using ICP-MS. In: *Plasma Source Spectrometry: Developments and Applications* (eds. G. Holland & S.D. Tanner). Royal Society of Chemistry, London.
140. Vos, G., Movens, J.P.C. & Delft, W.V. (1987) Arsenic, cadmium, lead and mercury in meat, livers and kidneys of cattle slaughtered in the Netherlands during 1980–1985. *Food Additives and Contaminants*, **4**, 73–88.
141. Asami, T. (1991) Cadmium pollution of soils and human health in Japan. In: *Human and Animal Health in Relation to Circulation Processes of Selenium and Cadmium* (ed. J. Lag), pp. 115–25, Norwegian Academy of Science and Letters, Oslo.
142. Buchet, J.P., Lauwerys, R. and Roels, H. (1990) Renal effects of cadmium. *Lancet*, **336**, 699–70.
143. Cook, M.E. (1991) Kadmium – Produktion, Eigenschaften, Aussichten, *Metal*, **45**, 278–81.
144. European Community (1992) *Community Action Programme, Directive No. 89/677/ EEC.*
145. Bothwell, R. (1983) Cadmium production declines 30 per cent. *Engineering and Mining Journal*, **184**, 126–7.
146. European Community (1992) *Community Action Programme, Directive No. 89/677/ EEC.*
147. Ysart, G., Miller, P., Crews, H. *et al.* (1999) Dietary exposure estimates of 30 elements from the UK Total Diet Study. *Food Additives and Contaminants*, **16**, 391–403.
148. NHMRC (1978) *Report on Revised Standards for Metals in Food*. National Health and Medical Research Council, Canberra.
149. Dabeka, R.W., McKenzie, A.D. & Lacroix, G.M.A. (1987) Dietary intake of lead, cadmium, arsenic and fluoride by Canadian adults: a 24 hour duplicate diet study. *Food Additives and Contaminants*, **4**, 89–102.
150. Urieta, I., Jalón, M. & Eguileor, I. (1996) Food surveillance in the Basque Country (Spain). II. Estimation of dietary intake of organochlorine pesticides, heavy metals, arsenic, aflatoxin M, iron and zinc through the Total Diet Study, 1990/91. *Food Additives and Contaminants*, **13**, 29–52.
151. Gems/Food (1991) *Summary of 1986–1988 monitoring data*. WHO, Geneva.
152. Muller, M., Anke, M., Illing-Gunther, H. & Thiel, C. (1998) Oral cadmium exposure of adults in Germany. 2: Market basket calculations. *Food Additives and Contaminants*, **15**, 135–41.
153. Hovinga, M.E., Sowers, M. & Humphrey, H.E.B. (1993) Environmental exposure and lifestyle predictors of lead, cadmium, PCB, and DDT levels in great lakes fish eaters. *Archives of Environmental Health*, **48**, 98–104.
154. Urieta, I., Jalón, M. & Eguileor, I. (1996) Food surveillance in the Basque Country (Spain). II. Estimation of the dietary intake of organochlorine pesticides, heavy metals, arsenic, aflatoxin M_1 iron and zinc through the Total Diet Study, 1990/91. *Food Additives and Contaminants*, **13**, 29–52.
155. Rayment, G.E., Best, E.K. & Hamilton, D.J. (1989) Cadmium in fertilisers and soil

amendments. *Chemistry International Conference*, Brisbane, 28 August–2 September. Royal Australian Chemical Institute, Canberra.

156. Bennet-Chambers, M., Davies, P. & Knott, B. (1999) Cadmium in aquatic ecosystems in Western Australia: a legacy of nutrient-deficient soils. *Journal of Environmental Management*, **57**, 283–95.
157. McLaughlin, M., Smart, M., Maier, N., Freeman, K., Williams, C. & Tiller, K. (1993) Cadmium accumulation in potatoes – occurrence and management. *Proceedings of the 7th National Potato Research Workshop*, Ulverstone, Tasmania, pp. 208–214.
158. Simpson, J. & Curnow, W. (1988) Cadmium accumulations in Australian agriculture: national symposium. *Bureau of Rural Resources Proceedings*, No. 2, Canberra.
159. Kjellström, T., Lind, B., Linnman, L. & Elinder, C.G. (1975) Variations of cadmium content in Swedish wheat and barley. *Archives of Environmental Health*, **30**, 321–8.
160. Wiseman, R. (1994) Cadmium, a modern day problem. *Rural Research*, **162**, 32–5.
161. Boukhars, L. & Rada, A. (2000) Plant exposure to cadmium in Moroccan calcareous soils treated with sewage sludge and wastewaters. *Environmental Technology*, **21**, 641–52.
162. Asami, M.O. (1984) Pollution of soil by cadmium. In: *Changing Metal Cycles and Human Health* (ed. J.O. Nriagu), pp. 95–111. Springer Verlag, Berlin.
163. Davies, B.E. & Ballinger, R.C. (1990) Heavy metals in soils in North Somerset, England, with special reference to contamination from base metal mining in the Mendips. *Environmental Geochemistry and Health*, **12**, 291–300.
164. Matthews, H. & Thornton, I. (1982) Seasonal and species variation in the content of cadmium and associated metals in the pasture plants at Shipman. *Plant and Soil*, **66**, 181–93.
165. Elliott, P., Arnold, R., Cockings, S. *et al.* (2000) Risk of mortality, cancer incidence, and stroke in a population potentially exposed to cadmium. *Occupational and Environmental Medicine*, **57**, 94–7.
166. Bustueva, K.A., Revich, B.A. & Bezpalko, L.E. (1994) Cadmium in the environment of three Russian cities and in human hair and urine. *Archives of Environmental Health*, **49**, 284–8.
167. Hoffmann, K., Becker, K., Friedrich, C., Helm, D., Krause, C. & Seifert, B. (2000) The German environmental survey 1990/1992 (GerES II): cadmium in blood, urine and hair of adults and children. *Journal of Exposure Analysis and Environmental Epidemiology*, **10**, 126–35.
168. Dudka, S., Piotrowska, M. & Terelak, H. (1996) Transfer of cadmium, lead, and zinc from industrially contaminated soil to crop plants: a field study. *Environmental Pollution*, **94**, 181–8.
169. MacIntosh, D.L., Spengler, J.J., Ozkaynak, H., Tsai, L. & Ryan, B. (1996) Dietary exposure to selected metals and pesticides. *Environmental Health Perspectives*, **104**, 202–9.
170. Queirolo, F., Stegen, S., Restoovic, M. *et al.* (2000) Total arsenic, lead, and cadmium levels in vegetables cultivated at the Andean villages of northern Chile. *Science of the Total Environment*, **255**, 75–84.
171. Lyon, T.D.B. & Lenihan, J.M.A. (1977) Kitchen boilers as source of lead and cadmium. *Lancet*, **i**, 423.
172. De Laeter, J.R., Ware, L.J., Taylor, K.R. & Rosman, K.J.R. (1976) The cadmium content of rural tank water in Western Australia. *Search*, **7**, 444–5.
173. Rosman, K.J.R., Hosie, D.J. & De Laeter, J.R. (1977) The cadmium content of drinking water in Western Australia. *Search*, **8**, 85–6.
174. Hoffman, C.M. (1968) Trace metals in illicitly alcholic beverages. *Journal of the Association of Official Analytical Chemists*, **51**, 580–6.
175. Ysart, G., Miller, P., Crews, H. *et al.* (1999) Dietary exposure estimates of 30 elements from the UK Total Diet Study. *Food Additives and Contaminants*, **16**, 391–403.
176. EC (1997) *Food Science and Techniques. Report on Tasks for Scientific Cooperation, Dietary Exposure to Cadmium.* **EUR 17527**. Office for Official Publications of the European Community, Luxembourg.

177. Ostergaard, K. (1977) Cadmium in cigarettes. *Acta Medica Scandinavica*, **202**, 193–7.
178. Nwankwo, J.N., Elinder, C.G., Piscator, M. & Lind, B. (1977) Cadmium in Zambian cigarettes: an interlaboratory comparison in analysis. *Zambian Journal of Science and Technology*, **2**, 1–4.
179. Bourgoin, B.P., Bommer, D., Powell, M.J., Willie, S., Edgar, D. & Evans, D. (1992) Instrumental comparison for the determination of cadmium and lead in calcium supplements and other calcium-rich matrices. *Analyst*, **117**, 19–22.
180. Hammer, D.I., Finklea, J.F. & Creason, J.P. (1971) Cadmium exposure and human health effects. In: *Trace Substances in Environmental Health* (ed. D.D. Hempill), pp. 269–88. University of Missouri Press, Columbia, Mo.
181. Gross, S.B., Yeagere, D.W. & Middendorf, M.S. (1976) Cadmium in liver, kidney and hair of humans, foetal through old age. *Journal of Toxicology and Environmental Health*, **2**, 153–67.
182. Friberg, L. & Vahter, M. (1983) Assessment of exposure to lead and cadmium through biological monitoring of a UNEP/WHO global study. *Environmental Research*, **30**, 95–123.
183. Carruthers, M.M. & Smith, B. (1979) Evidence of cadmium toxicity in a population living in a zinc-mining area. Pilot study of Shipman residents. *Lancet*, **i**, 663–7.
184. Kazantsis, G., Lam, T.H. & Sullivan, K.R. (1988) Mortality of cadmium-exposed workers. A five year update. *Scandinavian Journal of Work and Environmental Health*, **14**, 220–23.
185. Kazantsis, G., Lam, T.H. & Sullivan, K.R. (1988) Mortality of cadmium-exposed workers. A five year update. *Scandinavian Journal of Work and Environmental Health*, **14**, 220–3.
186. Piscator, M. (1981) Carcinogenicity of cadmium – Review. *Third International Conference on Cadmium*, Miami, Florida, 3–5 February 1981.
187. Sandstead, H.H. & Klevay, L.M. (1975) Cadmium–zinc interactions: implications for health. *Geological Society of America Special Papers*, **155**, 73–83.
188. Friberg, L., Piscator, M. & Norberg, G. (1971) *Cadmium in the Environment*. CRC Press, Cleveland, Ohio.
189. Analytical Methods Subcommittee (1975) Sample preparation. *Analyst*, **100**, 761–3.
190. Tinggi, U. (1998) Cadmium levels in peanut products. *Food Additives and Contaminants*, **15**, 789–92.
191. Muñoz, P., Macho, M.L. & Eguilfor, I. (1991) Analysis of lead, cadmium, mercury and arsenic for dietary intake calculation in the Basque Country. In: *Strategies for Food Quality Control and Analytical Methods in Europe*, volume 2 (eds. W. Baltes, T. Eklund, R. Fenwick, W. Pfannhausser, A. Ruiter & H.P. Their), pp. 703–8. Behr Verlag, Hamburg.
192. Ysart, G., Miller, P., Crews, H. *et al.* (1999) Dietary exposure estimates of 30 elements from the UK Total Diet Study. *Food Additives and Contaminants*, **16**, 391–403.
193. Caspar, S.G. & Yess, N.J. (1996) US Food and Drug Administration survey of cadmium, lead and other elements in clams and oysters. *Food Additives and Contaminants*, **13**, 553–60.

Chapter 7
The packaging metals: aluminium and tin

Metal pick-up by food from cooking utensils and containers has probably occurred ever since metals began to be used for domestic purposes rather than, solely, to make weapons. Pewter plates and bronze cauldrons, no doubt, contributed their share of lead and copper to meals consumed by our medieval forebears, but their effects on human health would not have been suspected at the time. It was only in the nineteenth century, when canning began to be introduced as a way of preserving food, that suspicions were raised about the safety of the use of metal containers to hold food. The public began to hear for the first time of what was then known as 'tin poisoning', though the metal responsible for gastric upsets in those who consumed contaminated canned food was most probably lead.

When the 'new' metal, aluminium, began to be used for making saucepans and other kitchen utensils, it, too, came under suspicion and there were moves in some countries to have its use in this way prohibited by law because of its suspected toxicity. Though both tin and aluminium have weathered the storms and continue to be used in enormous quantities, especially for food and beverage containers, every now and again, doubts are raised among consumers as to whether these are indeed acceptable metals to use in food preparation and preservation. This question will be considered here, in the light of recent observations and reports.

7.1 Aluminium

Aluminium is the most common metal in the Earth's crust. Because of its reactive nature, it does not occur as a free element but only in combination with oxygen, silicon, fluoride and other elements as silicates and other compounds. Many of its naturally occurring compounds are insoluble and thus aluminium in soil is relatively immobile and its concentration in natural waters, both fresh and marine, is normally very low. However, under acid conditions, the solubility of aluminium can be increased. As a consequence of environmental changes resulting from population growth and intensified agriculture and industrialisation, the pH of surface waters has been decreasing, leading to increased mobilisation and solubilisation of soil aluminium[1]. The resulting increase in uptake of the element by plants and animals, and thus into the human food chain, is a cause of considerable concern to environmentalists and to health authorities today[2].

There is evidence that the first use of aluminium was in China in about AD 300 when a copper-rich aluminium alloy (aluminium bronze) was produced by thermal reduction of copper and aluminium minerals. In Europe, however, though the substance known as alum (impure potassium aluminium sulphate) had been known and used for many hundreds of years as a mordant in dyeing, it was not until 1807 that the metal itself was discovered, and named, by the English chemist Humphrey Davy. He gave it its -ium ending, in keeping with the names he had given to two other metals he had discovered, sodium and potassium. The alternative spelling and pronunciation, aluminum, is widely used, especially in America. Elemental aluminium was first produced in pure form, though in minute quantities, by Oerstead in Denmark in 1825. Commercial extraction techniques were developed in France thirty years later by Delville, who exhibited the new metal at the Paris Exhibition of 1855. The cost of extracting the metal at that time was about £130/kg.

It was only after 1889 when, with the increasing availability of electricity, a less expensive extraction method was developed by Bayer in Germany that the possibility of commercial exploitation of aluminium began to be seriously considered. Further developments in extraction of alumina (aluminium oxide) from its ores, and improvements in processing, reduced the cost of production to about £0.44/kg by the end of the nineteenth century. By the mid-twentieth century, aluminium was well on its way to becoming the world's second most important metal[3].

The growth rate of aluminium use during the past century has been greater than for any other metal. Though there have been some ups and downs in aluminium usage in recent decades, as a result of economic slumps and surges, it is expected that annual rates of increase in production will continue to be between 1 and 3% for the foreseeable future[4]. Demand is currently met by increasing primary production from bauxite ore, as well as by secondary production using scrap metal. Today recycling is a major source of the metal. Total world production today is about 15 million tonnes[5].

7.1.1 *Chemical and physical properties of aluminium*

Aluminium has an atomic weight of 27 and is number 13 in the periodic table. It is a light metal with a density of 2.7 and a melting point of 660.4°C. It is a soft, ductile, silver-white metal, with good electrical and heat conductivity. It is extremely resistant to corrosion, though its alloys are less so. It has an oxidation state of +3. Aluminium is a very reactive metal and can ignite in air if mixed with the oxides of other metals and exposed to heat. This is the basis of the thermite reaction which was used in incendiary bombs during World War II.

The principal inorganic compounds of aluminium are the oxide (Al_2O_3), hydroxide {$Al(OH)_3$}, sulphate {$Al_2(SO_4)_3$}, fluoride (AlF_3) and chloride ($AlCl_3$). Alums are double salts of aluminium with the general formula $M_2SO_4.R_2(SO_4)_3$, where M represents Na, K or NH_4, and R is Al (or Cr). Aluminium also forms organic compounds, some of which are highly reactive and become hot and fume in air.

7.1.2 *Production and uses*

Though aluminium compounds make up about 7% of the Earth's crust, mainly as clays in soils and silicates in rocks, it is commercially extracted only from a few types of ores. Bauxite (hydrated aluminium oxide) is the principal ore, with cryolite (sodium aluminium fluoride) used in smaller amounts. Three countries currently account for about 60% of the world's production of bauxite, Australia, Guinea and Jamaica. China and Russia are also major producers.

For over a century the Bayer process for the production of alumina (aluminium oxide) from bauxite and the Hass–Heroult process for the subsequent production of metallic aluminium have been in use[6]. The first step in the production of the metal involves the conversion of the ore into aluminum hydroxide by treating the bauxite with sodium hydroxide. The hydroxide is then converted into the oxide by heating to about 150°C. The alumina is then mixed with fused cryolite and fluorspar (CaF_2) at a high temperature. An electric current is passed through this cryolite bath to electrolyse the dissolved alumina. Oxygen is formed at the carbon anode and aluminium collects as a metal pad at the cathode.

Aluminium produced in this way is usually cast into ingots. These can be reheated and rolled and pressed into a variety of forms, as plate, sheet, wire and foil. At this stage the aluminium contains impurities, mainly iron and silicon, plus smaller amounts of zinc, gallium, titanium and vanadium. It is purified by fractional crystallisation and other means.

Aluminium metal has many industrial applications because of its lightness, electrical conductivity, corrosion resistance and other useful properties. Much of the world's production is used in the electrical industry, but there are increasing applications in construction and automobile engineering, aircraft and ship manufacture, building construction, and in the manufacture of domestic and industrial appliances and other goods. In most modern households aluminium will be found in cooking and food storage utensils, foil and takeaway containers, cookers, refrigerators, freezers, air conditioners and other appliances. Another major non-construction use of aluminium is in cans, especially for beverages.

In many of its applications, aluminium is used in alloy form, combined with other metals. These include copper, zinc, chromium, magnesium, nickel, titanium, iron and silicon. Alloying increases the strength and improves other qualities of aluminium, especially as regards casting and machining. There are numerous non-metallurgical uses of aluminium and its compounds. Alumina is used extensively in the refractory, abrasive and ceramics industries. Smaller amounts of aluminium compounds are used as flame retardants, catalysts, and adsorbents in a variety of industries. The food industry uses them as food additives, in baking powder, processed cheese, manufactured meat and other products. Aluminium is also widely used in pharmaceutical products, from toothpaste to antiperspirants and products such as antacids. Aluminium sulphate and other compounds are used almost universally as a coagulant for particle sedimentation in water treatment plants.

7.1.3 *Aluminium in food and beverages*

A great deal of information is available on levels of aluminium in foods and in the diets of different populations. However, some of the earlier data in the literature are limited and their accuracy suspect because of the inadequate analytical techniques used to produce them[7]. Even some more recent information may not be reliable because of continuing analytical problems[8].

Aluminium is ubiquitous in the physical environment and is present in every food and beverage. Levels in plant foods reflect the aluminium content of the soil and water where they were grown. All plants accumulate some of the element during growth. There are, in addition, a small number of others, such as tea and some herbs and spices, that are able to take up relatively high levels of the metal. A Spanish study found that levels of aluminium in 17 types of spices and aromatic herbs contained from 3.74 to 56.50 μg/g (dry weight) of the metal[9]. The average value for aluminium levels in dried leaves of *Camellia sinensis*, from which tea is brewed, is reported to be 50–1500 mg/kg[10]. The highest level recorded was 30 000 mg/kg[11].

7.1.3.1 *Aluminium in fresh foods*

There can be considerable variation in levels of aluminium in fresh foods, depending on a variety of environmental and geographical factors. Some of this may be due to soil contamination and failure to wash the products adequately. In the absence of gross contamination, the range of aluminium levels that can be expected in most fresh vegetables, as reported from a number of countries, is about 0.5 to 3.0 mg/kg. However, it was found that levels in vegetables available for sale in Sweden could vary by a factor of 10[12]. An Australian study found a range of < 0.1–18.5 mg/kg in root vegetables and 0.6–26.0 in lettuce[13]. Levels of aluminium in animal products generally are similar to those in plant foods, with reports of about 0.3 mg/kg in beef to 4.0 mg/kg in bacon in the US[14]. Similar data have been reported from other countries, with the exception of organ meat and offal, which may contain 0.1–53.0 mg/kg[15].

7.1.3.2 *Aluminium in processed foods*

The major source of aluminium in the diet in many countries is processed food. Aluminium is added either directly in legally permitted additives or adventitiously through the use of aluminium cooking utensils and containers. A considerable number of aluminium-containing food additives and processing aids are used internationally and there are differences between legislation relating to them and the types approved in different countries. Those permitted for use under food legislation in one country, Australia, are listed in Table 7.1 Additional substances, including aluminium, in its own right, as a silver metallic surface colour for decorating confectionery, are permitted in other countries, including the UK[16]. British law also permits use of aluminium silicates (officially numbered E554, E556, E557 and E559) in a limited number of foods at certain maximum levels[17].

Additives can have the most dramatic effects in increasing aluminium levels in manufactured foods, with cereal products particularly affected[18]. This is illustrated by data in Table 7.2 from the UK TDS of 1991[19]. The high level of aluminium in

Table 7.1 Aluminium-containing food additives permitted for use in Australia.

Additive	Use
Aluminium lakes	in synthetic colours
Sodium aluminosilicates	anti-caking agent flavourings premixes dried milk (in hot-drink vending machines) beverage whiteners
Sodium aluminium phosphate	acid aerator (in baking compounds/powders, e.g. in cakes)
Fosetyl aluminium/aluminium phosphide	agricultural residue found on certain fruits, vegetables, nuts, spices etc.
Aluminium stearate	processing aid
Aluminium sulphate	desiccating preparation
Sodium aluminosilicate	lubricant, anti-stick agent (e.g. in table salt)

Adapted from Allen, J.L. & Cumming, F.J. (1998) *Aluminium in the Food and Water Supply: an Australian Perspective. Reserch report* No. 202. Urban Water Research Association of Australia, Melbourne.

Table 7.2 Aluminium in UK foods.

Food group	Mean concentration (mg/kg, fresh weight)
Bread	3.7
Miscellaneous cereals	78
Carcass meat	0.49
Offal	0.35
Meat products	3.2
Poultry	0.33
Fish	5.5
Oils and fats	1.2
Eggs	0.27
Sugar and preserves	3.6
Green vegetables	1.8
Potatoes	2.2
Other vegetables	3.2
Canned vegetables	1.1
Fresh fruit	0.57
Fruit products	1.0
Beverages	1.7
Milk	<0.27
Dairy products	0.64
Nuts	11

Adapted from Ysart, G., Miller, P., Crews, H. *et al.* (1999) Dietary exposure estimates of 30 elements from the UK Total Diet Study. *Food Additives and Contaminants*, **16**, 391–403.

'Miscellaneous cereals' reflects use of aluminium-containing additives in some bakery products[20]. A US study found 86.0 mg/kg of aluminium in a commercially manufactured chocolate cake[21], while a similar type of cake analysed by the Australian Government Analytical Laboratory contained 224.8 mg/kg of the metal[22].

Additives used in several other types of foods, especially snack foods, can also contribute significant amounts of aluminium to the diet. In the US levels of up to 274 mg/kg have been reported in chewing gum, with as much as 4 mg in a single stick[23]. Chewing gum sold in the UK appears to contain less aluminium, with only 1.1 mg/kg reported by MAFF[24]. Among snack foods, certain types of extruded foods, such as cheese snacks, were found to be rich in aluminium in an Australian study[25].

7.1.3.3 Aluminium in infant formulas

Because of concern that infants and young children can be particularly at risk of possible adverse effects of a high intake of aluminium, and evidence that some ingredients used in infant formulas and feeds may contain relatively high levels of the metal, authorities in many countries pay particular attention to intakes by this group of consumers. There have been a number of reports of higher than normal levels of aluminium in infant formulas[26], particularly those for low-birth-weight infants[27]. Within different types of formulas, the highest levels of aluminium have been found in soy-based products[28]. However, in recent years there has been a reduction in the use of soy in infant foods and levels of aluminium in them have been reported to be decreasing[29]. Overall dietary exposure of infants to aluminium in the US has been estimated to be on average 0.7 mg/day[30]. Mean daily intakes by Australian infants are reported to be 0.048 mg on whey-based formulas, 0.408 mg on soy-based formulas, and 0.205 mg when infant cereal is consumed, compared to an intake of 0.104 mg by infants consuming breast milk. All these intakes are below the WHO/FAO PTWI[31].

7.1.3.4 Aluminium in beverages

Certain types of beverages can be a significant source of dietary aluminium. Water normally contains very low levels, since most aluminium compounds are only slightly soluble in water, except under acid conditions. However, high levels of aluminium have been reported in some natural surface waters, with concentrations up to 20 mg/litre in a few places[32]. In the Kii Peninsula of Japan, where aluminium levels in drinking water are high, an unusual form of dementia occurs among local residents. The condition, which has symptoms similar to those of Parkinson's disease, is believed by some investigators to be linked with an excess of aluminium in drinking water and the diet, though there is also evidence that a chronic low intake of calcium and magnesium, as well as a high intake of manganese, may be involved[33].

7.1.3.5 Aluminium in domestic water

Domestic water which is delivered from a central source through a reticulation system by municipal and other authorities is commonly treated with coagulants to remove turbidity and improve clarity and colour. The most commonly used coagulant is aluminium sulphate (alum). It has been shown that use of alum at normal accepted levels can increase aluminium levels in water threefold, from about 0.2 to 0.6 mg/litre[34]. Concentrations of aluminium in tap-water in several countries, including the US, UK, Sweden, Canada and France, have been reported to range from 10 to 2670 μg/litre, the latter value being for alum-treated water in the US[35]. Levels found in reticulated domestic water in eight Australian cities ranged from < 20–36 μg/litre in Sydney to < 20–360 μg/litre in Melbourne. The nationally weighted means for total aluminium were estimated to be 81 μg/litre for alum-treated water, to 66 μg/litre for non-alum treated water (in Sydney where ferric chloride, and not alum, is used)[36].

These Australian figures, and many of those reported in other countries, are generally in line with guidelines published by the Joint Expert Committee on Food Additives of WHO[37]. These recommend that the aluminium concentration in domestic water should not exceed 0.2 mg/litre. Interestingly the recommendation is made on aesthetic grounds, not on the basis of health – a practice followed both by the USEPA and the Australian NHMRC[38]. European Community guidelines[39] set a Maximum Permitted Concentration (MAC) of 220 μg/litre, with a preferred value of 50 μg/litre[40]. This MAC has been incorporated into law in England and Wales since 1989[41].

Addition of excessive amounts of aluminium sulphate to water in treatment plants has been known to increase aluminium levels in reticulated water considerably beyond acceptable levels. This occurred in the Camelford district of Cornwall, England, in July 1988 when 20 tonnes of aluminium sulphate were accidentally added to the regional water supply[42]. Consumption of 100–300 mg aluminium in local tap-water 2–3 days after the spillage was reported to have caused gastrointestinal effects, joint pains, headaches and blurred vision[43]. Subsequent investigations of locals produced evidence of possible organic brain damage[44].

7.1.3.6 *Levels of aluminium in bottled waters and canned soft drinks*

Analyses of 43 different local and imported brands of bottled mineral waters available on the Italian market found levels between 2 and 25 μg aluminium/litre, with a mean value of 6 μg/litre for still waters and 11 μg/litre for carbonated waters[45]. A similar study in Australia of mineral waters sold in supermarkets found a range of < 0.005–0.447 mg/litre, with a mean of 0.12 (± 0.19) mg/litre in still waters[46].

Because many soft drinks are sold in aluminium cans, there is some concern that since they are the preferred drink of many young people, they could be a source of high levels of intake of aluminium. Leaching from cans could be expected to occur, especially since many of the drinks, including those that are cola-based, are acidic. However, this view is not supported by the (limited) data available in the international literature, which indicates that aluminium levels in both cola- and non-cola soft drinks range from < 0.1–3.2 mg/litre[47], though there is one Australian report of a level of 3.24 mg/litre in a widely consumed brand of cola drink[48]. An investigation in the UK concluded that aluminium levels in canned beverages were normally < 1 mg/kg[49].

Reported typical levels of aluminium in fruit-based drinks are 0.3–1.5 mg/litre[50]. A US study found levels of 5 mg/litre or less in fruit juices, with the exception of 23 mg/litre in very acid rhubarb juice[51]. Australian freshly made orange juice was reported to contain 0.03 mg aluminium/litre, in contrast to 0.015 mg/litre in commercially prepared juice[52].

7.1.3.7 *Aluminium in brewed tea*

Tea is said to be the second most commonly consumed beverage in the world, after water[53]. We have already seen that the tea plant, *Camellia sinensis,* is a natural accumulator of aluminium, sometimes to very high levels. Even after tea is brewed and the spent tea leaves strained away, the resulting beverage can contain significant levels of aluminium. There have been reports of as much as 40–100 mg/litre in infused tea[54]. However, more typical levels are 2–6 mg/litre[55]. Significant differences in levels in teas from different countries have been reported. These differences may owe more to the maturity of the leaves used, than to their place of origin[56].

7.1.3.8 *Aluminium in alcoholic beverages*

Though wooden barrels and glass bottles continue to be used, especially for traditional beers, aluminium casks and cans are today increasingly used in the brewing industry. Thus many beers might be expected to contain a significant amount of aluminium. However, though levels of up to 8–10 mg/kg have been reported in some UK beers, these were the exception. More than 60% of British beers tested in an extensive study had less than 0.2 mg aluminium/litre[57]. Similar levels were found in another UK study conducted by MAFF[58]. Canned beers in Australia have been reported to contain aluminium levels similar to those found in tap-water, around 0.15 mg/litre, with a range of 0.10–0.25 mg/litre[59]. A Scandanavian study has reported that aluminium levels in European beers range from < 0.1 to about 1.0 mg/litre[60].

7.1.4 **Dietary intake of aluminium**

A PTWI of 7 mg aluminium/kg body weight was established by JEFCA in 1989[61]. This is equivalent to 420 mg/week or 60 mg/day for a 60 kg adult. Recent estimates of dietary intakes of aluminium in a number of countries show that this PTWI is not normally exceeded. Daily adult intakes are reported to average 3.9 mg in the UK[62], 3.1 mg in Holland[63], 2.5–6.3 mg in Italy[64], 4.5–6.5 mg in Australia[65], 2.5 mg in Japan[66], and in the US 7–9 mg[67]. The more general figures published by FAO/WHO in 1989 give an adult intake of 6–14 mg/day and 2–6 mg/day for children[68].

Reported intakes of aluminium from food and beverages have been falling in several countries in recent years. Some of these falls are most probably artefacts due to improvements in analytical techniques, but more recent data point to real reductions. In the US there is evidence that a decrease in the use of aluminium compounds in processed cheese has had a significant effect on dietary levels of the metal[69]. In the UK the fall in aluminium intakes for adults from 11 mg/day in 1994 to 3.4 mg/day in 1997 has been attributed to the lower aluminium concentrations in bread and other cereal products resulting from a decrease in use of aluminium-containing additives.

7.1.4.1 High consumers of aluminium

There are a number of circumstances in which higher than normal intakes of aluminium may occur, for example where there is a high proportion of aluminium-rich foods in the diet. This was found to be the case in an Australian study which looked at aluminium intakes and found that the foods and beverages which contributed most to the total daily aluminium intakes of all population groups aged two years and over were cakes and tea[70]. Using data from the National Nutrition Survey (NNS) of 1995, and concentrating on 95th percentile intakes, aluminium levels ingested by 'high consumers' of tea and cakes were estimated. These were, for the youngest group of 2–11 years of age, 10 mg/kg bw, 6 mg/kg bw for 12–18 years old, 5 mg/kg bw for 19–64 years old and 4 mg/kg bw for those over 65 years old. Though aluminium intakes by the adults were high compared to average intakes by the whole population, only the youngest group was found to exceed the PTWI, with the 12–18-year-olds not far off. As the authors of the Australian report point out, though these intakes are high, they represent only a small proportion of the whole population. Only 5–7% of 2–11-year-olds were reported to consume tea, according to the NNS data, and the high intakes reflect only 5% (the 95th percentile) of that group.

7.1.4.2 Adventitious contributions of aluminium to the diet

Accidental and other non-food sources of aluminium can also be a source of increased intake. It is difficult, however, to quantify the contribution made to the diet by such means. Several studies have found that use of aluminium cooking utensils, as well as aluminium foil, can increase the amount of the metal in some, but not all, foods[71]. In countries where aluminium utensils are widely used, such as India, significant levels of the metal have been found in most domestically prepared foods[72].

Uptake depends, to a certain extent, on the acidity of the food. Potatoes, for instance, showed no increase when boiled in an aluminium saucepan, compared to a glass pan. Stewed tomatoes, in contrast, with 0.14 mg/kg after cooking in glass, had 15.5 mg/kg after use of an aluminium saucepan. Levels in cabbage likewise increased from 0.34 to 90.8 mg/kg. Increases in aluminium in meat stored and cooked in aluminium foil also occurred, though to a lesser extent. Raw frozen beef contained 0.39 mg/kg; when frozen in foil, the level went up to 0.47 mg/kg. Cooking in foil increased it to 5.7 mg/kg, and it reached 7.4 mg/kg when reheated in the foil.

Uptake of aluminium by beverages stored in aluminium cans has been shown to occur, sometimes to a significant extent. Beer in an all-tinplate can contained 0.15 mg aluminium/kg, compared to 0.38 mg/kg in a can with aluminium ends. Fresh orange juice in an aluminium can contained 12.4 mg/kg, compared to 1.3 mg/kg in fresh juice. Aluminium in water boiled in an aluminium pan increased in 20 minutes from 0.05 to 8.08 mg/kg, though only at a low pH of 3.0[73]. A UK study found that rhubarb, a highly acid food, took up more than 100 mg/kg from an aluminium cooking pan. Even such a non-acidic product as sponge flan cases, prepacked in aluminium foil, were found to have 1150 mg aluminium/kg, compared with 13 mg/kg in unpacked cases[74].

It has to be pointed out that though aluminium can be contributed to the diet from such adventitious sources, this route of dietary intake is probably neither consistent nor of major significance[75]. However, this is unlikely to reassure everyone concerned about the use of aluminium in contact with food. There has been opposition to the use of the metal in catering for many years, and it is still widespread today. As early as 1886, shortly after the metal began to be used extensively for making cooking utensils, claims were made that food cooked in them was toxic[76]. The resulting controversy was vigorous. It was still going on forty years later when British manufacturers of aluminium felt obliged to issue a report in defence of the metal[77]. Almost immediately their report was attacked by a Dr Spira, who argued strongly in support of the anti-aluminium case[78]. The controversy rumbled on for another four decades, until, in 1971, the American Medical Association entered the field with a statement that evidence was lacking to support the view that the small amounts of aluminium ingested in the diet, even when aluminium cooking utensils were used, had any ill effects on human health[79]. This view was substantially repeated by the USFDA some years later[80].

Even now, three decades later, the controversy refuses to go away. Reports in the medical literature, as well as in the popular media, of connections between aluminium and a variety of diseases, and even the commercial promotion of filtering and other devices to remove aluminium from domestic water, makes it an issue of continuing interest to the community. This concern is not confined only to the non-scientific public but is even expressed by some food scientists[81].

7.1.5 Aluminium absorption

In spite of the widespread distribution of aluminium in the environment, as well as its presence in all diets, little is normally absorbed from food or beverages. Why this is so is still not fully understood[82]. It is clear that the chemical form of the element is an important factor in controlling absorption. Aluminium hydroxide, as well as the phosphate, is very insoluble and is poorly absorbed from the gut, in contrast to the soluble and easily absorbed citrate. Both vitamin D and the parathyroid hormone may help to moderate absorption of aluminium. Body stores of iron also appear to play a part, since there is increased absorption in people with low ferritin levels. There is evidence that age and other conditions affect absorption, with increased aluminium absorption in those suffering from Alzheimer's disease[83].

The percentage of aluminium in ingested food that is absorbed from the gastrointestinal tract varies widely between individuals[84]. Uptakes as high as 7% have been reported in healthy young men[85]. In contrast, absorption of aluminium in the elderly Alzheimer's patients was only 0.06–0.1%[86]. Other studies have found that absorption in general accounts for upwards of 1% of total ingested aluminium[87]. Once absorbed, aluminium accumulates in the bone, kidney, liver and brain. In blood, aluminium is bound mainly to transferrin and albumin[88]. Most of the absorbed aluminium is excreted in urine, with a small amount in faeces and bile[89].

The GI tract is a major barrier to aluminium absorption. When this is by-passed, as in patients undergoing total parenteral nutrition (TPN), a high body burden of the metal can result. This has been shown in patients on dialysis where aluminium-treated domestic water was used in the dialysate. The aluminium was directly transferred across the dialysing membrane, with serious consequences for the patient. For this reason, care is taken today that only aluminium-free water is used during dialysis[90].

The kidneys are normally able to excrete absorbed aluminium efficiently, but in kidney failure this does not occur. In order to prevent build-up of high levels of serum phosphate, a symptom of the disease, chronic renal failure patients are prescribed aluminium hydroxide-containing gels to limit absorption of phosphate from the GI tract. Unfortunately, as a consequence, these patients, with inadequate kidney function and a high aluminium intake, can accumulate a high body burden of the metal. Aluminium can also be absorbed in excess, even by healthy persons, if they are given large oral loads of the metal[91]. In such circumstances, urinary excretion increases, but some storage also takes place. Tissues most affected are liver, kidney, spleen and bone. Some of the extra aluminium may also be stored in the brain.

7.1.5.1 *Metabolic consequences of high aluminium absorption*

The effects of high aluminium intake, especially with regard to its possible connection with Alzheimer's disease, have been the subject of considerable controversy in recent years. There is no doubt that aluminium is a neurotoxin. As has been noted above, chronic renal failure patients who have undergone dialysis with aluminium-containing water have frequently developed severe neurological problems[92]. Accumulation of aluminium in the brain has also been associated with Alzheimer's disease[93]. High concentrations of aluminium have been reported in the neurofibrillary tangles found in the brains of Alzheimer's patients[94], but whether the metal is a cause, or a consequence, of the disease is still debated[95]. High intakes have also been associated with osteomalacia in adults and defective bone mineralisation in children[96], as well as with several other conditions, such as metabolic alkalosis and bowel obstruction.

Whether aluminium-related conditions such as have been described here can be caused by intake of aluminium in normal foods and beverages by individuals who do not have chronic kidney failure, are not on TPN or dialysis, and do not consume large amounts of aluminium-containing pharmaceutical products, is far from clear. A recent Australian study has shown that intake of aluminium in treated water as well in ordinary food and beverages, including tea, was minimal[97]. Some evidence which suggests that there is an association between aluminium in drinking water and Alzheimer's disease (AD) has been produced by a number of epidemiological studies[98], while others have failed to do so[99]. However, as has been pointed out by Doll[100] among others, epidemiological studies of this kind have many confounding variables and inherent design weaknesses which make interpretation of the results difficult. According to MAFF, evidence supporting the view that a reduction in the aluminium intake of the general population would be likely to reduce the incidence of AD is too tentative to justify changes in the use of aluminium sulphate in water treatment[101].

7.1.6 *Analysis of foodstuffs for aluminium*

As has been mentioned, many of the analytical data relating to aluminium in food and beverages published some years ago were unreliable, because of inadequate equipment and procedures available to analysts. The principal difficulty they experienced was due to the ubiquitous nature of aluminium and the high risk of contamination throughout the various steps in the analytical process. Newer techniques have helped overcome these problems and enabled greater accuracy, reliability and sensitivity to be achieved in the determination of aluminium in dietary components and other materials in recent years. However, the problem of contamination can still occur and render results unreliable unless the greatest care is taken to isolate samples from the environment. For this reason, anyone embarking on aluminium determination in food and beverages for the first time would be well advised to read the paper by Savory and Wills on the subject, even though it was written back in 1989[102].

Wet acid digestion, preferably in a closed-circuit microwave system, is the preferred sample preparation method. Dry ashing can be used, though unless it is carried out at low temperatures, loss of aluminium chloride can occur[103]. Graphite furnace AAS provides an accurate and relatively easy method for determining aluminium levels in foods and beverages, especially when relatively small numbers of samples are involved. Background correction using a deuterium lamp, or Zeeman background correction, is desirable[104]. ETA-AAS, following microwave digestion of samples in high-pressure bombs, was used to study aluminium and other metals in commercial canned seafoods by Tahán and colleagues, who describe their method in detail[105].

ICP-MS was used to analyse aluminium and other trace elements in samples collected in the UK Total Diet Study[106]. ICP-AES has been successfully used to determine aluminium in the US TDS[107], as well as in many small-scale investigations, such as a study of trace elements in breakfast cereals[108]. A considerable advantage of using ICP rather than GF is that it helps to overcomes matrix problems caused by the refractive nature of aluminium.

If available, and the cost is appropriate to the nature of the investigation, NAA should be considered for the analysis of aluminium in dietary samples[109]. Its use eliminates the need for extensive sample preparation, and thus reduces the likelihood of contamination. However, in samples with a high phosphorus content, a preliminary separation step may be necessary to remove this element as it can cause interference.

7.2 Tin

Tin is one of the ancient metals. There is evidence of tin production in Anatolia, in Turkey, in the Early Bronze Age[110]. It was valued by early metallurgists because of its ability to form alloys with distinctive and very useful properties when mixed with other metals. Soft and vulnerable copper was converted into tough and resistant bronze by the addition of tin; lead was converted into rigid, and lighter, pewter. A bronze containing 10% tin was produced in Cornwall in the Middle Bronze Age, some 4500 years ago. Though a tin bracelet was found on the Greek island of Lesbos, dating from about 3000 BC, and grains of metallic tin have been recovered from the wreck of a Phoenician ship in the Mediterranean[111], it is only in relatively recent times that the metal has been used in its own right. For the past two centuries considerable quantities of tin have been used to plate utensils used to store and process food.

Even before tinplate was developed, the metal had an important place in the kitchen and on the dining table. Bronze cooking and storage utensils, and eating implements, were common, at least in wealthy homes. Pewter plates and dishes were only replaced when porcelain and glazed earthenware were introduced into common use in the seventeenth century. Undoubtedly, during those earlier centuries, some food must have been contaminated with tin leached from bronze and pewter culinary equipment, but there is no evidence that our medieval forebears suffered any major ill effects from intake of the metal in this way. Indeed, the culinary use of tin has stood the test of time well, from a practical as well as a health point of view. Only in some exceptional cases has tin resulted in serious poisoning. Even today, in spite of a few expressions of concern, it is still generally accepted as safe for use with food.

7.2.1 Chemical and physical properties of tin

Tin has the symbol Sn, after its Latin name *stannum*. It is element number 50 in the periodic table, with an atomic weight of 118.7. Tin is normally a mixture of three crystalline forms and, consequently, its specific gravity is not fixed but lies between 5.8 and 7.3, depending on the proportion of each form present. The metal is soft, white and lustrous and is easily rolled into wire and foil, and extruded into tubes. It has a low melting point of 231.9°C and is highly resistant to corrosion. Oxidation states of tin are +2 and +4. It forms two series of compounds, stannous and stannic. Among its inorganic compounds are stannous chloride ($SnCl_2$), stannic chloride ($SnCl_4$), stannic oxide (SnO_2) and sodium metastannate (Na_2SnO_3). Tin also forms a number of organic compounds, the organotins. These are important industrially, especially the alkyl and phenyl compounds.

7.2.2 Production and uses of tin

Though tin is widely distributed in small amounts in most soils, it is commercially produced in quantity in only a few places in the world. Tin deposits are associated with certain types of granite and occur in only a few known localities such as Malaysia, China, Bolivia, Cornwall (UK), Saxony–Bohemia and Nigeria. In all these areas it is found mainly as the mineral cassiterite or tinstone (stannic oxide, SnO_2). Much of the South-east Asian ore occurs in alluvial deposits, known as placer deposits, from which it is extracted by washing away the associated sand and gravel[112]. Mined vein ore in granite rock is much more difficult to extract and requires that the ore is heated with powdered charcoal in a reverberatory furnace.

Until well towards the end of the twentieth century, about 60% of the world production of tin took place in what is known as the South-east Asian Tin Belt, extending from Burma (present day Myanmar) and Thailand to Malaysia and Indonesia. But in recent years China has become the world's major producer, with Brazil, Peru and Bolivia following closely behind[113]. Small-scale production is undertaken in some other countries, such as Australia and Zambia, but this is usually intermittent and depends on world prices. The ancient Cornish tin mines have not been in commercial use for some years. Interestingly, the US, which uses more tin than any other country, has little tin production of its own.

Primary production of tin worldwide is about 250 000 tonnes per annum, with an increasing amount of secondary tin being produced each year from scrap. Most tin is used to produce tinplate, which is sheet steel coated with tin to prevent rusting. Today the plate is usually applied by electrodeposition, though dipping in molten metal is still used on a small scale in some specialist industries, such as the manufacture of copper cooking utensils. Most of the tinplate is used for making food containers ('tin cans'). Tin coatings are also used in the manufacture of engineering and electrical components and for other applications where resistance to corrosion is important.

Important alloys of tin include bronze (copper and tin) which was one of the earliest alloys to be used. Solder (mainly lead and tin) melts more easily than pure tin and is widely used to join metals together. Pewter (lead and tin), and various related alloys with antimony, copper and other metals, which are known by such names as Monel and Britannia metal, are used today largely for ornamental purposes. An alloy of tin, with lead and antimony, was formerly widely used as type metal in printing works.

About 5% of total tin production is used in the chemical industry. Tin compounds are employed in the manufacture of certain types of glass and enamels, in dyeing and printing of textiles, and as a reducing agent in certain chemical processes. The pharmaceutical industry uses them in a number of ways, such as tin fluoride in toothpaste. Organic compounds of tin are used as stabilisers in PVC plastics, as well as in rubber paints. Other organic compounds are used as fungicides and other agrochemicals. An important, though highly controversial, use is that of organotin compounds as antifouling agents on boats and ships.

7.2.3 Tin in food and beverages

The normal level of tin in food and beverages is low, except where they have been in contact with the metal in cans and other tinplated containers, as is shown in Table 7.3, which is based on data collected in the UK TDS in 1994[114]. Similar levels were reported in subsequent studies, for example in the 1997 TDS, which found that levels of tin in all food were below 0.1 mg/kg, except for canned vegetables and fruit products, which include canned fruit[115]. Data from other countries confirm the UK finding of low levels of tin in most foods, apart from those preserved in cans. A French study found that tin levels in fresh foods were 0.03 ± 0.03 mg/kg, compared to 76.6 ± 36.5 mg/kg in food preserved in unlacquered cans, and 3.2 ± 2.3 in lacquered cans[116]. Similar levels have been reported in Japanese food[117].

Table 7.3 Tin in UK foods.

Food group	Mean concentration (mg/kg fresh weight)
Bread	0.03
Miscellaneous cereals	0.02
Carcass meat	0.02
Offal	0.02
Meat products	0.31
Poultry	< 0.02
Fish	0.44
Oils and fats	0.02
Eggs	< 0.02
Sugar and preserves	0.02
Green vegetables	0.02
Potatoes	< 0.02
Other vegetables	0.02
Canned vegetables	44
Fresh fruit	0.03
Fruit products	17
Beverages	0.02
Milk	< 0.02
Dairy products	0.31
Nuts	0.03

Adapted from Ysart, G., Miller, P., Crews, H. *et al.* (1999) Dietary exposure estimates of 30 elements from the UK Total Diet Study. *Food Additives and Contaminants*, **16**, 391–403.

As is clear from the above studies, and from many others reported in the literature, levels of the metal in food are in every case related to storage in tinplated cans or contact with tinplated culinary utensils[118]. The tinplate is normally little affected by the food with which it is in contact, but can dissolve, usually very slowly, if certain substances are present which accelerate corrosion[119]. These substances, which include residual and nitrate oxygen, are known as cathode depolarisers. Acid conditions can also encourage corrosion. It is to prevent this that insides of cans are commonly coated with lacquer, generally a thermosetting resin polymerised on the surface by heat[120].

Uptake of tin by food depends on the nature of the foods as well as whether cans are lacquered or not. Meat and meat products are often contained in lacquered cans and are not normally aggressive towards the tin. Canned meats seldom contain more than trace amounts of the metal. In contrast, tomatoes, which often have a high nitrate content, and are often sold in unlacquered cans, have been found to contain more than 50 mg/kg of tin. Highly coloured fruits, such as blackcurrants and raspberries, contain anthocyanins which are highly aggressive towards tinplate, even when preserved in lacquered cans, and have been found to accumulate up to 100 mg/kg of tin[121]. In a study of tin in infant foods in the UK, levels of up to 33 mg/kg were found in some samples which were either sold in cans or contained fruit ingredients which may have been stored previously in cans[122]. In a similar study of infant foods in New Zealand, levels of up to 18 mg/kg of tin were detected in some canned products[123].

Considerably higher levels of tin have, on occasion, been found in individual cans, and in a few cases in whole batches of cans. Tins of Italian peaches, on sale in Germany, were found to contain 400 mg tin/kg. A high level of corrosion of the tinplate was attributed to high levels of nitrate in the water used to process the fruit[124]. A spectacular example of tin uptake by food stored in cans was seen in foods stored in cans which had been taken to the Arctic in 1837 and remained there unopened for a century, with 783 mg/kg in veal and 2440 mg/kg in carrots[125].

7.2.3.1 Organotin compounds in food

In addition to inorganic tin, food may also contain organic tin compounds, which pose a much greater risk to health than do the inorganic compounds. While it is highly unlikely that tin in canned food is in anything other than the inorganic form, there is evidence that certain foods may also contain organotin compounds. This does not appear to be as a result of natural conversion of inorganic to organic forms, as is the case for mercury in marine organisms. It is most probably the result of increasing industrial use of organotin compounds. Dibutyltin salts are used as stabilisers in PVC plastics, and trialkyl- and triaryl-tin compounds are used as biocides. Tributyltin (TBT) has been used extensively until recently as a marine antifouling agent, though legislation has been introduced in many countries to limit its use.

An Indian study has revealed the high levels of accumulation of organotin compounds in freshwater food chains, largely as a result of the discharge of untreated domestic sewage into rivers. High concentrations of mono-, di- and tri-butyltin were detected in dolphin, several species of fish and various invertebrates in the River Ganges, pointing to high levels of biomagnification of these compounds in living organisms[126]. High levels of organotin compounds have also been detected in municipal wastewater and sewage sludge in Switzerland[127].

Because of the use of TBT-containing antifouling paints on the majority of the world's deep-ocean ships, as well as on smaller vessels, many harbours and coastal waters are now heavily contaminated with organotin compounds. This has resulted in contamination of many marine organisms, such as oysters in parts of Australia[128] and various molluscs in rivers in the UK and elsewhere[129]. In the early 1980s TBT was detected in Atlantic farmed salmon in the UK, apparently as a result of the use of antifouling paints on netting cages used to contain the fish. It was also detected in freshwater trout as a result of spill from a wood treatment plant which used TBT as a timber preservative[130]. Because of concern about the danger to the consumer's health, the use of TBT-based antifouling paints has been effectively banned in the UK since 1987[131]. Most other countries now have legislation to restrict the use of these products.

7.2.4 Dietary intakes of tin

Apart from the dramatic findings in the case of the food stores of the ill-fated Franklin expedition, even the highest of the levels of tin reported in UK, as in several other countries, including France[132] and Japan[133], seldom exceed the UK's legal limit of 200 mg/kg for tin in food[134]. Total tin intake in the UK in 1997 was reported to be 1.9 mg/day for the mean and 6.3 mg/day for 97.5th percentile adult consumers, with 94% of this intake accounted for by consumption of canned foods and fruit products[135]. These intakes are similar to those reported in other countries, such as the US and France, which were reported to have intakes of 3.5 and 2.7 mg/day of tin respectively[136]. A lower intake of 0.644 mg/day of tin by Japanese adults has been reported[137]. Since the PTWI for tin established by the WHO is 2 mg/kg bw, equivalent to 120 mg/day for a 60 kg person, tin intakes in the UK, and in the other countries for which we have consumption data, is not a cause for concern.

7.2.5 Absorption and metabolism of tin

Tin ingested in food is poorly absorbed and is excreted mainly in faeces. At most about 1% of ingested tin enters the blood. Absorption appears to be related to the chemical form of the element, with Sn(II) being four times more readily absorbed than Sn(IV) compounds. Absorbed tin is rapidly excreted initially, but small amounts may be retained in kidney, liver and bone, with little in soft tissue. Excretion is mainly in urine, with a small amount also in bile[138]. Tin does not appear to have any biological role in the body, though it has been shown to be capable of inducing the enzyme haemoxygenase and, in rats at least, to stimulate growth under certain conditions[139].

Inorganic tin is generally considered to be non-toxic. High intakes of tin in food can irritate the GI tract and can cause nausea and vomiting in some individuals. These are short-term effects and usually occur at concentrations above 200 mg/kg[140]. However, there is considerable individual variation in response to tin. Levels as high as 500 mg/litre in fruit juice have failed to cause gastric upset in some volunteers. An intake of 200 mg/kg over an extended period of time failed to cause toxicity in others[141]. In rats, however, prolonged intake of low levels of inorganic tin have been associated with growth retardation and histological changes in the liver. Tin may also interfere with the metabolism of iron and zinc. Rats fed a tin-enriched diet developed anaemia. They also lost more zinc in faeces and retained less in bones than did controls[142].

Organotin compounds are much more toxic than are inorganic forms of the element. Triethyl- and trimethyl-tin, like their lead equivalents, are dealkylated rapidly to toxic trialkyl derivatives and are then metabolised more slowly to less toxic di- and mono-derivatives. All of these and other related compounds, especially TBT, are genotoxic[143]. Triethyltin interferes with cellular metabolism by uncoupling oxidative phosphorylation, causing mitochondrial damage and resulting in cerebral oedema leading to neurotoxicity. Organotin compounds can also stimulate catecholamines, leading to hyperglycaemia, changes in blood pressure and immunotoxicity[144].

7.2.6 *Analysis of foodstuffs for tin*

Problems can be experienced when analysing tin in foods and beverages. Graphite furnace AAS, with deuterium lamp or Zeeman background correction, is an effective analytical procedure, and has helped overcome some of the difficulties experienced by some earlier analysts. However, high ashing temperatures are required and matrix interference can be severe with different tin compounds. The use of pyrolytically coated graphite tubes and *in situ* coating is recommended, allowing lower atomisation temperatures and higher sensitivity[145].

Total tin concentrations in foods and beverages are routinely determined as part of multielement surveys in a number of countries. In the UK this is performed using ICP-MS, following acid digestion in pressurised vessels using microwave heating[146]. A similar procedure has been followed in France[147]. Techniques for speciation of organotin compounds using gas chromatography with mass spectrometry are available[148]. Different techniques which employ liquid chromatography–mass spectrometry interfacing systems to determine levels of organic tin have been applied to the analysis of food and beverages[149].

References

1. Gerhardsson, L., Oskarsson, A. & Skerfving, S. (1994) Acid precipitation – effects on trace elements and human health. *Science of the Total Environment*, **153**, 237–45.
2. Martin, R.B. (1994) Aluminium: a neurotoxic product of acid rain. *Accounts of Chemical Research*, **27**, 204–10.
3. Tylecote, R.F. (1976) *A History of Metallurgy*, pp. 149–50. The Metal Society, London.
4. Roskill Report (1998) *The Economics of Bauxite and Alumina.* http://www.roskill.co.uk/bauxite.html
5. Bureau of Mines (1997) *Mineral Year Book*. US Department of the Interior. US Government Printing Office, Washington, DC.
6. Mahadevan, H. & Ramachandran, T.R. (1996) Recent trends in alumina and aluminium production technology. *Bulletin of Materials Science*, **19**, 905–20.
7. Allen, J.L. & Cumming, F.J. (1998) *Aluminium in the Food and Water Supply: an Australian Perspective. Research Report No. 202.* Water Services Association of Australia, Melbourne, Vic.
8. Pennington, J.A.T. (1987) Aluminium content of foods and diets. *Food Additives and Contaminants*, **5**, 161–232.
9. Lopez, F.F., Cabrera, C., Lorenzo, M.L. & Lopez, M.C. (2000) Aluminium levels in spices and aromatic herbs. *Science of the Total Environment*, **257**, 191–7.
10. Hampton, M.G. (1992) Tea consumption. In: *Tea Cultivation to Consumption*, (eds. K.C. Wilson & M.N. Clifford), pp. 459–81. Chapman & Hall, London.
11. Ravichandran, R. & Parthiban, R. (1998) Aluminium content in South Indian teas and their bioavailability. *Journal of Food Science and Technology*, **35**, 349–51.
12. Jorhem, L. & Haegglund, G. (1992) Aluminium in foodstuffs and diets in Sweden. *Zeitschrift für Lebensmittel Untersuchung und Forschung*, **194**, 38–42.
13. Allen, J.L. & Cumming, F.J. (1998) *Aluminium in the Food and Water Supply: an Australian Perspective. Research Report No. 202.* Water Services Association of Australia, Melbourne, Vic.
14. Pennington, J.A.T. & Jones, J.W. (1988) Aluminum in American diets. In: *Aluminum in Health, a Critical Review* (ed. H.J. Gitelman), pp. 67–100. M. Dekker, New York.

15. Allen, J.L. & Cumming, F.J. (1998) *Aluminium in the Food and Water Supply: an Australian Perspective. Research Report No. 202.* Water Services Association of Australia, Melbourne, Vic.
16. Statutory Instrument (1995a) *The Colours in Food Regulations* (SI No. 3124). HMSO, London.
17. Statutory Instrument (1995b) *The Miscellaneous Food Additives Regulations* (SI No. 1384). HMSO, London.
18. Pennington, J.A.T. (1987) Aluminium content of foods and diets. *Food Additives and Contaminants*, **5**, 161–232.
19. Ysart, G., Miller, P., Crews, H. *et al.* (1999) Dietary exposure estimates of 30 elements from the UK Total Diet Study. *Food Additives and Contaminants*, **16**, 391–403.
20. Ministry of Agriculture, Fisheries and Food (1998) *Lead, arsenic and other metals in food. Food Surveillance Paper No. 53.* The Stationery Office, London.
21. Allen, J.L. & Cumming, F.J. (1998) *Aluminium in the Food and Water Supply: an Australian Perspective. Research Report No. 202.* Water Services Association of Australia, Melbourne, Vic.
22. Allen, J.L. & Cumming, F.J. (1998) *Aluminium in the Food and Water Supply: an Australian Perspective. Research Report No. 202.* Water Services Association of Australia, Melbourne, Vic.
23. Lione, A. & Smith, J.C. (1982) The mobilization of aluminium from three brands of chewing gum. *Food and Cosmetic Toxicology*, **22**, 265–8.
24. Ministry of Agriculture, Fisheries and Food (1993) *Aluminium in Food. Food Surveillance Paper No. 39.* HMSO, London.
25. Allen, J.L. & Cumming, F.J. (1998) *Aluminium in the Food and Water Supply: an Australian Perspective. Research Report No. 202.* Water Services Association of Australia, Melbourne, Vic.
26. Koo, W.W.K., Kaplan, L.A. & Krug-Wispe, S.K. (1988) Aluminium contamination of infant formulas. *Journal of Parenteral and Enteral Nutrition*, **12**, 170–3.
27. Bouglé, D., Foucault, P., Voirin, J. & Duhamel, J.F. (1989) Concentrations en aluminium des formules pour prématurés. *Archives Pédiatriques*, **46**, 768.
28. Baxter, M.J., Burrell, J.A. & Massey, R.C. (1990) The aluminium content of infant fermula and tea. *Food Additives and Contaminants*, **7**, 101–7.
29. Coni, E., Bellomonte, G. & Caroli, S. (1993) Aluminium content of infant formulas. *Journal of Trace Elements and Electrolytes in Health and Disease*, **7**, 83–6.
30. Pennington, J.A.T. & Schoen, S.A. (1995) Estimates of dietary exposure to aluminium. *Food Additives and Contaminants*, **12**, 119–28.
31. Allen, J.L. & Cumming, F.J. (1998) *Aluminium in the Food and Water Supply: an Australian Perspective. Research Report No. 202.* Water Services Association of Australia, Melbourne, Vic.
32. Driscoll, C. (1989) The chemistry of aluminium in surface waters. In: *The Environmental Chemistry of Aluminum* (ed. G. Sposito). CRC Press, Boca Raton, Florida.
33. Yase, Y., Yoshida, S., Kihira, T., Wakayama, I. & Komoto, J. (2001) Kii ALS dementia. *Neuropathology*, **21**, 105–9.
34. Miller, R.G., Kopfler, F.C., Kelty, K.C., Stober, J.A. & Ulmer, N.S. (1984) The occurrence of aluminum in drinking water. *Journal of the American Water and Wastewater Association*, **76**, 84–91.
35. Pennington, J.A.T. (1987) Aluminium content of foods and diets. *Food Additives and Contaminants*, **5**, 161–232.
36. Allen, J.L. & Cumming, F.J. (1998) *Aluminium in the Food and Water Supply: an Australian Perspective. Research Report No. 202.* Water Services Association of Australia, Melbourne, Vic.
37. WHO (1989) *Evaluation of Certain Additives and Contaminants. 33rd Report of the Joint Expert Committee on Food Additives. WHO Technical Series 776.* World Health Organization, Geneva.
38. Allen, J.L. & Cumming, F.J. (1998) *Aluminium in the Food and Water Supply: an*

Australian Perspective. Research Report No. 202. Water Services Association of Australia, Melbourne, Vic.
39. EC (1986) Directive relating to the quality of water intended for human consumption (80/778/EEC). *Official Journal of the European Communities* L229/11–29. European Community, Brussels.
40. Plessi, M. & Monzani, A. (1995) Aluminium determination in bottled mineral waters by electrothermal atomic absorption spectrometry. *Journal of Food Composition and Analysis*, **8**, 21–26.
41. Statutory Instrument (1991) *The Water Supply (Water Quality) (Amendment) Regulations 1991.* (SI No. 1837). HMSO, London.
42. Lowermoor Incident Advisory Group (1991) *Water Pollution at Lowermoor, North Cornwall, 2nd Report.* HMSO, London.
43. Rowland, A., Grainger, R., Smith, R.S., Hicks, N. & Hughes, A. (1990) Water contamination in North Cornwall: a retrospective cohort study into the acute and short-term effects of the aluminium sulphate incident in July 1988. *Journal of the Royal Society of Health*, **110**, 166–72.
44. Altmann, P., Cunningham, J., Dhanesha, U., Ballard, M., Thompson, J. & Marsh, F. (1999) Disturbance of cerebral function in people exposed to drinking water contaminated with aluminium sulphate: retrospective study of the Camelford water incident. *British Medical Journal*, **319**, 807–12.
45. Plessi, M. & Monzani, A. (1995) Aluminium determination in bottled mineral waters by electrothermal atomic absorption spectrometry. *Journal of Food Composition and Analysis*, **8**, 21–6.
46. Allen, J.L. & Cumming, F.J. (1998) *Aluminium in the Food and Water Supply: an Australian Perspective. Research Report No. 202.* Water Services Association of Australia, Melbourne, Vic.
47. Allen, J.L. & Cumming, F.J. (1998) *Aluminium in the Food and Water Supply: an Australian Perspective. Research Report No. 202.* Water Services Association of Australia, Melbourne, Vic.
48. Walton, J., Hams, G. & Wilcox, D. (1994) *Bioavailability of aluminium from drinking water: co-exposure with foods and beverages. Research Report No. 83* Urban Water Research Association of Australia, Melbourne, Australia.
49. PIRA International (1994) Literature survey and critical review of published data relating to leaching of aluminium from containers into foods and beverages (unpublished). Quoted in MAFF (1998) *Lead, Arsenic and Other Metals in Food. Food Surveillance Paper No. 52.* The Stationery Office, London.
50. Aikoh, H., & Nishio, M.R. (1996) Aluminium content of various canned and bottled beverages. *Bulletin of Environmental Contamination and Toxicology*, **56**, 1–7.
51. Pennington, J.A.T. (1987) Aluminium contents of foods and diets. *Food Additives and Contaminants*, **5**, 161–232.
52. Plessi, M. & Monzani, A. (1995) Aluminium determination in bottled mineral waters by electrothermal atomic absorption spectrometry. *Journal of Food Composition and Analysis*, **8**, 21–6.
53. Record, I. (1997) Tea and cancer. *Perspective in Food and Nutrition*, **6**, 7.
54. Coriat, A.-M. & Gilliard, R.D. (1986) Beware the cup that cheers. *Nature*, **321**, 570.
55. Baxter, M.J., Burrell, J.A. & Massey, R.C. (1990) The aluminium content of infant formula and tea. *Food Additives and Contaminants*, **7**, 101–7.
56. Fairweather-Tait, S.J., Faulks, R.M., Fatemi, S.J.A. & Moore, G.R. (1987) Aluminium in the diet. *Human Nutrition: Food Sciences and Nutrition*, **41F**, 183–92.
57. Williams, D. (1997) Aluminium in beer. http//www.breworld.com/the_brewer/9603/br3.html
58. Ministry of Agriculture, Fisheries and Food (1993) *Aluminium in Food. Food Surveillance Paper No. 39.* HMSO, London.
59. Allen, J.L. & Cumming, F.J. (1998) *Aluminium in the Food and Water Supply: an Australian Perspective. Research Report No. 202.* Water Services Association of Australia, Melbourne, Vic.

60. Jorhem, L. & Haegglund, G. (1992) Aluminium in foodstuffs and diets in Sweden. *Zeitschrift für Lebensmittel Untersuchung und Forschung*, **194**, 38–42.
61. Pennington, J.A.T. & Schoen, S.A. (1995) Estimates of dietary exposure to aluminium. *Food Additives and Contaminants*, **12**, 119–28.
62. Ministry of Agriculture, Fisheries and Food (1993) *Aluminium in Food. Food Surveillance Paper No. 39*. HMSO, London.
63. Ellen, G., Egmond, E., Van Loon, J.W., Sahertian, E.T. & Tolsma, K. (1990) Dietary intakes of some essential and non-essential trace elements, nitrate, nitrite and *N*-nitrosamines, by Dutch adults: estimated via a 24-hour duplicate portion study. *Food Additives and Contaminants*, 7, 207–21.
64. Gramiccioni, L., Ingrao, G., Milana, M.R., Santorini, P. & Tomassi, G. (1996) Aluminium levels in Italian diets and in selected foods from aluminium utensils. *Food Additives and Contaminants*, **13**, 767–74.
65. Allen, J.L. & Cumming, F.J. (1998) *Aluminium in the Food and Water Supply: an Australian Perspective. Research Report No. 202*. Water Services Association of Australia, Melbourne, Vic.
66. Shimbo, S., Hayase, A., Murakami, M. *et al.* (1996) Use of food composition database to estimate daily dietary intake of nutrient or trace elements in Japan, with reference to limitation. *Food Additives and Contaminants*, **13**, 775–86.
67. Pennington, J.A.T. (1987) Aluminium content of foods and diets. *Food Additives and Contaminants*, **5**, 160–232.
68. Pennington, J.A.T. & Schoen, S.A. (1995) Estimates of dietary exposure to aluminium. *Food Additives and Contaminants*, **12**, 119–28.
69. Pennington, J.A.T. (1987) Aluminium content of foods and diets. *Food Additives and Contaminants*, **5**, 161–232.
70. Allen, J.L. & Cumming, F.J. (1987) *Aluminium in the Food and Water supply: an Australian Perspective. Research Report No. 202*. Water Services Association of Australia, Melbourne, Vic.
71. Greger, J.L., Goetz, W. & Sullivan, D. (1985) Aluminium levels of foods cooked and stored in aluminium pans, trays, and foil. *Journal of Food Protection*, **48**, 772–7.
72. Neelam, M., Bamji, M.S. & Kaladhar, M. (2000) Risk of increased aluminium burden in the Indian population: contribution from aluminium cookware. *Food Chemistry*, **70**, 57–61.
73. Samsahl, K. & Wester, P.O. (1977) Metal contamination of food during preparation and storage: development of methods and some preliminary results. *Science of the Total Environment*, **8**, 165–77.
74. Ministry of Agriculture, Fisheries and Food (1998) Lead, arsenic and other metals in food. *Food Surveillance Paper No. 53*. The Stationery Office, London.
75. Pennington, J.A.T. & Jones, J.W. (1988) Aluminum in American diets. In: *Aluminum in Health, a Critical Review* (ed. H.J. Gitelman), pp. 67–100. M. Dekker, New York.
76. Glaister, G. & Allison, A. (1913) Aluminium in food. *Lancet*, **i**, 843.
77. Burns, J.H. (1932) *Aluminium and Food. A Critical Examination of the Evidence Available as to the Toxicity of Aluminium. BNFMRA Research Reports, External Series No. 162*. British Non-Ferrous Metals Research Association, London.
78. Spira, L. (1933) *The Clinical Aspects of Chronic Poisoning by Aluminium and its Alloys*. Bale & Danielsson, London.
79. American Medical Association Council on Food and Nutrition (1951) Aluminium in food. *Journal of the American Medical Association*, **146**, 477.
80. FDA (1971) *Safety of Cooking Utensils Fact Sheet*. United States Food and Drug Administration, Washington, DC.
81. Neelam, M., Bamji, S. & Kaladhar, M. (2000) Risk of increased aluminium burden in the Indian population: contribution from aluminium cookware. *Food Chemistry*, **70**, 57–61.
82. Alfrey, A.C. (1989) Physiology of aluminium in man. In: *Aluminium and Health: a Critical Review* (ed. H.J. Gitelman), pp. 101–24. M. Dekker, New York.
83. Moore, P.B., Day, J.P., Taylor, G.A., Ferrier, I.N., Fifield, L.K. & Edwardson, J.A. (2000)

Absorption of aluminium-26 in Alzheimer's disease, measured using accelerator mass spectrometry. *Dementia and Geriatric Cognitive Disorders*, **11**, 66–9.
84. Priest, N.D., Talbat, R.J., Austin, J.G. *et al.* (1996) The bioavailability of ^{26}Al-labelled aluminium citrate and aluminium hydroxide in volunteers. *Biometals*, **9**, 221–8.
85. Gormican, A. & Catli, E. (1971) Mineral balance in young men fed a fortified milk-based formula. *Nutrition and Metabolism*, **13**, 364–77.
86. Neelam, M., Bamji, S., & Kaladhar, M. (2000) Risk of increased aluminium burden in the Indian population: contribution from aluminium cookware. *Food Chemistry*, **70**, 57–61.
87. Exley, C., Burgess, E., Day, J.P., Jeffrey, E.H., Melethil, S. & Yokel, R.A. (1996) Aluminium toxicokinetics. *Journals of Toxicology and Environmental Health*, **48**, 569–84.
88. Day, J.P., Barker, J., Evans, L.J.A. *et al.* (1991) Aluminium absorption studied by ^{26}Al tracer. *Lancet*, **337**, 1345.
89. De Voto, E. & Yokel, R.A. (1994) The biological speciation and toxicokinetics of aluminium. *Environmental Health Perspectives*, **102**, 940–51.
90. Fernandez-Martin, J.L., Canteros, A., Alles, A., Massari, P. & Cannata-Andia, J. (2000) Aluminum exposure in chronic renal failure in Iberoamerica at the end of the 1990s: overview and perspectives. *American Journal of the Medical Sciences*, *320*, 96–9.
91. Priest, N.D. (1993) The bioavailability and metabolism of aluminium compounds in man. *Proceedings of the Nutrition Society*, **52**, 231–40.
92. Platts, M.M., Goode, G.C. & Hislop, J.S. (1977) Composition of the domestic water supply and the incidence of fractures and encephalopathy in patients on home dialysis. *British Medical Journal*, **2**, 657–60.
93. Crapper, D.R., Krishnan, S.S. & Dalton, A.J. (1973) Brain aluminium distribution in Alzheimer's disease and experimental neurofibrillary degeneration. *Science*, **180**, 511–13.
94. McLachlan, D.R. (1995) Aluminium and the risk of Alzheimer's disease. *Environmetrics*, **6**, 233–75.
95. Doll, R. (1993) Review: Alzheimer's disease and environmental aluminium. *Age and Ageing*, **22**, 138–53.
96. Bougle, D., Sabatier, J.P., Bureau, F. *et al.* (1998) Relationship between bone mineralisation and aluminium in the healthy infant. *European Journal of Clinical Nutrition*, **52**, 431–5.
97. Stauber, J.L., Davies, C.M., Adams, M.S. & Buchanan, S.J. (1998) *Bioavailability of Aluminium in Alum-treated Drinking Water and Food. Research Report* No. *203*. Water Services Association of Australia, Melbourne, Vic.
98. Forbes, W.F. & MaCainey, C.A. (1992) Aluminium in dementia, *Lancet*, **340**, 668–9.
99. Martyn, C.N., Barker, D.J.P., Oxmond, C., Harris, E.C., Edwardson, J.A. & Lacey, R.F. (1989) Geographical relation between Alzheimer's disease and aluminium in drinking water. *Lancet*, **14**, 59–62.
100. Doll, R. (1993) Review: Alzheimer's disease and environmental aluminium. *Age and Ageing*, **22**, 138–93.
101. Ministry of Agriculture, Fisheries and Food (1998) *Lead, Arsenic and Other Metals in Food. Food Surveillance Paper No. 52*. The Stationery Office, London.
102. Savory, J. & Willis, M.R. (1989) Analytical techniques for the analysis of aluminium. In: *Aluminium and Health: a Critical Review* (ed. H.J. Gitelman), pp. 1–27. M. Dekker, New York.
103. Jorhem, L. & Haegglund, G. (1992) Aluminium in foodstuffs and diets in Sweden. *Zeitschrift für Lebensmittel Untersuchung und Forschung*, **194**, 38–42.
104. Tinggi, U., Reilly, C. & Patterson, C. (1989) Aluminium in milk and soya products. *Proceedings of the Nutrition Society of Australia*, **14**, 132.
105. Tahán, J.E., Sanchez, J.M., Grenadillo, V.A., Cubillán, H.S. & Romero, R.A. (1995) Concentration of total Al, Cr, Cu, Fe, Hg, Na, Pb, and Zn in commercial canned seafood determined by atomic spectrometric means after mineralisation by microwave heating. *Journal of Agriculture and Food Chemistry*, **43**, 910–15.
106. Ysart, G., Miller, P. Crews, H. *et al.* (1999) Dietary exposure estimates of 30 elements from the UK Total Diet Study. *Food Additives and Contaminants*, **16**, 391–403.

107. Pennington, J.A.T. & Shoen, S.A. (1995) Estimates of dietary exposure to aluminium. *Food Additives and Contaminants*, *12*, 119–28.
108. Booth, C.K., Reilly, C. & Farmakalidis, E. (1996) Mineral composition of Australian ready-to-eat breakfast cereals. *Journal of Food Composition and Analysis*, **9**, 135–47.
109. Reis, M.F., Abdulla, M., Parr, R.M., Chatt, A., Dang, H.S. & Machado, A.A.S.C. (1994) Trace element contents in food determined by neutron activation analysis and other techniques. *Biological Trace Element Research*, **43**, 481–7.
110. Adriaens, A. (1996) Elemental composition and microstructure of early bronze age and medieval tin slags. *Mikrochimica Acta*, **124**, 89–98.
111. Tylecore, R.F. (1976) *A History of Metallurgy*, pp. 149–50. The Metal Society, London.
112. Schwartz, M.O., Rajah, S.S., Askury, A.K. & Putthapiban, P. (1995) The Southeast Asian Tin Belt. *Earth Science Reviews*, **38**, 95–286.
113. Newman, P. (1996) Tin enters free market era. *Metall*, **50**, 616–19.
114. Ysart, G., Miller, P. Crews, H. *et al.* (1997) Dietary exposure estimates of 30 elements from the UK Total Diet Study. *Food Additives and Contaminants*, **16**, 391–403.
115. Ministry of Agriculture, Fisheries and Food (1999) *MAFF UK – 1997 Total Diet Study – Aluminium, Arsenic, Cadmium, Chromium, Copper, Lead, Mercury, Nickel, Tin and Zinc. Food Surveillance Information Sheet No. 191.* http://www.foodstandards.gov.uk
116. Biego, G.H., Joyeux, M., Hartemann, P. & Debry, G. (1999) Determination of dietary tin intake in an adult French citizen. *Archives of Environmental Contamination and Toxicology*, **36**, 227–32.
117. Shimbo, S., Hayase, A., Murakami, M. *et al.* (1996) Use of food composition database to estimate daily dietary intake of nutrient or trace elements in Japan, with reference to limitation. *Food Additives and Contaminants*, **13**, 775–86.
118. FAO (1986) Guidelines for can manufacturers and food canners. Prevention of metal contamination of canned foods. *FAO Food and Nutrition Paper No. 36*. Food and Agricultural Organization, Rome.
119. Sherlock, J.C. & Smart, G.A. (1984) Tin in foods and the diet. *Food Additives and Contaminants*, **1**, 277–82.
120. Monier-Williams, G.W. (1949) *Trace Elements in Food*. Chapman & Hall, London.
121. Britton, S.C. (1975) *Tin Versus Corrosion*. International Tin Research Institution Publication No. 510. ITRI, Greenford, Middlesex, UK.
122. Newman, P. (1996) Tin enters free market era. *Metall*, **50**, 616–19.
123. Vannort, R.W. & Cressey, P.J. (1997) *Assessment of Selected Pesticides and the Elements Cadmium, Lead, Tin, Iodine and Fluorine in Infant Formulas and Weaning Foods. A Report for the Ministry of Health*. Report No FW97/01. Institute of Environmental Science and Research Limited, Christchurch, New Zealand.
124. Board, P.W. (1973) The chemistry of nitrate-induced corrosion of tinplate. *Food Technology Australia*, **25**, 16–17.
125. Bartsiokas, A. (2000) The Franklin expedition and lead poisoning. *European Journal of Oral Sciences*, **108**, 78–9.
126. Kannan, K., Senthilkumar, K. & Sinha, R.K. (1997) Sources and accumulation of butyltin compounds in Ganges river dolphin, *Platanista gangetica*. *Applied Organometallic Chemistry*, **11**, 223–30.
127. Fent, K. (1996) Organotin compounds in municipal wastewater and sewage sludge: contamination, fate in treatment process and ecological consequences. *Science of the Total Environment*, **185**, 151–9.
128. Batley, G.E., Fuhua, C., Brockbank, C.I. & Flegg, K.J. (1989) Accumulation of tributyltin in Sydney rock oysters, *Saccostrea commercialis*. *Australian Journal of Marine and Freshwater Research*, **40**, 49–54.
129. Thain, J.E. & Waldock, M.J. (1986) The impact of tributyltin (TBT) in antifouling paints on molluscan fisheries. *Marine Science and Technology*, **18**, 193–202.
130. Ministry of Agriculture, Fisheries and Food (1998) *Cadmium, Mercury and Other Metals in Food. Food Surveillance Paper No. 53*. The Stationery Office, London.

131. Statutory Instrument (1987) *The Control of Pollution (Antifouling Paints and Treatments) Regulations 1987* (SI No. 783). The Stationery Office, London.
132. Ministry of Agriculture, Fisheries and Food (1999) *MAFF UK – 1997 Total Diet Study – Aluminium, Arsenic, Cadmium, Chromium, Copper, Lead, Mercury, Nickel, Tin and Zinc. Food Surveillance Information Sheet No. 191.* http://www.foodstandards.gov.uk
133. Shimbo, S. Hayase, A., Murakami, M. *et al.* (1996) Use of food composition database to estimate daily dietary intake of nutrient or trace elements in Japan, with reference to limitation. *Food Additives and Contaminants*, **13**, 775–86.
134. Statutory Instrument (1992) *The Tin in Food Regulations 1992* (SI No. 496). The Stationery Office, London.
135. Newman, P. (1996) Tin enters free market era. *Metall*, **50**, 616–19.
136. Ministry of Agriculture, Fisheries and Food (1999) *MAFF UK – 1997 Total Diet Study – Aluminium, Arsenic, Cadmium, Chromium, Copper, Lead, Mercury, Nickel, Tin and Zinc. Food Surveillance Information Sheet No. 191.* http://www.foodstandards.gov.uk
137. Shimbo, S. Hayase, A., Murakami, M. *et al.* (1996) Use of food composition database to estimate daily dietary intake of nutrient or trace elements in Japan, with reference to limitation. *Food Additives and Contaminants*, **13**, 775–86.
138. WHO (1973) *Trace Elements in Human Nutrition*. Technical Report Series No. 532, 38–9. World Health Organization, Geneva.
139. Schwartz, K., Milne, D.B. & Vineyard, E. (1970) Growth effects of tin compounds in rats maintained in a trace element-controlled environment. *Biochemistry and Biophysics Research Communications*, **40**, 22–9.
140. Thain, J.E. & Waldock, M.J. (1986) The impact of tributylin (TBT) in antifouling paints on molluscan fisheries. *Marine Science and Technology*, **18**, 193–202.
141. De Groot, A.P., Feron, V.J. & Til, H.P. (1973) Toxicology of tin compounds. *Food and Cosmetic Toxicology*, **11**, 19–30.
142. Greger, J.L. & Johnson, M.A. (1981) Effect of dietary tin on zinc retention and excretion by rats. *Food and Cosmetic Toxicology*, **19**, 163–6.
143. Baldwin, D.R. & Marshall, W.J. (1999) Heavy metal poisoning and its laboratory investigation. *Annals of Clinical Biochemistry*, **36**, 267–300.
144. Walker, A.W. (1988) *Clinical and Analytical Handbook*, 3rd edn. SAAS Trace Element Laboratories. Guildford, UK.
145. Tsalev, D.L. (1983) *Atomic Absorption in Occupational and Environmental Health Practice*, Volume 2, pp. 204–206. CRC Press, Boca Raton, Florida.
146. Martyn, C.N., Barker, D.J.P., Oxmond, C., Harris, E.C., Edwardson, J.A. & Lacey, R.F. (1989) Geographical relation between Alzheimer's disease and aluminium in drinking water. *Lancet*, **14**, 59–62.
147. Ministry of Agriculture, Fisheries and Food (1999) *MAFF UK – 1997 Total Diet Study – Aluminium, Arsenic, Cadmium, Chromium, Copper, Lead, Mercury, Nickel, Tin and Zinc. Food Surveillance Information Sheet No. 191.* http://www.foodstandards.gov.uk
148. Schramel, P., Wendler, I. & Angerer, J. (1997) The determination of metals (antimony, bismuth, lead, cadmium, mercury, palladium, platinum, tellurium, thallium, tin and tungsten) in urine samples by inductively coupled plasma–mass spectrometry. *International Archives of Occupational Health*, **69**, 219–23.
149. Careri, M., Mangia, A. & Musci, M. (1996) Applications of liquid chromatography mass spectrometry interfacing systems in food analysis: pesticide, drug and toxic substance residues. *Journal of Chromatography*, **727**, 153–84.

Chapter 8
Transition metals: chromium, manganese, iron, cobalt, nickel, copper and molybdenum

Chemists classify the metals in terms of their electronic structure, either as *representative* or *transition* metals. The former have all their valence electrons in one shell, whereas the transition metals have their valence electrons in more than one shell. Their electronic structure gives the transition metals considerable chemical variety and flexibility: they are able to form complexes and to bind particular molecules and functional groups and to have more than one stable oxidation state[1].

Metals with incomplete underlying electron shells are usually coloured, with the colour dependent on the oxidation state. The colour can be modified by the nature of non-metallic elements or groups with which the metal is combined. These colours are the basis for the many colorimetric analytic methods which were formerly widely used for transition metals. Because they are usually easy to detect, quantitatively as well as qualitatively, the biological significance of several of the transition metals has been recognised for some time and much is known about their functions and properties in foods and diets.

The transition metals are subdivided by chemists into three groups: the main transition or *d* block elements, the lanthanide elements, and the actinide elements[2]. Here we are interested principally in the main transition group. This contains elements in which the *d* shells only are partially filled. Scandium, with an outer electron configuration of $4s^2 3d$, is the lightest of the group. The eight following elements: titanium, vanadium, chromium, manganese, iron, cobalt, nickel and copper, have partly filled $3d$ shells, either in the ground state of the free atom (all except copper) or in one of their more important ions (all except scandium). This group of elements makes up the first transition series, numbers 21–29 in the periodic table.

The next element, number 30, zinc, is sometimes considered to belong to the transition metals, since it shares some of their properties, but, strictly speaking, it is not. Its outer electron configuration is $3d^{10}4s^2$ and it forms no compounds in which the $3d$ shell is ionised. It also differs from the transition metals in that, with its filled outer shell, it does not form coloured compounds. However, zinc is usually associated with the transition metals in textbooks, and that convention will be followed here.

Two other characteristics of transition metals are of particular interest in relation to the present study. The first is their strong metallic quality: they are mainly hard, tough metals, with high melting points, and are capable of conducting heat and electricity well. They can also form useful alloys among themselves and with other metals. This has meant that the transition metals have long been used for construction of equipment for the food processing and catering industries.

The second characteristic is of interest from the point of view of chemical and biological activity. Transition metals, as they go from left to right across the periodic table, show only small variations in their physical and chemical properties. This is because, generally, succeeding elements in the series differ in electron configuration by one electron in the next to outer valence shell, rather than in the outer valence shell. The interchangeability of several of the transition elements in their biological roles, for example as coenzymes, may be accounted for by this characteristic.

We will be concerned here mainly with metals of the first transition series, with particular attention to those of most significance in relation to food. These are, in ascending order of their atomic numbers, chromium, manganese, iron, cobalt, nickel and copper, as well as molybdenum, from the second transition series. In the next chapter we will look at the remaining transition elements from all three series, as well as their close relation, zinc.

8.1 Chromium

Chromium is widely distributed in the Earth's crust, though at low levels. It has a concentration of about 250 μg/kg of soil[3]. It is a metal about which, in the context of human biology, we did not know a great deal until very recently. We now know that its low environmental concentration bears little relation to its importance to health. Its toxicity to industrial workers has been recognised for many years[4], but it was only in the 1950s that its essentiality to human life began to be appreciated[5]. Today chromium is attracting increasing attention in the medical field, especially because of evidence for its important role in carbohydrate and lipid metabolism[6].

One of the principal reasons why progress has been slow in establishing the vital role of chromium in animal and human metabolism was the difficulty formerly experienced in determination of the element. In 1977 Eric Underwood, in his pioneering study of trace elements, noted that reliable data on the distribution of chromium in the animal body were sparse and sporadic because of the absence of satisfactory methods for its analysis[7]. Though that situation has now been remedied, much still remains to be learned about the role of chromium in food and nutrition.

8.1.1 *Chemical and physical properties of chromium*

Chromium has an atomic weight of 52 and is element number 24 in the periodic table. It is a heavy metal, with a density of 7.2, and a high melting point of approximately 1860°C, the exact temperature depending on the crystal structure of the sample in question. It is a very hard, white, lustrous and brittle metal and is extremely resistant to corrosion. Chromium can occur in every one of nine oxidation states, from −2 to +6, but only the ground 0, and the +2, +3 and +6 states, are of practical significance. The most stable and important oxidation state is Cr(III), which gives a series of chromic compounds, including the oxide Cr_2O_3, chloride $CrCl_3$, and the sulphate $Cr_2(SO_4)_3$. Industrially important Cr(III) salts are the acetate, citrate and chloride.

In higher oxidation states almost all chromium salts are oxo-forms and are potent oxidising agents. The Cr(VI) compounds include the chromates CrO_4^{2-} and the dichromate $Cr_2O_7^{2-}$. Important compounds of this type include lead, zinc, calcium and barium chromates, as well as sodium and potassium chromates and dichromates, each of which is very soluble. From the biological point of view, hexavalent chromium is of particular importance because of its toxicity due to its oxidising properties.

The relatively unstable Cr(II) chromous ion is rapidly oxidised to the Cr(III) form. Few of the other forms of chromium, such as Cr(IV) or Cr(V), are known, or if they are, they do not appear to be of biological significance.

8.1.2 Production and uses of chromium

Through chromium is widely distributed in small amounts in rocks and soils, its only commercial ore is chromite, $FeOCr_2O_3$, which can contain up to 55% chromic oxide. Chromite is found in commercial quantities in a few parts of the world, especially in South Africa and Russia. Smaller deposits are worked in the US, Zimbabwe and a few other countries. The chromite ore is reduced in a furnace with carbon to produce ferrochrome, a carbon-containing alloy of chromium and iron. Pure chromium metal is produced from this, either by electrolytic treatment or by conversion into sodium dichromate by treatment with hot alkali and oxygen followed by reaction with aluminium.

Chromium has many industrial uses. More than half of the metal produced worldwide is destined for the metallurgical industries, where it is used for the production of stainless steels of various kinds, in alloys with iron, nickel and cobalt. Its high resistance to oxidation has led to its extensive use as an electroplated coating agent and in stainless steel. Another 30% is used in the production of refractory materials for lining furnaces and kilns, and the remainder in the chemical industry[8]. Chemical uses of chromium include production of tanning agents, pigments, catalysts and timber preservatives. Some chromium is used as an additive in water to prevent corrosion in industrial and other cooling systems. It is probably this use that accounts for a significant amount of industrial chromium emission[9]. Chromium salts are also found in some household detergents, as well as in pottery glazes, paper and dyes.

8.1.3 Chromium in food and beverages

As was noted in the UK Report on Dietary Reference Values published in 1991[10], because of inadequate analytical techniques and possible chromium contamination, values for levels of the metal in foods and diets published prior to 1980 should be viewed with caution. Certainly, as a perusal of the earlier literature shows, some high levels were reported, which subsequent investigations have failed to confirm. Though analytical procedures have undoubtedly improved during the past two decades, and much more reliable data are now published, the determination of chromium in biological materials is still considered a challenge by some analysts[11].

Chromium levels are generally very low in foods and beverages, as can be seen in Table 8.1. In many of the food groups analysed by the MAFF scientists who prepared the data used in the table, chromium was found at or below the limit of detection (LOD) of 0.1 mg/kg. The best food sources of the metal are meat, whole grains, legumes and nuts, though they are unlikely to contain more than about 0.5 mg chromium/kg. One of the best sources of the metal is brewer's yeast, which can contain up to 5 mg/kg, though levels in different samples can vary widely[12]. Chromium-enriched yeast is widely used as a nutritional supplement for humans as well as animals[13]. There have also been reports of high levels of chromium in spices, such as black pepper, and in raw sugar[14].

Though water may contain chromium, especially if it is affected by emissions from industry, it is unlikely to make a significant contribution to dietary intake of the metal. If any chromium is present in drinking water, it is likely to be Cr(VI), which is far more soluble than hydrated Cr(III) oxide. WHO drinking water standards, which have been adopted by the EU, set a maximum permitted level of 50 μg/litre for Cr(VI)[15]. Total chromium levels in drinking water in several European cities range from 1.0 to 5.0 μg/litre, with a mean of 2.0 μg/litre. A person drinking 1.5 litre/day would have a daily intake of between 1.5 and 7.5 μg of chromium from this source alone[16].

Table 8.1 Chromium in UK foods.

Food group	Mean concentration (mg/kg fresh weight)
Bread	0.1
Miscellaneous cereals	0.1
Carcass meat	0.2
Offal	0.1
Meat products	0.2
Poultry	0.2
Fish	0.2
Oils and fats	0.4
Eggs	0.2
Sugar and preserves	0.2
Green vegetables	0.2
Potatoes	0.1
Other vegetables	0.1
Canned vegetables	0.1
Fresh fruit	<0.1
Fruit products	<0.1
Beverages	<0.1
Milk	0.3
Dairy products	0.9
Nuts	0.7

Adapted from Ysart, G., Miller, P., Crews, H. *et al.* (1999) Dietary exposure estimates of 30 elements from the UK Total Diet Study. *Food Additives and Contaminants*, **16**, 391–403.

Increased levels of chromium in agricultural soils, and thus potentially in crops, can occur through the application of sewage sludge, as well as by atmospheric deposition of fumes produced by industry. Several countries have introduced legislation to limit such sources of pollution. In the UK the Department of the Environment has set a guideline limit of 400mg/kg for the chromium content of sludge-amended soils. This is about ten times greater than the mean chromium content of most agricultural soils[17].

8.1.3.1 Adventitious chromium in foods

Canned and other processed foods may include adventitious chromium, taken up from cooking utensils and storage vessels, including cans. Stainless steel, which is widely used in food processing equipment and fittings, may contain chromium, as well as a number of other metals. All of these can migrate from the steel into food, the amount transferred depending particularly on the length of contact time between the food and the metal[18]. The temperature at which processing is carried out is also important, as is the pH of the food. The level of pick-up by acid foods from stainless steel is seen in red cabbage, pickled in 4% acetic acid (vinegar), which contained 72 μg chromium/kg, compared to about 10 μg/kg in the fresh vegetable[19].

The 'passivation treatment', in which chromic acid or another chromium compound is used to increase the lacquer adherence and resistance to oxidation of tinplate, can contribute significantly to the level of pick-up of chromium by canned food. It has been calculated that if all the chromium on the inner surface of a 454-gram can migrated into the contents, it would result in a final concentration of about 400 μg/kg[20]. Lacquering actually protects the contents of cans from this chromium. It has been found, for example, that while fresh fruit and vegetables contain 0.009 mg/kg chromium, this can increase to 0.018 mg/kg when they are canned[21].

8.1.3.2 Dietary intakes of chromium

Reported levels of chromium in foods, and hence in daily diets, have decreased over recent decades due to improved analytical procedures, instrumentation and prevention of chromium contamination[22]. Today's reported intakes are generally much lower than the 0.30 mg/day or higher estimated in the 1970s[23]. In contrast daily mean adult intake in, for example, the UK, as estimated from the 1997 TDS, is 0.1 mg/day, with 0.17 mg/day for the 97.5th percentile[24]. Similar findings have been reported in recent years, including the US (0.2 mg)[25], Japan (0.89 mg)[26] and Denmark (0.5 mg)[27].

8.1.3.3 Recommended intakes of chromium

Chromium is included among the nutrients for which, in the UK, a range of 'Safe Intakes' (SI), rather than an RNI has been established on the grounds that a DRV could not be set because of insufficient reliable data on human requirements. The SI is judged to be a range of intakes at which there is no risk of deficiency or of undesirable effects. For chromium the SI is, for adults, above 25 μg/day, and for children and adolescents 0.1–1.0 μg/day. COMA, in its 1999 report, noted that the WHO had not set a PTWI for chromium and that current estimates of intake in the UK, 'did not warrant any major concerns in terms of toxicity, or deficiency'[28].

In the US, because there was inadequate data for establishing an RDA, an estimated safe and adequate daily dietary intake (ESADDI) of 50–200 μg/day of chromium has been established for adults[29]. Since the intakes of many people, in the US and elsewhere, are below the lower range of the ESADDI, the tentative nature of these recommendations was clearly pointed out in the 10th (1989) edition of the US *Recommended Dietary Allowances* and the advice given that

> *until more precise recommendations can be made, the consumption of a varied diet, balanced with regard to other essential nutrients, remains the best assurance of an adequate and safe chromium intake.*

It is to be expected that these more precise recommendations will emerge from the review of dietary reference values currently ongoing in the US[30].

8.1.4 Absorption and metabolism of chromium

The mechanism of absorption of chromium from food in the GI tract is not yet well understood. It appears to involve an active process, not simple diffusion[31]. Various factors affect absorption. Uptake is inversely related to the amount of the metal in food. Phytate significantly reduces absorption, while it is enhanced by ascorbic acid[32]. There also appears to be competition between chromium and other metals, including zinc, iron, manganese and vanadium, for uptake[33]. Overall, the rate at which chromium is absorbed into the intestine is low, ranging from 0.5 to 3%. Studies using rats have shown that Cr(VI) is absorbed at a faster rate than is Cr(III)[34]. Absorbed chromium is transported in blood bound to transferrin, with both metals competing for the same binding sites[35]. The adult human body appears to contain between 4 and 6 mg of chromium. The metal is stored to a limited extent in organs, including kidneys, spleen and testes. Absorbed chromium is excreted mainly in urine, with a small amount in faeces. The chromium content of body tissues decreases with age[36].

There is strong evidence that chromium is an essential nutrient for humans, though this is disputed by some investigators[37]. Its essentiality was first recognised in the 1960s through observations of patients maintained on TPN who developed impaired glucose tolerance[38]. It was found that they could be returned to normal by addition of small amounts of chromium to the infusion fluid[39]. The chromium appears to function in an organic complex which potentiates the action of insulin. The metal has been found to be an integral part of a dinicotinic acid–glutathione complex, the so-called glucose tolerance factor, GTF[40]. Though this complex has been extracted from brewer's yeast, and other apparently biologically active chromium complexes have been detected in other organisms, none of them has yet been fully characterised[41].

There is evidence that chromium deficiency is associated with elevated blood glucose levels and that chromium fed to elderly persons improves glucose tolerance, as well as blood lipid levels[42]. Chromium appears to participate in lipoprotein metabolism, and a low chromium status may result in decreased levels of high-density cholesterol (HDLC). Improved chromium nutrition may reduce risk factors associated with cardiovascular disease as well as diabetes mellitus[43].

There is considerable interest, especially among athletes and sports nutritionists, in the possibility that chromium can help to increase lean body mass and decrease percentage body fat. Numerous studies have been conducted to investigate the relationship of the metal with exercise and fitness. However, findings have been equivocal. Nevertheless, the use of chromium supplements has gained in popularity among athletes and others, in the belief that it will compensate for losses caused by exercise, increase lean body mass (LBM) and reduce fat[44].

8.1.5 Chromium toxicity

Trivalent chromium has a low order of toxicity and then only at high levels of intake. The hexavalent form is, in contrast, extremely toxic. Intakes of 1–2 grams/day can cause kidney and liver necrosis[45]. There is no evidence, however, that the chromium normally present in food is a danger to health, whether the metal is present naturally or has migrated into the food from stainless steel cooking equipment.

Concern has been raised by reports of toxic effects of ingestion of certain chromium-containing supplements. Chromium picolinate, a widely available supplement, much used by some athletes and others, has been reported to cause clastogenesis (chromosome breakdown). It is, however, unclear whether it is the Cr(III) or the picolinate that is responsible for the toxicity[46]. There is an urgent need for closer scrutiny of the possible toxic effects of this and other chromium supplements[47].

8.1.6 Analysis of foodstuffs for chromium

Because of the usually very low levels of chromium in foods and beverages, and the considerable danger of contamination of samples, chromium has traditionally been considered by many analysts to be a difficult element to determine[48]. Today many of the analytical problems encountered have been overcome by the development of new equipment, especially closed microwave digestion and highly sensitive analytical instruments, but difficulties can still be encountered. Apart from ensuring absolute cleanliness and use of a suitable work station during sample preparation, care has to be taken to avoid loss through volatilisation, especially if dry ashing is used. A problem can be encountered if acid digestion is used with samples containing appreciable levels of silicates that can interfere with recovery of the metal. Release of the chromium can be achieved by adding hydrofluoric acid to the digestion mixture[49].

A problem encountered when using GFAAS for chromium is inadequate background correction. The deuterium lamp is of little use for this purpose because its intensity at the 357.9 nm line is low compared to the hollow cathode lamp and will not eliminate the non-specific absorption. A high-intensity tungsten-iodide lamp, or Zeeman background correction, can be used to overcome the problem[50]. Where available, NAA is the analytical method of choice, since it avoids extensive sample preparation[51].

8.2 Manganese

Until relatively recently manganese, if considered at all by toxicologists and other medical and biological investigators, was included among the scientific curiosities, of little consequence to human health. They might have heard of 'manganese madness' in workers exposed to fumes of the metal, but that was probably the extent of their knowledge. We now know that there is much more to manganese and its relation to human health than its industrial toxicity.

8.2.1 Chemical and physical properties of manganese

Manganese is element number 25 in the periodic table, with an atomic weight of 54.94. Its chemical and physical properties are, in many respects, similar to those of iron, which it immediately precedes in the first transition series. Manganese, however, is harder, more brittle and refractory, with a lower melting point, 1247°C. Like iron, it is a reactive metal which dissolves in dilute, non-oxidising acids. It ignites in chlorine to form $MnCl_2$ and reacts with oxygen at high temperatures to form Mn_3O_4. It also combines directly with boron, carbon, sulphur, silicon and phosphorus.

Manganese has several oxidation states, the most characteristic of which, especially in solution and in biological complexes, is the divalent Mn(II). This forms a series of manganous salts with all the common anions. Most of these are soluble and crystallise as hydrates. Mn(II) also forms a series of chelates with EDTA and other agents.

The other oxidation state, Mn(III), is important *in vivo*. Ingested Mn(II) is thought to be converted into Mn(III) which is the form of the element that binds to transferrin[52]. Mn(IV) compounds are of little importance, except for the oxide MnO_2, pyrolusite, which occurs in nature and is one of the chief ores of the metal.

Mn(VI) is found only in the manganate ion (MnO_4^{2-}). Similarly Mn(VII) is best known in the permanganate ion (MnO_4^-). Potassium permanganate is a common and widely used compound, traditionally known to many as the wine-red aqueous solution, Condy's fluid. Permanganates are powerful oxidising agents. Manganese forms several organo-compounds. The best known of these is methylcyclopentadienyl manganese tricarbonyl, which is used as an anti-knock additive in petroleum fuels.

8.2.2 Production and uses of manganese

Manganese is a ubiquitous constituent of the environment, comprising about 0.1% of the Earth's crust. It is the twelfth most common element and the second most common heavy metal, after iron. Ores of manganese occur in a number of substantial deposits, especially as pyrolusite, the Mn(IV) oxide. Major commercially exploited deposits occur in Africa, Russia, Canada and Australia. The metal is produced by roasting the ore, followed by reduction by aluminium.

The principal industrial use of manganese is in the manufacture of steels. Manganese steel alloys are particularly strong. The metal is also used to make other types of alloys that are used in a variety of applications, such as in electrical accumulators. Manganese compounds are used to make pigments and glazes, as well as domestic and pharmaceutical products. A recent use has been as a replacement of alkyllead additives in petroleum fuels and of metallic lead in shotgun cartridges.

8.2.3 Manganese in food and beverages

Manganese is present in all plant and animal tissues, usually at low levels of concentration. There does not seem to be much variation in levels in foods between countries. Data in Table 8.2 are typical of reported levels in several other countries, such as the US[53]. Interestingly, in contrast to iron and in spite of the chemical similarities between the two metals, cereals and cereal products are among the richest sources of manganese, with low levels in meat and offal. An interesting and, in the national context, important finding in the UK is that tea can be a major source of manganese in the diet. Dry tea leaves can contain between 350 and 900 mg/kg, and the beverage itself 7.1–38.0 mg of the metal/kg[54].

Table 8.2 Manganese in UK foods.

Food group	Mean concentration (mg/kg fresh weight)
Bread	8.0
Miscellaneous cereals	6.8
Carcass meat	0.14
Offal	2.8
Meat products	1.4
Poultry	0.17
Fish	1.1
Oils and fats	0.02
Eggs	0.31
Sugar and preserves	1.5
Green vegetables	2.0
Potatoes	1.9
Other vegetables	1.6
Canned vegetables	1.8
Fresh fruit	2.0
Fruit products	2.2
Beverages	2.7
Milk	0.03
Dairy products	0.27
Nuts	15

Adapted from Ysart, G., Miller, P., Crews, H. *et al.* (1999) Dietary exposure estimates of 30 elements from the UK Total Diet Study. *Food Additives and Contaminants*, **16**, 391–403.

8.2.4 *Dietary intake of manganese*

The mean dietary intake of manganese by adults in the UK is 4.5 mg/day[55]. About half of this is believed to be from tea[56]. Daily intakes in the US were reported to be somewhat lower, averaging 2.85 and 2.21 mg for 25–30-year-old men and women respectively over the years 1982–91[57]. Similar figures have been reported for Belgium and Spain. It is generally believed that these levels of intake are adequate and that there is no need for concern about possible deficiencies of manganese[58]. However, in theory at least, it is possible that in a diet in which wholemeal cereals and tea are not consumed, manganese intakes could be less than optimal.

8.2.4.1 Dietary requirements for manganese

There is no evidence that dietary manganese deficiency ever occurs in the general population. Deficiency has been induced in animals fed a specially designed, low manganese diet, with symptoms of impaired growth and skeletal abnormalities. There have been a small number of reports of apparent deficiency in humans who consumed a restricted diet with very low levels of manganese[59]. There is at present inadequate information on which to base a DRV. It has not been possible to define requirements on the basis of a number of short-term balance studies[60]. Health authorities in a number of countries have, on the assumption that current population intakes are adequate, adopted a pragmatic approach to determining recommendations. In the UK an SI has been set of above 1.4 mg/day for adults and 16 μg/day for infants and children. The US ESADDI is 2.0 to 5.0 mg/day.

8.2.5 *Absorption and metabolism of manganese*

There are many gaps in our knowledge of the metabolism of manganese. It is known that uptake from food is low, with considerable variation between individuals. An average retention rate of 3% has been reported, with a range of 0.6–9.2% in adult men[61]. Absorption occurs along the length of the small intestine. The process appears to be an active one and is related to the manganese status of the body[62]. Absorbed manganese leaves the blood quickly and is concentrated in liver and other organs. Excretion is mainly in bile in faeces, with only a small amount in urine. Manganese is an essential nutrient for humans. Its known biochemical functions are as activator and constituent of enzymes. Those it activates include hydrolases, kinases, decarboxylases and transferases[63]. Most of these can also be activated by other metals, especially magnesium, with the exception of the manganese-specific glycotransferases. Enzymes of which manganese is an essential cofactor include arginase, hexokinase, superoxide dismutase and xanthine oxidase.

The effects of manganese deficiency in humans are far from clear. Evidence from animal studies suggest that the metal is required for normal bone and cartilage formation, and for protecting cells against oxidative damage by free radicals. Very little else is known about its role in human metabolism or its clinical importance. It may be that certain types of stress accentuate the effects of dietary manganese insufficiency. An example of this is ethanol toxicity, which increases superoxide production and can be counteracted by manganese superoxide dismutase, thus placing an increased demand on manganese stores[64].

8.2.6 Manganese toxicity

Though manganese has long been recognised as a toxic element to industrial workers and miners, it is considered to be one of the least toxic through oral intake. This has been attributed to the fact that when excess is consumed absorption is very low and what is absorbed is efficiently excreted via the bile and kidneys[65]. However, this may not be true in the case of people exposed to high levels of environmental pollution by manganese. Severe neurological damage has been observed in inhabitants of Groote Eylandt, an island in the Gulf of Carpentaria off the north coast of Australia which contains one of the world's richest manganese mines[66]. About a thousand Aboriginal people of the Angurugu tribe live there, literally on top of deposits of high-grade manganese ore deposits. All over the island are opencast mines and piles of crushed ore. A layer of black pyrolusite dust covers everything. The soil contains upwards of 103 000 mg/kg of manganese. Fruit and vegetables eaten by the local people are rich in manganese: bananas contain 31 mg/kg, yams 720 mg/kg and paspalum, 240 mg/kg.

It has been estimated that islanders consume 100–200 mg/day of the metal. Up to 2% of them are affected by motor neurone defects and cerebellar dysfunctions. A distinctive symptom of the condition is an awkward, high-stepping gait, which has earned them the name of 'bird people'. Whether this condition is due to an oral intake of manganese, or is a form of the fume fever which is responsible for the 'manganese madness' that can affect industrial workers, has not been determined[68]. It has been suggested that the 'bird people' have a genetic predisposition which is triggered by the high environmental level of manganese[69]. Whatever may be the ultimate cause of the illness suffered by the inhabitants of Groote Eylandt, a major role for chronic manganese poisoning, as a result of excessive dietary intake, cannot be ruled out.

8.2.7 Analysis of foodstuffs for manganese

Manganese is relatively easily determined in food by GFAAS, following acid digestion. For low concentrations, background correction is essential. A procedure using a nitric/sulphuric acid digestion mixture and employing Zeeman background correction has been described. No problems with matrix interferences or losses of analyte were encountered[70].

While the use of background correction is strongly recommended, various preconcentration procedures of samples can also be effective when using AAS to determine manganese in foods. A method in which manganese in the digest was complexed with cupferron and then adsorbed on activated carbon has been described. A detection limit for manganese of 1.8 μg/litre was achieved by the method[71].

8.3 Iron

Iron is the second most abundant metal, after aluminium, and the fourth most abundant element in the Earth's crust. The Earth's core is believed to consist of a mass of iron and nickel, and, to judge from the composition of many meteorites that have been analysed, iron is also abundant elsewhere in the universe. The importance of the metal in the physical world is matched by its importance to human life. Civilisation, as we know it, would not exist without iron. We use more of it than of any other metal. The Iron Age has been followed by the age of steel and technology, in which the metal plays an ever-increasing role.

The human body, like nearly all other living organisms, has, during the course of evolution, come to depend on iron as the linchpin of its existence. The release and utilisation of energy from food depend on the iron-containing enzymes of tissues. Iron is needed for the synthesis of blood pigments, as well as for many other essential activities of cells. Out of all living organisms, only certain bacteria, of the genera *Lactobacillus* and *Bacillus,* appear to possess no iron-containing enzymes. In these organisms, the functions of iron are replaced by other transition metals[72]. Nevertheless, in spite of its abundance, iron deficiency is probably the most common nutritional deficiency in the world, even in the best-fed countries[73].

8.3.1 *Chemical and physical properties of iron*

Iron, which has the chemical symbol Fe after its Latin name *ferrum,* is element number 26 in the periodic table. It has an atomic weight of 55.85 and a density of 7.86. Pure iron is a white, lustrous metal which melts at 1528°C. It is not particularly hard and is ductile and malleable. It is a good conductor of both heat and electricity. The physical properties of iron are significantly affected by the presence of even small amounts of other metals and carbon, an effect which accounts for one of the most significant characteristics of the metal, its ability to be transformed into various types of steel. Iron is a chemically active element. In moist air the metal rapidly oxidises to a hydrous oxide which we call rust. Unlike the patina on copper, rust does not provide permanent protection for the underlying metal since it readily flakes off, exposing the underlying untarnished surface.

Iron forms two series of compounds, ferrous Fe(II) and ferric Fe(III). Other oxidation states also occur, but are not of biological importance. Three oxides are known, FeO, Fe_2O_3, and Fe_3O_4, representing the Fe(II) and the Fe(III) as well as the mixed Fe(II)–Fe(III) oxide which occurs in nature as the mineral magnetite. Iron dissolves in acids to form salts. With non-oxidising acids, in the absence of air, Fe(II) salts are obtained. Many iron salts, as well as the hydroxides, are insoluble in water.

One of the most important of the chemical properties of iron, from the point of view of human biology, is its ability to form coordination compounds with organic molecules, especially the porphyrin nucleus. The most important of these are the haem pigments, found in haemoglobin and myoglobin, the compounds responsible for oxygen transport in the body. Alternation between the metal's two oxidation states, Fe(II) and Fe(III), is essential to the role of these compounds in oxidation/reduction reactions involving other iron porphyrin proteins active in respiration, the cytochromes[74]. Another characteristic of such organically complexed iron, of considerable biological importance, is the fact that, compared to inorganic iron salts, organic iron compounds are highly bioavailable, and easily absorbed from food.

8.3.2 *Production and uses of iron*

Large-scale production of iron, especially for making weapons, seems to have begun in Asia Minor and elsewhere in the Middle East during the first millennium BC. Though iron is sometimes found in the metallic state in meteorites, and many soils are rich in the metal, it has always been extracted from a number of different ores. These include haematite, Fe_2O_3, magnetite, Fe_3O_4, limonite, FeO(OH), and siderite, $FeCO_3$. Extraction of the metal is a highly skilled process, involving reduction of the ore by smelting in a blast furnace with various forms of carbon.

Though iron itself is widely used in a number of different forms, particularly as the almost pure wrought iron, it is mainly employed to make various kinds of steel. Steel is iron containing between 0.1 and 1.5% carbon, as well as different amounts of other metals. The amount of carbon and the types of metal added to the original cast iron depend on the quality of the steel required and the functions for which it will be used. These uses are so numerous and so well known that they will not be discussed further here, except to remark that whether in the form of cast iron, or as a component of steel or other alloys, iron plays a major part in the construction of food processing equipment, containers and other utensils used for food. It is through the use of such equipment that much of the food contamination by iron comes about.

8.3.3 Iron in food and beverages

Iron is present in all foods and vegetables, as it is in all biological materials. Concentrations in different foods can vary greatly, depending on the kind of food and the processes to which it has been exposed. Overall a range of about 0.1 mg/kg or less at the lower end, to about 100 mg/kg at the upper end, can be expected in foods usually consumed in a western-style diet. Table 8.3 gives the mean concentrations determined in different food groups examined in the 1994 UK TDS[75]. Similar results have been reported in similar studies carried out elsewhere, including the US[76].

Table 8.3 Iron in UK foods.

Food group	Mean concentration (mg/kg fresh weight)
Bread	21
Miscellaneous cereals	32
Carcass meat	21
Offal	69
Meat products	23
Poultry	7.1
Fish	16
Oils and fats	0.5
Eggs	20
Sugar and preserves	9.4
Green vegetables	11
Potatoes	8.1
Other vegetables	7.5
Canned vegetables	12
Fresh fruit	2.7
Fruit products	3.2
Beverages	0.4
Milk	4.1
Dairy products	12
Nuts	34

Adapted from Ysart, G., Miller, P., Crews, H. *et al.* (1999) Dietary exposure estimates of 30 elements from the UK Total Diet Study. *Food Additives and Contaminants*, **16**, 391–403.

One of the foods with the highest concentration of iron is animal offal, especially liver. Other animal products, in particular red meat, are also rich in iron. This iron is organically bound haem iron, which is easily absorbed. Plant foods are generally poor sources of iron and what is present is in the inorganic form. However, in contrast to most other plant foods, dark green vegetables, such as spinach, are relatively rich in iron. Generally cereals are low in iron, but breakfast cereals, fortified voluntarily, and flour, to which the addition of iron is required by law, are relatively rich in the metal.

8.3.3.1 *Iron in canned foods*

Particularly high levels of iron are, on occasion, found in canned foods.Uptake of the metal, especially by certain kinds of fruit and vegetables, in this way can be a problem. Several factors, such as the pH of the contents, whether or not nitrate is present, as well as the standards of hygiene used in the canning plant, determine the level of iron accumulated in foods in tinplated cans. Some fruits, especially the tropical pawpaw (papaya), naturally contain such high levels of nitrate that they cannot be stored in this way without excessive iron uptake. Other fruits, temperate as well as tropical, also present problems. One study found iron levels of 5.5 mg/kg in canned grapefruit, 5.0 mg/kg in canned pears and 1300 mg/kg in canned blackcurrants[77].

The age of the can and the length of storage are also important in relation to iron uptake. It has been shown that the iron content of a variety of canned foods increased by an average of 34.8% after 280 days' storage and 42.8% after a full year. An extreme case was a can of anchovies stored for four years, with 5800 mg/kg of iron. Many countries now require canned food to carry a 'use by' date to overcome the problem of spoilage by metal pick-up with excessive storage time[78].

8.3.3.2 *Iron fortification of food*

Iron levels in foods can be significantly increased by fortification. As has been noted above, deliberate addition of iron to certain foodstuffs during processing was initially introduced in some countries as a health intervention to prevent iron deficiency as a result of wartime food restrictions. In 1941 legislation was introduced in the US which required the enrichment of flour and bread with iron, and certain other nutrients, to the level found in whole wheat[79]. The UK and other countries adopted a similar programme about the same time. It is still in force in the UK today[80] and requires that flour, apart from wholemeal, must contain not less than 1.65 mg iron/100 grams, as ferric ammonium citrate, ferrous sulphate or iron powder[81].

Several other foods besides flour and bread are required by UK law to be fortified with iron. Infant formulas must contain iron at a minimum of 0.12 mg/100 kJ and a maximum of 0.36 mg/kJ, and follow-on formulas a minimum of 0.25 mg/100 kJ and a maximum of 0.50 mg/kJ[82]. Processed products used as substitutes for meat, officially described as 'foods that simulate meat', must also be fortified to a mean concentration of 20 mg/100 gram protein. In addition, UK legislation allows for the voluntary fortification with iron of several different types of processed foods, such as breakfast cereals, infant weaning foods, meal replacements and other diet foods. The amounts of iron which can be added are related to the recommended dietary intakes for iron and range from 17 to 100% of the RDA[83].

When the requirement for fortification with iron first became mandatory, white flour was chosen as the vehicle because white bread provided up to one third of food energy of poorer families in the UK and it was believed that this would reduce the level of iron deficiency anaemia which was then common. Besides the officially sanctioned forms of iron, many other forms have been proposed. There is considerable debate about their relative effectiveness in meeting the requirements of the legislation[84]. A major problem with all of them is their bioavailability, as is shown in Table 8.4. Unfortunately, some of the more bioavailable forms, such as ferrous sulphate, can cause problems, including enhanced vitamin degradation, oxidative rancidity, colour

Table 8.4 Bioavailability of different forms of iron.

Form of iron	Relative bioavailability (%)
Ferrous sulphate	100
Ferrous lactate	106
Ferrous fumarate	101
Ferrous succinate	92–123
Ferrous gluconate	89
Iron-glycine	85
Ferrous citrate	74
Ferrous saccharate	74
Ferrous tartrate	62
Ferrous pyrophosphate	39
Haemoglobin iron	37–118
Ferric sulphate	34
Ferric citrate	31
Ferric orthophosphate	31
Iron(III) EDTA	30–290
Sodium iron pyrophosphate	15
Reduced elemental iron	13–90
Carbonyl iron	5–20

After Fairweather-Tait, S.J. (1997) Iron fortification in the UK: how necessary and effective is it? In: *Trace Elements in Man and Animals – TEMA 9* (eds. P.W.P. Foscher, M.R. L'Abbé, K.A. Cockell & R.S. Gibson), pp. 392–6. NRC Research Press, Ottawa, Canada.

changes, off-flavours and precipitates in the foods to which they are added. To avoid such problems, less well-absorbed but less damaging forms are used.

There is, in fact, considerable opposition among some health professionals to the continuation of the practice of compulsory fortification of food with iron. Iron overload, which may be exacerbated as a result of fortification of staple foods, is a real problem for some people[85]. It is also argued that, because of the use of some forms of iron with low bioavailability, fortification fails to produce any worthwhile result. Reduced iron powder, in particular, has been singled out for criticism[86].

The problem of bioavailability of the different forms of iron and their suitability for use in food fortification is not, however, clear cut. In the case of reduced iron in particular, it has been pointed out that human studies on its absorption have produced inconsistent and conflicting results[87]. Its bioavailability relative to ferrous sulphate has been reported to range from 13 to 148%. Such variations can be attributed to such factors as differences in particle size, porosity, surface area and manufacturing processes. When tested under carefully controlled conditions, using bread enriched with the stable isotope ^{58}Fe, it has been shown that hydrogen-reduced iron powder is soluble under gastric conditions and is available for uptake into the mucosal cells of the GI tract, to an extent comparable to that of iron(II) sulphate[89].

The levels of iron required by legislation to be added to fortified foods have caused controversy in some countries. In 1971 the USFDA proposed that levels in refined wheat products should be increased from about 2.6 to 8.8 mg/100 grams. The proposal met with strong opposition because of possible danger to the health of some people[90]. The proposal was withdrawn. In Sweden, where an increase in levels of iron fortification up to 6.5 mg/100 grams had been approved in 1970, legislation requiring fortification of flour was totally withdrawn in 1995[91]. It is anticipated that similar moves will be made to remove mandatory iron fortification of cereal products in other countries within the next decade[92].

Unintentional fortification of foods can occur through the use of iron cooking and processing equipment. The significance of this adventitious uptake is highlighted by two recent studies. An investigation of the incidence of iron deficiency anaemia (IDA) in Brazilian infants found that the use of iron cooking pots to prepare their food significantly improved their iron nutritional status[93]. Another study, in South Africa and Zimbabwe, found that use of a traditional beverage, a low-alcohol 'sweet' beer, which was made in non-galvanised iron vessels and cast iron pots, helped prevent IDA in women of childbearing age[94]. The finding of the protective effect of iron enrichment of traditional beer in the second study is of interest, since, in contrast, home-brewed beer has been indicated as a contributing factor in iron overload in several other studies, in Southern Africa in particular[95].

8.3.4 Dietary intake of iron

In spite of the wide-scale distribution of iron in the environment and in most foods, iron deficiency is probably one of the most common nutritional problems, even in prosperous communities. Indeed, IDA remains one of the few deficiency disorders observed commonly in Western industrialised countries[96]. One reason for this is that we live in a world in which the energy demands made on humans are continually decreasing and where, as a consequence, if excessive body weight is to be avoided, a decreased intake of dietary energy is essential. The requirements of most nutrients including iron, however, are not reduced and adjustments have to be made to the composition of meals to provide the right balance of energy and nutrients. This is done by increasing the nutrient density of food, that is, the amounts of essential nutrients per unit energy. Unfortunately, many people, rather than adjusting their diets in this way, have a high intake of 'empty calories', in so-called 'junk foods' which are rich in energy, fat and sugar, but low in iron, among other essentials. A second, and probably more important, reason for an inadequate intake of iron is that though the metal may be present in foods consumed, it is not there in a readily bioavailable form and thus the amount absorbed from the GI tract is inadequate for the body's needs.

8.3.5 Iron absorption and metabolism

The uptake of iron is a complex process. Its initial absorption in the GI tract depends on the bioavailability of the particular form of the element present in the food consumed as well as on the overall composition of the diet. Iron occurs in foods in two different forms, as organically complexed haem iron and as inorganic compounds. The iron in cereals, vegetables and fruits and other plant foods is inorganic, or non-haem iron, and has a far lower level of bioavailability than does the iron in meat and other animal products. In addition, other components of a meal can either help or hinder iron absorption. Bioavailability will be hindered by phytate, fibre, tannin and oxalate and enhanced by ascorbic and other organic acids, such as citrate[97].

A diet with adequate sources of energy and relatively rich in inorganic iron can fail to meet nutritional needs for the element, if the food consumed is also rich in dietary fibre or tannin. It has been shown that one cup of tea can reduce absorption of iron from 11.5 to 2.5% from a meal consisting mainly of bread[98]. On the other hand, citric acid in fruit can increase iron absorption from a mixed meal sevenfold[99].

Iron absorption is apparently regulated by the existing iron status of the body. If this is low, the absorption mechanism can be stimulated to increased activity. When iron stores are high, absorption is slowed down. The pH of the gut also has an effect, with food iron mainly in the more readily absorbed ferrous state under acid conditions.

Once the iron in ingested food has been absorbed and has entered the blood, only small amounts are lost from the body, except when bleeding occurs. There is no physiological mechanism for secreting iron, so iron homeostasis depends on its absorption. Thus the healthy individual with a good store of iron is able to maintain a balance between the small normal losses and the amounts of the element absorbed from food. Normally only a very small amount of iron, about 1 mg/day, needs to be absorbed. The metal first enters the intestinal mucosal cells where it is bound into ferritin, an iron-storage protein. This is a large molecule from which the iron can be readily mobilised when required. Some of the incoming iron may be transferred directly by a transport protein, transferrin, to bone marrow and other tissues to be used in the synthesis of haemoglobin and myoglobin[100].

Loss of iron from the body normally occurs through the ageing of cells in the intestinal mucosa. As these are lost from the gut and excreted they carry with them iron which had been incorporated into ferritin. This accounts for the adult body's approximately 1 mg/day iron loss which must be made up by absorption from the diet. Iron released from the blood's red cells, as these are broken down in the spleen after about 120 days of life, is recirculated in plasma and re-utilised. The kidneys are normally very effective in preventing any loss of this circulating iron.

8.3.5.1 Functions of iron

Iron is an essential nutrient for all living organisms, with the exception of certain bacteria. An adult human contains about 4 grams of iron, of which nearly 3 grams are contained in the two metalloproteins, haemoglobin and myoglobin. Both molecules are able to bind to oxygen in a reversible manner. Haemoglobin is packed within the red blood cells and is used to carry oxygen from lungs to tissues. Myoglobin is confined to muscle cells, where it provides an oxygen reserve.

Iron is also an essential component of a variety of enzymes, such as the cytochromes and catalase. In these the iron atoms, present in the ferrous and ferric states, interchange with gain or loss of an electron, as part of the electron chain responsible for the redox reactions necessary for release of energy in cellular catabolism and synthesis of large molecules. These enzymes account for about 4% of the functional iron compounds of the body.

In addition to its major functions in oxygen transport and as a cofactor in many enzymes, iron also plays an important role in the immune system. Though the mechanisms involved are complex, there is good evidence that an abnormal iron nutritional status can lead to impaired immune function, with serious consequences for health[101].

8.3.5.2 Iron deficiency anaemia

Iron deficiency ultimately results in failure of the body to produce new blood cells to replace those which are constantly being destroyed at the end of their normal life span. Gradually the number of blood cells falls and, with this, the amount of haemoglobin in the blood. The cells become paler in colour and smaller in size. These undersized cells are unable to carry sufficient oxygen to meet the needs of tissues, so energy release is hindered. This is what is known as microcytic hypochromic anaemia, or, simply, as iron deficiency anaemia (IDA)[102]. Those suffering from IDA show symptoms of chronic tiredness, persistent headache, and, in many cases, a rapid heart rate on exertion. There may also be other functional consequences, including a decreased work capacity, a fall in intellectual performance, and a reduction in immune function.[103] There is concern that IDA in infancy and childhood may have serious consequences, such as morbidity and defects in growth and development in infancy, as well as impaired educational performance in schoolchildren[104].

8.3.6 *Recommended intakes of iron*

The UK RNI is 1.7–8.7 mg/day for both males and females from birth to ten years of age. This rises to 14.8 mg/day for females from years 11 to 50, when it is reduced to 8.7 for the post-childbearing years. Women with a high menstrual loss are recommended to increase iron intake by consuming a supplement. The RNI for males from years 11 to 18 is 11.3 mg/day, with a drop to 8.7 mg/day for later years[105].

8.3.7 *High intakes of iron*

Iron toxicity can occur as a result of ingestion of large amounts of iron compounds. This is most unlikely to be caused by iron in normal foods, but by accidental intake of a chemical substance. A lethal dose for adults is about 100 g, while in children, among whom most cases of iron poisoning occur, it is 200–300 mg/kg body weight[73]. The source of the iron in most childhood poisoning appears to be supplement pills used by their mothers which the children mistook for sweets[106].

A high intake of iron can be a serious problem for persons with the hereditary disorder of haemochromatosis. This condition, in which there is a gradual accumulation of iron in tissues and which can result in liver failure, occurs in nearly 1% of Europeans and more frequently in people of African origin. There is evidence that in some cases iron overload is caused by excessive intake of the metal as a result of consumption of foods and beverages prepared in iron cooking pots and containers[107]. Concern has been expressed by some investigators about the possibility that an excessive intake of supplementary iron, in addition to a high meat diet, could result in high levels of cellular oxidation and be linked to ageing, coronary artery disease and a variety of neurodegenerative and inflammatory disorders[108].

8.3.8 *Analysis of foodstuffs for iron*

Iron in foods is readily determined by AAS. Concentrations in most samples generally fall within the working range of the instruments. Both dry ashing and acid digestion can be used to prepare samples for analysis. For lower concentrations of iron GFAAS can be used directly on digests, without the need for preconcentration.

8.4 Cobalt

Cobalt is one of the most fascinating of the biological trace elements, with several unusual facets. Its name was derived from the German word for goblin, and was coined by the miners of the Ore Mountains of Thuringia and Saxony. They considered it a useless and evil metal, because its presence in ores prevented the extraction of other metals and, as a consequence, they named it after 'Kobold', an evil earth spirit[109]. Cobalt is a relatively rare element, making up about only 0.001% of the lithosphere. It occurs in animals, including man, in minute amounts, and the daily intake of the human body is not much more than 0.1 μg, yet it is an essential trace element. It is an integral part of vitamin B_{12}, or cobalamin, of which the body contains not much more than 5 mg. However, if there is less than this, the fatal illness pernicious anaemia can develop. Its role in vitamin B_{12} appears to be the main, if not sole, function of cobalt in humans.

8.4.1 Chemical and physical properties of cobalt

Cobalt has an atomic weight of 58.9, and an atomic number of 27. It is a hard, brittle, bluish-white metal, with magnetic properties similar to those of iron. It has a density of 8.9 and a melting point of 1490°C. The metal is fairly unreactive and dissolves only slowly in dilute acids. Like other transition metals, cobalt has several oxidation states, but only states +2 and +3 are of practical significance. Co(II) forms a series of simple and hydrated cobaltous salts with all common anions. Few cobaltic Co(III) salts are known, since oxidation state +3 is relatively unstable. However, numerous stable complexes of both the cobaltous and the cobaltic forms of the element are known.

8.4.2 Production and uses of cobalt

Cobalt occurs in the Earth's crust in association with other elements, especially arsenic. Smaltite, $CoAs_2$, cobaltite, CoAsS, and linnaeite, Co_3S_4, are three of its best-known ores. However, the chief commercial source of the element is the residue left after electrolytic refining of nickel, copper and lead. Major ore deposits are found in Morocco, Canada and the Democratic Republic of the Congo. Other deposits are also worked in Australia, Uganda and Zambia. World demand for cobalt has been increasing rapidly in recent decades and is estimated to reach 40 000 tonnes in 2002. The growth in demand has been largely due to increased use of the metal in chemical applications (primarily in portable rechargeable batteries) and in alloys. Whether demand will continue to increase in coming years is debatable[110].

A growing use of cobalt is in the production of high-strength 'superalloys'. The addition of cobalt to steel results in greater hardness and brittleness, improves the cutting power of high-speed tools, and alters the magnetic properties. Combined with aluminium and nickel it produces an alloy which is used to make strong permanent magnets for radios and electronic devices. When alloyed with chromium and tungsten it is used to make hard-edged metal-cutting and surgical instruments. Cobalt is also used in the manufacture of glass and glazing for pottery. Some of its salts produce an intense and beautiful blue and are used in cerulean, cobalt green and cobalt yellow pigments and paints. An important use is in the form of its artificially produced isotope, cobalt-60 (^{60}Co), which is a source of γ-rays used in cancer treatment. Cobalt is also used in the pharmaceutical industry. Cobalt–chromium alloys are employed in some metallic prostheses.

8.4.3 Cobalt in food and beverages

Information on levels of cobalt in foods and diets throughout the world is not very plentiful. Data from the UK indicate that levels in most foods are very low, at < 0.1 mg/kg, as is seen in Table 8.5. The highest levels are found in offal meat and nuts. It has been reported that another good source of cobalt is yeast extract, sold under such brand names as Marmite in the UK and Vegemite in Australia.

Table 8.5 Cobalt in UK foods.

Food group	Mean concentration (mg/kg fresh weight)
Bread	0.02
Miscellaneous cereals	0.01
Carcass meat	0.004
Offal	0.06
Meat products	0.008
Poultry	0.003
Fish	0.01
Oils and fats	0.003
Eggs	0.002
Sugar and preserves	0.03
Green vegetables	0.009
Potatoes	0.02
Other vegetables	0.006
Canned vegetables	0.01
Fresh fruit	0.004
Fruit products	0.005
Beverages	0.002
Milk	0.002
Dairy products	0.004
Nuts	0.09

Adapted from Ysart, G., Miller, P., Crews, H. *et al.* (1999) Dietary exposure estimates of 30 elements from the UK Total Diet Study. *Food Additives and Contaminants*, **16**, 391–403.

Surprisingly, though in general dairy products are low in cobalt, an Italian study found that the mean concentration range in cheese was 1.6–32 μg/kg, with the highest levels in curd and hard cheeses. Some samples contained as much as 46 μg/kg. It was estimated that cheese contributed approximately 10% of the total cobalt intake in Italy[111].

Average daily intake in the UK is about 0.3 mg. This is about the same as estimated daily intakes in the US (150–600 μg)[112] and Japan (300 μg)[113], but considerably higher than that reported in Finland (13 μg)[114] and Belgium (26 μg)[115]. The use of cobalt-containing food additives has been known to increase dietary levels of the metal excessively. Formerly cobalt salts were sometimes added to beer to improve foaming qualities[116]. The practice is no longer in use following fears that the resulting excessive consumption of cobalt was responsible for an epidemic of cardiac failure in heavy drinkers, some of whom were ingesting the equivalent of 8 mg of cobalt sulphate each day. An investigation concluded that the cobalt alone was not responsible for the epidemic, and that two other factors, a high alcohol intake and a dietary deficiency of protein and/or thiamin, played a part in the incident[117].

8.4.4 Recommended intakes of cobalt

There are no recommended intakes for cobalt itself, but only for the cobalt-containing vitamin B_{12}, cobalamin. This is a large, complicated molecule in which a single atom of cobalt sits at the centre of a porphyrin-like corrin ring. Neither animals, including humans, nor higher plants are able to synthesise vitamin B_{12} and it has to be obtained preformed in food. Neither elemental cobalt nor any of its compounds can meet this requirement. The vitamin is made by bacteria, particularly soil bacteria, which have the ability to incorporate cobalt into the corrin ring structure. The UK adult RNI for vitamin B_{12} is 1.5 μg/day, which is equivalent to about 0.06 μg of cobalt.

8.4.5 *Absorption and metabolism of cobalt*

Reports by different investigators on levels of absorption of cobalt from the diet present a confusing picture, with estimates ranging from 5 to 45%[118]. The variation may be due to nutritional and other factors which affect levels of absorption, such as body iron status[119] as well as the chemical form of the element in food[120]. The human body contains little more than 1 mg of cobalt, with about a fifth of this stored in the liver[121]. Excretion is mainly in urine.

No functions of cobalt in human nutrition, other than as an integral part of vitamin B_{12}, have been clearly established. The vitamin plays a very important role in the body, especially in cells where active division is taking place, such as in blood-forming tissues of bone marrow. Deficiency of the vitamin is responsible for the development of pernicious anaemia when abnormal cells known as megablasts develop. This leads to potentially fatal macrocytic anaemia. In addition to the anaemia, the nervous system is also seriously affected.

There are some indications that cobalt may be required for synthesis of thyroid hormone, according to a report which noted an inverse relationship between cobalt levels in food, water and soil in some areas with the level of incidence of goitre in animals and humans[122]. It has also been claimed that cobalt, rather than chromium, is the metal in GTF, the Glucose Tolerance Factor[123].

8.4.6 *Toxicity of cobalt*

Cobalt, like certain other hard metals, such as tungsten, can cause lung fibrosis and asthma in industrial workers exposed to its dust and fumes[124]. Concern has been expressed that cobalt in metal implants, for example in hip replacement prostheses, may eventually be redistributed in the body, with possible effects on health. However, the evidence for this is slight[125]. Another medical use of cobalt has caused problems. When cobalt salts have been used as a non-specific bone marrow stimulus in the treatment of certain refractory anaemias, high doses have been reported to cause serious toxic effects, including goitre, hypothyroidism and heart failure[126].

Apart from such high and non-ingested intakes of chromium, there is no evidence that cobalt, at least at the levels normally found in the diet, has any toxic effects. Health authorities have not felt it necessary to set limits for cobalt in food. However, there may be problems with high levels in some dietary supplements and health foods. In the UK a Working Party on Dietary Supplements and Health Foods of MAFF warns that cobalt could have undesirable effects above a chronic dose of 300 mg/day[127].

8.4.7 *Analysis of foodstuffs for cobalt*

Because of its low levels of concentration, the determination of chromium in foods and beverages presents some difficulties. AA is normally the method of choice, though FAAS is insufficiently sensitive. GFAAS, with background correction, preceded by a preconcentration, is recommended[128]. Sensitivity can be improved considerably by the use of pyrolytically coated graphite tubes[129].

8.5 Nickel

Nickel, like cobalt, owes its name to its refractory nature and to the problems this gave to medieval Saxon miners. It occurred in a type of copper ore that was very difficult to refine and was known as 'kupfer-nickel' after a local bad earth spirit which plagued the miners[130]. The problems associated with the extraction of nickel from its ores have been overcome and the metal is now of major industrial consequence, but it still remains, for the biologist, something of a problem. Though there is good evidence that it is an essential nutrient for animals, it has not been shown to have a clearly defined biochemical function in humans[131]. Moreover, while its toxicity for industrial workers is well known, it is not certain that it can have similar consequences when consumed in food.

8.5.1 *Chemical and physical properties of nickel*

Nickel is element number 28 in the periodic table with an atomic weight of 58.71. It is a tough, silver-white metal with a density of 8.9 and a melting point of 1453°C. The metal is very adaptable to the requirements of fabricators and can be drawn, rolled, forged and polished. It has, in addition, good thermal and electrical conductivities. It readily forms alloys with other metals, especially iron. Though highly resistant to attack by air and water, it can be dissolved in dilute acids.

The chemistry of nickel is relatively complex. Like other transition metals it has several oxidation states, including Ni(0), Ni(I) and Ni(II), but only the divalent form is commonly encountered. One important exception is nickel dioxide, NiO_2, in which it is in the tetravalent state. Ni(II) forms a number of binary compounds with various non-metals, such as carbon and phosphorus. It binds many substances of biological interest, including amino acids and proteins. Nickel forms an extensive series of inorganic compounds, many of which are green in colour and are soluble in water. When finely divided nickel metal is exposed to carbon monoxide, nickel carbonyl, $Ni(CO)_4$, is formed. This is a colourless volatile liquid that decomposes at high temperature to form CO and free nickel metal. This reaction is the basis of the Mond method for separation of nickel from other metals.

8.5.2 *Production and uses of nickel*

Nickel occurs naturally in combinations with arsenic, antimony, sulphur and other elements. Its principal ores include kupfer-nickel and garnerite, a magnesium–nickel silicate, as well as certain varieties of the iron mineral pyrrhotite, which contains up to 5% nickel. One of the first commercial sources of the metal was the Pacific Island of New Caledonia, where a silicate containing 10% nickel occurs near the surface. An even richer source, and one of the world's major deposits, is at Sudbury, in Ontario, which was discovered during the cutting of the Canadian Pacific Railway[132]. Elemental nickel, alloyed with iron, is found in some meteorites. There is evidence that the Earth's core may contain considerable amounts of the metal.

Extraction and refining of nickel from its ore are complex. Usually the ore is converted by heating to Ni_2S_3 and then to NiO, which is reduced to the metal by hydrogen. Pure high-grade nickel is produced by the Mond process, in which the crude metal is reacted with carbon monoxide to produce gaseous nickel carbonyl. Heating to 200°C reduces the $Ni(CO)_4$ to 99.9% pure metal.

The major use of nickel metal is in the production of high-quality, resistant alloys with iron, copper, aluminium, chromium, zinc and molybdenum. There are said to be some 3000 different nickel alloys[132]. Nickel-containing steels are highly corrosion-resistant and are used in the manufacture of food-processing equipment. Nickel alloys are also used in storage batteries, automobile and aircraft parts, and electrodes. Nickel compounds are important to the chemical industry, for example as catalysts in hydrogenation of fats. They are also used in pigments in lacquers, paints, pottery and cosmetics. Nickel sulphate is used as a mordant in dyeing and fabric printing.

8.5.3 Nickel in food and beverages

Nickel occurs in small amounts in most soils. Plants may contain between 0.5 and 3.5 mg of the metal/kg[134]. Levels in most foods are low, at less than 0.2 mg/kg. Data for levels in different foods in the UK are given in Table 8.6. There are exceptions to the generally low levels in foods. Some food plants, such as tea, have naturally high levels of the metal. A UK study found that tea leaves contain 3.9–8.2 mg/kg, and instant tea 14–17 mg/kg[135]. Certain culinary herbs also contain relatively high levels. The cacao bean, from which cocoa and chocolate are made, may contain as much as 10 mg/kg. Nuts are also a good source of the element[136].

Table 8.6 Nickel in UK foods.

Food group	Mean concentration (mg/kg fresh weight)
Bread	0.10
Miscellaneous cereals	0.17
Carcass meat	0.04
Offal	0.06
Meat products	0.06
Poultry	0.04
Fish	0.08
Oils and fats	0.03
Eggs	0.03
Sugar and preserves	0.28
Green vegetables	0.11
Potatoes	0.10
Other vegetables	0.09
Canned vegetables	0.25
Fresh fruit	0.03
Fruit products	0.06
Beverages	0.03
Milk	<0.02
Dairy products	0.02
Nuts	2.5

Adapted from Ysart, G., Miller, P., Crews, H. *et al.* (1999) Dietary exposure estimates of 30 elements from the UK Total Diet Study. *Food Additives and Contaminants*, **16**, 391–403.

Nickel levels in processed foods can be increased by pick-up from cooking equipment and containers. As seen in Table 8.6, higher than average levels are found in canned fruit in the UK. Similar findings have been reported from the US[137]. Nickel migration in dairy products and other foods from stainless steel containers has been reported[138].

8.5.4 *Dietary intakes and requirements*

Dietary intake of nickel is normally very low. The mean total adult exposure to nickel in the UK has been reported to be 0.12 mg/day, with a range up to 0.21 mg/day[139]. These results are of the same order as those reported in other countries, including Finland (0.13 mg)[140] and the US (0.17 mg)[141]. The human requirement for nickel is unknown and no RDA or DRV has been set for the element. It has been estimated that requirements may be as low as 5 μg/day[142]. There is no evidence that levels of intake in the UK or elsewhere fail to meet the body's needs and symptoms of deficiency in humans are unknown.

8.5.5 *Absorption and metabolism of nickel*

Nickel is poorly absorbed from food. Between 3 and 6% of dietary intake is believed to be retained by the body. Absorption appears to be enhanced by iron deficiency. Nickel is transported in the blood principally bound to albumin. It seems to be fairly evenly distributed among tissues and is not accumulated in any particular organ[143]. Excretion is mainly in faeces, with a small amount in urine.

Nickel is recognised as an essential nutrient for animals, and possibly also for humans, though its essentiality has still to be proven definitively[144]. No clear biochemical function for the element has been identified in humans, though it is possible that it serves as a cofactor or structural component of certain metalloenzymes similar to the nickel-containing enzymes that have been identified in plants and micro-organisms. These include urease, hydrogenase and carbon monoxide dehydrogenase. Other enzymes, including carboxylase, trypsin and coenzyme A, are activated by nickel.

Signs of nickel deficiency have not been described for humans, though they are well recognised in animals. Rats and other animals deprived of nickel show depressed growth, reduced reproductive rates and alterations of serum lipid and glucose levels[145]. It has been suggested that because the nickel content of some human diets can be lower than that inducing such changes in animals, the metal should be considered a possibly limiting nutrient under specific human conditions[146].

8.5.6 *Nickel toxicity*

In contrast to industrial nickel, which can cause cancer of the respiratory tract and dermatitis[147], dietary nickel appears to be relatively non-toxic to humans. It has been estimated that the level of toxicity in humans is about 250 mg[148]. However, there is increasing concern at the possibility of allergic reactions to nickel in foods. Nickel is a common sensitising agent with a high prevalence of allergic contact dermatitis. An oral dose of 0.6 mg of nickel sulphate has been shown to produce a positive skin reaction in some individuals with nickel allergy[149]. Interestingly, it has been reported that nickel eczema sufferers have a significantly lower intake of nickel-rich foods, such as chocolate, nuts, beans and porridge oats, than do non-sufferers, and that serum nickel levels correlate with intakes of these foods[150].

8.5.7 *Analysis of foodstuffs for nickel*

The method of choice for nickel analysis is GFAAS, with adequate background correction. The procedure has been used in the study of trace elements in the diet by the Swedish National Food Administration[151]. Samples were prepared by dry ashing, followed by solution in nitric acid. Deuterium lamp background correction was found to give satisfactory results, though it has been claimed that the Zeeman method is superior[152]. The determination of nickel using GFAAS directly on a slurry of wheat flour in a nitric acid/hydrogen peroxide mixture has been described[153].

8.6 Copper

Copper was probably among the first of the metals to be discovered. This was because it is relatively easily extracted from its ores by simple metallurgical processes, and, moreover, is often found ready for use in its elemental, metallic state. Its ores, because of their relative abundance and distinctive colours, were easily recognised even by unskilled prospectors. However, though it could be used for making ornaments and utensils for holding food and drink, copper was not good for making weapons or large structures of any kind until its ability to form a tough alloy with tin was discovered. The discovery of bronze ushered in a great period of human development, the Bronze Age, which lasted for many hundreds of years until the copper alloy was replaced by the more useful iron. In China and India, brass, an alloy of copper and zinc, has played a major role in the lives of millions of people for thousands of years. Copper and its alloys are still of importance, not least in the food industry, in today's world.

8.6.1 Chemical and physical properties of copper

Copper has an atomic weight of 63.54 and its atomic number is 29. It has a density of 8.96 and a melting point of 1083°C. It is a tough though soft and ductile metal, second only to silver as a conductor of heat and electricity. These are the properties which have made copper so valuable to the electrical industry, as well as to the manufacture of utensils of many kinds. Copper is resistant to corrosion in the sense that on exposure to air it is superficially oxidised. Copper-sheathed roofs and domes, in the damp, sulphur-polluted air of many cities, take on a patina of green hydroxo-carbonate and hydroxo-sulphate, verdigris, which protects the metal from corrosion.

Copper has two stable oxidation states, +1 (cuprous) and +2 (cupric). There is some evidence that it may also occur in a third oxidation state (+3) in some crystalline compounds and complexes, but, if so, its existence is transitory. Copper forms a wide range of cuprous and cupric inorganic compounds. Most cuprous compounds are readily oxidised to the cupric form. The ability of copper to form complexes with amines and other ligands is a distinctive characteristic of the metal and accounts for many of its biological properties. These complexes, which are coloured, also provide the basis for several colorimetric methods for determination of the element, such as Fehling's and Benedict's tests.

8.6.2 Production and uses of copper

Copper ores are widely distributed throughout the world. The most important are malachite, azurite, chalcopyrite, cuprite and bornite. Most are sulphides, but oxides, carbonates, arsenides and chlorides also occur. Many of the ores are mixed with other metals, including zinc, cadmium and molybdenum. The metal is relatively easily extracted by roasting the ores in air to form the oxide, and smelting.

World production of copper depends very much on industrial needs. In recent years there has been a dramatic slump in demands by industry, as copper began to be replaced by cheaper metals, as well as by other materials such as fibre optics, especially in the electrical industry. Even today, however, the manufacture of electrical cables and associated equipment accounts for about half of the world production. Copper pipes and fittings are still extensively used in plumbing, though plastic has been replacing them in recent years. Another industry in which copper is no longer used as extensively as in the past is the manufacture of cooking and other food-processing equipment. Copper sheeting is still used for roofing buildings and sheathing boat hulls. A major use continues to be as a component of alloys with zinc and other metals. Brass utensils are widely used, especially for domestic purposes, in some countries. Copper-containing bronzes of various kinds, such as phosphor-bronzes, are widely used to make high-quality machines and industrial fittings. Other industrial applications include plating and circuit board production, and the manufacture of vehicle parts such as brake pads.

Copper salts have important pharmaceutical and agricultural uses. Bordeaux mixture, a fungicide used to prevent blight in potatoes and grapes, was one of the earliest and most important to be developed. Copper-containing insecticides and wood preservatives, such as cupro-arsenate, are widely available. Copper algicides are used in water treatment, and copper salts to clear underground pipes blocked by tree roots. In agriculture copper salts are used as growth enhancers, especially of pigs[154].

8.6.3 *Copper in food and beverages*

Copper is present in all foods, with concentrations in most ranging from about 0.05 to 2.0 mg/kg. There are some exceptions, such as offal and nuts, which can contain considerably more. Copper levels in the different food groups in the UK are shown in Table 8.7[155]. Similar levels have been reported in other countries, including the US[156], Italy[157] and Spain[158]. The best food sources of copper, apart from animal offal and nuts, are cereals, especially wholegrain, with about 1–4 mg/kg and up to 10 mg/kg in certain ready-to-eat breakfast cereals, especially those based on wheat bran[159].

8.6.3.1 *Copper in drinking water*

Apart from food, a significant source of copper can be domestic water supplies. Even though copper has been replaced in plumbing fittings in many modern homes, tap water can still deliver a significant amount to consumers. Levels of naturally occurring copper can vary considerably, depending on the nature of the soil and the rock where the water originates. The WHO standard for drinking water is 50 μg/litre, but this can easily be exceeded by water which has passed through copper pipes. A study of metal uptake from domestic plumbing in an Australian city found that daily consumption of 1.5 litres of water could give an intake of 80 μg of copper when the water was drawn from the cold tap. This increased to 12.3 mg when water from the hot system, which had a copper tank and immersion heater, was used[160]. This instance, was, in fact, an extreme exception, and most domestic water supplies are unlikely to have such high concentrations of copper or other metals.

Most countries have guidelines for copper in water, both at the treatment plant and the tap. These range from a conservative 50 μg/litre in France and Sweden (in line with WHO recommendations), to 1.0 mg/litre in the US, Norway and Finland, and to a maximum of 3.0 mg/litre in Germany and the UK[161]. It is interesting that even in the UK, with a higher allowable total copper level than the WHO, a study of copper and zinc in urban children found that levels in water used to prepare their food had no discernible effect on their intakes[162]. However, drinking of domestic water which was

Table 8.7 Copper in UK foods.

Food group	Mean concentration (mg/kg fresh weight)
Bread	1.6
Miscellaneous cereals	1.8
Carcass meat	1.4
Offal	40
Meat products	1.5
Poultry	0.73
Fish	1.1
Oils and fats	0.05
Eggs	0.59
Sugar and preserves	1.5
Green vegetables	0.84
Potatoes	1.3
Other vegetables	0.91
Canned vegetables	1.5
Fresh fruit	0.94
Fruit products	0.73
Beverages	0.10
Milk	0.05
Dairy products	0.45
Nuts	8.5

Adapted from Ysart, G., Miller, P., Crews, H. *et al.* (1999) Dietary exposure estimates of 30 elements from the UK Total Diet Study. *Food Additives and Contaminants*, **16**, 391–403.

contaminated with more than 1.6 mg copper/litre is reported to have caused poisoning in children in the US[163].

8.1.3.2 Adventitious copper in food

A non-food source of copper which can have a significant effect on dietary intake is the copper utensils used in cooking and food processing. Copper was once a favoured metal for making saucepans and other cooking equipment, because of its high heat conductivity. However, as the author of the present study has shown, a meal of meat and vegetables cooked in copper utensils can have more than twice the level of the metal found in the same meal cooked in stainless steel or aluminium equipment[164].

Though today copper saucepans are more often seen in a decorative than in a practical role, they can still be used, on occasion, for preparing or storing food and beverages. As a result poisoning can occur, as was the case with 15 Australian children who drank lime cordial which had been stored overnight in an old urn. The drink contained 300 mg/litre of copper and within minutes of drinking it the children developed symptoms of acute copper poisoning, with nausea and vomiting. The source of the copper was the urn's electrical heating element[165]. Chronic copper poisoning of infants whose feed was prepared using water taken from a hot-water tap has been reported[166].

There is some evidence that the endemic illness Indian childhood cirrhosis (ICC), in which there is an excessive accumulation of copper in the liver, is associated with the use of brass utensils to prepare the food of infants[167]. There have been reports that high intakes of copper in infant feeds are also associated with other forms of cirrhosis in children. A study in Austria found evidence of a connection between cow's milk containing 63.3 mg/kg copper and what is known as endemic Tyrolean childhood cirrhosis[168], though this is disputed[169].

8.6.4 Dietary intakes of copper

Copper intakes in different countries are generally similar at about 1–3 mg/day. The levels reported seem to have remained relatively constant in recent years. The UK TDS of 1974 found an intake of 1.4 mg/day in adults, with an upper range of 3.0 mg/day[170]. Three years later, the mean intake was still 1.4 mg/day and the 97.5th percentile 3.2 mg/day. In Finland adult male copper intake was estimated to be 1.7 mg/day[171]. Mean adult daily intakes reported in a number of other countries are 1.0 mg in Japan[172], and 1.2 ± 0.6 in Denmark (males)[173]. In other European countries reported copper intakes are very similar[174]. In the US intakes for both male and female adults varied little over a nine-year period, with, for 25–30-year-old males, 1.21 mg in 1982/83, 1.10 mg in 1986/87 and 1.16 mg in 1990/91[175].

8.6.4.1 Recommended and safe intakes of copper

Expert committees in many countries have experienced difficulties in trying to establish recommended levels of intake of copper, mainly because of absence of clarity about its metabolic role and, to some extent, because it is not only an essential element, but also toxic at high levels of intake. These problems are reflected in official recommendations in different countries. In the 1989 review of the US RDAs, it was noted that because of the uncertainty about quantitative human requirements, it was not possible to establish an RDA for copper. Instead a safe and adequate range of dietary intake for adults (ESADDI) of 1.5–3 mg/day was recommended.

The UK's expert panel on dietary intakes also noted the inadequacy of data, but felt that though it was unable to establish an Estimated Average Requirement (EAR) or a Lower Reference Nutrient Intake (LNRI), there were sufficient data for infants and biochemical changes in adults on which to base RNIs[176]. An RNI of 1.2 mg/day was proposed for adults, with an additional 0.3 mg for women who were lactating. Recommendations for infants were 0.3 mg/day, and for children 0.4–0.7 mg/day. These are similar to the European Community Population Reference Intake (PRI)[177], but lower than the WHO Normative Requirements[178]. The 2001 review of US dietary intakes recommends an RNI of 0.9 mg copper/day[179].

As available data show, typical western diets fail to meet even the lower limits of the US ESADDI, and just reach the UK and European recommendations. Surveys in the US have shown that over the period from 1974–1991, intakes never exceeded 80% of the lower end of the ESSADI[180]. Estimates in the UK based on the 1997 TDS, are below the PMTDI and approximately equal to the RNI. Therefore, according to MAFF (now DEFRA), copper intakes are not a concern in terms either of toxicity or of deficiency. However, if the US ESADDI is taken as appropriate, copper deficiency should be expected to be relatively common. In fact, overt symptoms of deficiency are rare in the UK as well as the US and most other countries for which information is available. Such findings suggest, according to Failla, either that the lower limit of the ESADDI is too high, that marginal copper deficiency may be common but is too difficult to diagnose, that water provides sufficient copper to meet additional requirements, or some combination of such factors[181]. The question of copper requirements is being reconsidered in a current review of Dietary Reference Intakes in the US, but the problem is still unresolved. A major reason for this is the absence of clear functional or other indices of copper status in humans[182].

8.6.5 *Absorption and metabolism of copper*

Levels of copper in the human body are under homeostatic control over a broad range of intakes[183]. Generally 30–60% of copper in food is absorbed across the intestinal mucosa, but efficiency of absorption varies with the dietary content present in the GI tract. The amount absorbed seems to be controlled by the level of copper already in the body, in a feedback type of mechanism. Several factors, besides total levels of copper in food and in the body, can affect absorption. Phytate has an inhibitory effect. There is also interaction between molybdenum and copper. Low levels of molybdenum in food can result in retention of copper by the body, while high levels provoke an increase in copper excretion. Molybdenum-induced copper deficiency is well known in farm animals[184]. Other metals, including iron and zinc, play a similar antagonistic role towards copper[185].

Absorbed copper is rapidly transferred to the liver, kidney, heart and brain. Some copper is lost in urine, but the main excretory route is via bile in faeces. Within tissues copper is complexed with proteins. The principal copper protein is caeruloplasmin (Cp). This is a blue coloured molecule, with a molecular weight of 150 000, and contains 8 Cu(I) and 8 Cu(II) ions. Cp accounts for about 3% of the total copper in the body. It is manufactured in the liver and is involved in the regulation of levels of copper in the body. It is an important, but not the only transporter of copper to cells of the body[186]. Cp plays an essential role in the oxidation of Fe(II) to Fe(III), an important step in iron transport and haemoglobin manufacture.

8.6.5.1 *Copper enzymes*

The adult human body contains about 100 mg of copper. Most of this is tightly bound to about 20 proteins, whose characteristics have been established. Several of these are enzymes, with copper serving as a catalytic cofactor. They function in a wide range of cellular and extracellular activities[187]. They include cytochrome *c* oxidase, involved in ATP production; superoxide dismutase (SOD), which functions in oxygen metabolism; caeruloplasmin, in iron transport; lysyl oxidase, necessary for maturation of extracellular matrix; as well as tyrosinase, which is involved in production of the pigment melanin. Copper enzymes are also involved in the synthesis of a range of neuroactive amines and peptides (such as catecholamines). Maintenance of all these enzymes and their activities at optimum levels requires that adequate supplies of copper are available for incorporation into the apocuproproteins.

Failure to meet the body's dietary requirements for copper could result in biochemical inadequacies in the cells and low levels of activity of copper enzymes. However, though copper deficiency is well recognised in farm animals, severe deficiency is rare in adult humans. In infants, especially those that are premature, and in young children who are hospitalised for long periods, copper deficiency is well recognised and leads to leucopenia and skeletal fragility[188], as well as susceptibility to respiratory tract and other infections[189]. Anaemia may develop if the deficiency is prolonged and severe. In adults there is growing evidence that copper deficiency can result in cardiovascular problems[190], for example in patients maintained on TPN[191]. Copper deficiency has also been associated with development of anaemia, hypopigmentation and various other symptoms, including abnormalities of glucose and cholesterol metabolism and heart function[192]. There are also indications that an inadequate copper status may be related to reduced immune response[193].

Two hereditary diseases involving copper occur in humans. Menke's 'steely hair syndrome' is characterised by an inability to absorb adequate copper from the diet. Its symptoms, as the name suggests, include changes in the appearance and structure of the hair, accompanied by serious defects in cellular metabolism. In contrast the other hereditary condition, Wilson's disease, results in accumulation of excessive amounts of copper in liver and brain. Both diseases can be fatal if not treated. Each disease results from the absence or dysfunction of unique copper-transporting proteins. These are adenosine triphosphatases, ATPases, which, in the case of Menke's disease, are required to transport copper across the placenta, the GI tract and the blood/brain barrier, and, in Wilson's, for carrying copper from the liver to the bile[194].

8.6.5.2 *Depression of copper absorption by zinc and other metals*

As with many other trace elements in the diet, interaction can occur between copper and several other metals, with consequences for copper status and metabolism. Excessive intakes of zinc, especially in the form of high-concentration dietary supplements, have been shown to inhibit intestinal absorption, hepatic accumulation and placental transfer of copper, as well as to induce clinical and biochemical symptoms of copper deficiency[195]. It has also been shown, at least in experimental animals, that a high intake of tin can result in a lowering of copper status through inhibition of copper absorption[196].

8.6.6 *Copper toxicity*

High oral intakes of copper salts can be toxic, though, since their effect is emetic, it is difficult to retain enough in the body to produce fatal results. Intakes of up to 0.5 mg/kg body weight/day, and an occasional intake of up to 10 mg/day, are considered to be safe for adults[197]. Chronic intakes can have long-term health effects, such as childhood cirrhosis. However, official data, in the UK and the US, indicate that overt toxicity from dietary sources of copper is extremely rare. High levels of copper in foods can, however, have undesirable consequences, apart from toxicity. Even a small amount can act as a catalyst for the oxidation of unsaturated fats and oils, bringing about undesirable changes in taste, odour and colour in certain foods. Such changes, while undesirable from the point of view of appearance and consumer acceptability, do not, however, cause a significant increase in toxicity or a decrease in nutritive value of the affected food.

8.6.7 *Analysis of foodstuffs for copper*

Today in most large-scale analytical programmes, such as the UK TDS, copper in food is determined with ICPMS or an equivalent multielement technique. However, in smaller studies, copper analyses are still frequently carried out by simpler procedures. Flame AAS is relatively easy to use and it is accurate, sensitive and suffers little matrix interference[198]. The current method of choice for small-scale studies is generally GFAAS, with deuterium-lamp[199] or Zeeman-mode background correction[200]. Sample preparation, by acid digestion, preferably in a closed, microwave-heated system, is recommended[201]. With very low levels of copper in the food sample, extraction and concentration following digestion may be necessary.

8.7 Molybdenum

Though molybdenum has been known to science for two centuries, following its isolation by Hjelm in 1782, its importance in animal and human life was not appreciated until relatively recently[202]. In 1930 it was found to act as a catalyst for nitrogen fixation in bacteria. In 1943 its role in the enzyme xanthine oxidase in rat liver was established. Then in 1967 its essentiality was recognised when the pathology of molybdenum deficiency was recognised in humans[203].

8.7.1 *Chemical and physical properties of molybdenum*

Molybdenum has an atomic weight of 95.94 and is number 42 in the periodic table. It is dull, silver-white in colour and is mealleable and ductile, with a high melting point of 2610°C. The metal is strongly resistant to acid. It undergoes oxidation when heated in air, but at lower temperatures is not easily affected by oxygen. The chemistry of molybdenum is complex. Like other transition metals, it has a number of oxidation states, from -2 to +6. Only the higher oxidation states are of practical significance, with the lower states occurring mainly in organometal complexes. The most stable oxidation state is Mo(VI). The trioxide, MoO_3, is a white solid which dissolves in alkaline solutions to form molybdates, $-MoO_4$. These tend to polymerise into polymolybdates of the form $X_6Mo(Mo_6O_{24}).4H_2O$. It seems that in blood and urine molybdenum exists mainly as the molybdate ion. Numerous complexes of Mo(V) are known, including cyanides, thiocyanates, oxyhalides and a variety of organic chelates. Molybdenum can form alloys with other metals. The chemistry of these, like that of its organic compounds, is extensive and complex.

8.7.2 *Production and uses of molybdenum*

Molybdenum is one of the rarer elements. Though some areas of high soil molybdenum are known, especially overlying deposits of certain types of shale, generally levels in agricultural soils are low. There are a number of molybdenum ores, the most important of which is molybdenite, MoS_2. Its only known major deposit is in Colorado in the US. To produce the metal, the ore is crushed, treated by a flotation process and roasted to form MoO_3, which is reduced to the metal with hydrogen. The major use of molybdenum is in the production of steel. Molybdenum steels are extremely hard and strong and are used extensively in heavy machinery and military hardware. Molybdenum is also used in the chemical and other manufacturing industries as a catalyst, as well as in paints and pigments and in pharmaceutical and agricultural products.

8.7.3 *Molybdenum in food and beverages*

Molybdenum is found in all foods and beverages, usually at low levels of less than 1 mg/kg. Typical levels in a range of foodstuffs are given in Table 8.8. In addition to nuts, animal offal is the only foodstuff that contains relatively high levels of the metal. Data similar to those reported in the UK have been found in the US TDS. The range of levels, expressed as µg/kg, in a variety of foods purchased in supermarkets throughout the US, were: cereals/cereal products 26–1170; meat/fish, 4–1290; milk/dairy products, 19–99; vegetables, 5–332. In the US study, cereals were found to make up almost 40% of total intake of the general population, with, in the case of infants, dairy products contributing nearly 30%[204].

Table 8.8 Molybdenum in UK foods.

Food group	Mean concentration (mg/kg fresh weight)
Bread	0.20
Miscellaneous cereals	0.23
Carcass meat	0.01
Offal	1.2
Meat products	0.12
Poultry	0.04
Fish	0.03
Oils and fats	0.02
Eggs	0.08
Sugar and preserves	0.03
Green vegetables	0.15
Potatoes	0.09
Other vegetables	0.06
Canned vegetables	0.31
Fresh fruit	0.01
Fruit products	0.009
Beverages	0.003
Milk	0.03
Dairy products	0.07
Nuts	0.96

Adapted from Ysart, G., Miller, P., Crews, H. *et al.* (1999) Dietary exposure estimates of 30 elements from the UKTotal Diet Study. *Food Additives and Contaminants*, **16**, 391–403.

There does not appear to be any particular type of food which, in the absence of extrinsic factors, has a naturally high level of molybdenum. However, environmental conditions, such as pollution from natural or human sources, can lead to high levels in certain plants. There have been reports that residents of parts of India and Armenia, where the soil is naturally rich in molybdenum, have abnormally high intakes of up to 15 mg/day[205]. High levels of molybdenum have been found in plants grown on soils treated with sewage sludge and with certain fertilisers[206].

8.7.4 Dietary intakes of molybdenum

Intake of molybdenum by adults in the UK averages 0.12 mg/day, with an upper level of 0.21 mg/day. The intakes are in the same range as those reported in a number of other countries, such as the US (120–240 μg)[207], Sweden (44–260 μg)[208], and Finland (120 μg)[209]. These are generally lower than the 250–1000 μg/day previously reported from some other countries. These higher figures may, however, reflect unreliable analytical procedures rather than true intakes.

8.7.5 Absorption and metabolism of molybdenum

Molybdenum in food is readily absorbed by the body, possibly by a passive mechanism. More than 80% is absorbed in the stomach and the remainder in the small intestine[210]. Retention rates are low and most of the absorbed molybdenum appears to be excreted in urine within 24 hours of intake. In blood, molybdenum is loosely bound to the red cells. Some is retained in the liver and kidneys. It is an essential constituent of several enzymes, including xanthine oxidase/dehydrogenase, which is necessary for the formation of uric acid. It is also found in aldehyde oxidase, which is involved in purine metabolism, as well as in sulphite oxidase. The latter is essential for the detoxification of sulphite arising from the metabolism of sulphur-containing amino acids, as well as from ingestion of bisulphite preservatives and the inhalation of sulphur dioxide.

There is evidence that in all molybdoenzymes, the metal forms part of a complex called the 'molybdenum cofactor'. Its chemical structure is not fully known, but is believed to be a small, non-protein structure containing a pterin nucleus. A consequence of the need of the body to incorporate molybdenum into a complex structure is a potential for genetic abnormalities. A rare lethal inborn error of metabolism which involves sulphite oxidase and xanthine oxidase activities is known to occur. The disorder is characterised by deranged cysteine metabolism caused by a lack of functioning molybdenum.

Dietary molybdenum deficiency has not been shown to occur in humans, though it has been recognised in experimental and farm animals. A patient receiving prolonged TPN acquired a syndrome described as 'acquired molybdenum deficiency' characterised by disturbances of sulphur metabolism. The patient suffered mental disturbances and lapsed into a coma. His condition was reversed by the administration of ammonium molybdate[211].

A close connection between the metabolism of molybdenum and of both sulphur and copper has been clearly recognised in farm animals. Sulphur and sulphur compounds can be shown to limit molybdenum absorption and retention in animals. Sulphur is, in fact, used to treat 'teart disease', which is the name given to molybdenum toxicity in cattle. A consequence of this disease is copper deficiency due to impaired copper availability caused by the molybdenum[212]. Evidence for similar effects in humans is not available.

8.7.6 Toxicity of molybdenum

Though molybdenum appears to be relatively non-toxic to humans, there is some evidence that high intakes of between 10 and 15 mg/day may be associated with altered metabolism of nucleotides and reduced copper absorption[213]. There is a report of a possible connection between regular intakes of vegetables containing high levels of molybdenum and a high incidence of gout in Armenia. It has been suggested that the molybdenum encourages increased activity of the enzyme xanthine oxidase which leads to an increased production of uric acid leading to gout[214]. In the UK a Safe Intake of 50–400 μg/day has been recommended. The US ESADDI is 75–250 μg/day. Both of these recommendations are based on the fact that usual intakes fall into these ranges and do not appear to result in either deficiency or toxicity.

There has been a report of acute human molybdenum toxicity caused by consumption of a dietary supplement. The supplement was described as 'chelated molybdenum' and contained 100 μg/tablet, of which one a day was consumed by the patient. Within three weeks he began to show severe psychotic symptoms, with strong audio and visual hallucinations. His condition was described as similar to 'lucor metallicum' or 'manganese madness'. Even though he had consumed the supplement for little over two weeks, a year later he still showed signs of the effects of molybdenum toxicity[215].

8.7.7 *Analysis of foodstuffs for molybdenum*

The generally very low levels of molybdenum in foods and beverages cause some problems for analysts. Acid digestion, followed by extraction and concentration of the element, is recommended. Subsequent analysis of the digest can be carried out effectively using GFAAS with appropriate background correction. Several different procedures for the determination of molybdenum, and other trace elements, in foods and other biological samples, are available. Both closed microwave and open acid digestion can be used for sample preparation. In those with a high silicon content which can bind molybdenum, digestion efficiency is sometimes improved by the addition of hydrogen fluoride[216]. Low-temperature digestion using a Teflon® tube has been shown to give excellent recoveries[217].

References

1. Shriver, F.D., Atkins, P.W. & Langford, C.H. (1994) *Inorganic Chemistry*, p. 64. Oxford University Press, Oxford.
2. Stillman, M.J. & Presta, A. (2000) Characterizing metal ion interactions with biological molecules – the spectroscopy of metallothionein. In: *Molecular Biology and Toxicology of Metals* (eds. R.K. Zalups & J. Koropatnick), pp. 1–70. Taylor and Francis, London.
3. Nriagu, J.O. & Pacyna, J.M. (1988) Quantitative assessment of worldwide contamination of air, water and soils by trace metals. *Nature*, **333**, 134–9.
4. Akasuka, K. & Fairhall, L.T. (1934) The toxicology of chromium. *Journal of Industrial Toxicology and Hygiene*, **16**, 1–10.
5. Schwarz, K. & Mertz, W. (1959) Chromium(III) and the glucose tolerance factor. *Archives of Biochemistry and Biophysics*, **85**, 292–303.
6. Anderson, R.A. (1995) Chromium, glucose tolerance, diabetes and lipid metabolism. *Journal of the Advancement of Medicine*, **8**, 37–50.
7. Underwood, E. (1977) *Trace Elements in Human and Animal Nutrition*, 4th edn., p. 258. Academic Press, New York and London.
8. Burrows, D. (1983) *Chromium: Metabolism and Toxicity*. CRC Press, Boca Raton, Florida.
9. Rowbotham, A.L., Levy, L.S. & Shuker, L.K. (2000) Chromium in the environment: an evaluation of exposure of the UK general population and possible adverse health effects. *Journal of Toxicology and Environmental Health – Part B: Critical Reviews*, **3**, 145–78.
10. Department of Health (1991) *Dietary Reference Values for Food Energy and Nutrients in the United Kingdom*, pp. 181–2. HMSO, London.
11. Tinggi, U., Reilly, C. & Patterson, C. (1997) Determination of manganese and chromium in foods by atomic absorption spectrometry after wet digestion. *Food Chemistry*, **60**, 123–8.
12. Mordenti, A., Piva, A. & Piva, G. (1997) The European perspective on organic chromium in animal nutrition. In: *Biotechnology in the Feed Industry. Proceedings of Alltech's 13th Annual Symposium* (eds. T.P. Lyons & K.A. Jacques), pp. 227–40. Nottingham University Press, Nottingham, UK.
13. Hegoczki, J., Suhajda, A., Janzso, B. & Vereczkey, G. (1997) Preparation of chromium enriched yeasts. *Acta Alimentaria*, **26**, 345–58.
14. Saner, G. (1986) The metabolic significance of dietary chromium. *Nutrition International*, **2**, 213–20.
15. EEC (1980) *Council Directive Relating to the Quality of Water Intended for Human Consumption*, **80/778**. European Economic Community, Brussels.
16. EEC (1980) *Council Directive Relating to the Quality of Water Intended for Human Consumption*, **80/778**. European Economic Community, Brussels.
17. Ministry of Agriculture, Fisheries and Food (1998) *Cadmium, Mercury aand Other Metals in Food. 53rd Report of the Steering Group on Chemical Aspects of Food Surveillance*. The Stationery Office, London.

18. Whitman, W.E. (1978) Interactions between structural materials in food plant, and foodstuffs and cleaning agents. *Food Progress*, **2**, 1–2.
19. Offenbachr, E.G. & Pi-Sunyer, F.X. (1983) Temperature and pH effects on the release of chromium from stainless steel into water and fruit juices. *Journal of Advances in Food Chemistry*, **31**, 89–92.
20. Hoare, W.E., Hedges, E.S. & Barry, B. (1965) *The Technology of Tinplate*. Edward Arnold, London.
21. Anderson, R.A. (1988) Chromium. In: *Trace Minerals in Food* (ed. K.T. Smith), pp. 231–47. Marcel Dekker, New York.
22. Anderson, R.A. (1988) Chromium. In: *Trace Minerals in Food* (ed. K.T. Smith), pp. 231–47. Marcel Dekker, New York.
23. Murthy, G.K., Rhea, U. & Peeler, J.R. (1971) Chromium in the diet of schoolchildren. *Environmental Science and Technology*, **5**, 436–42.
24. MAFF (1999) *MAFF – 1997 Total Diet Study – Aluminium, Arsenic, Cadmium, Chromium, Copper, Lead, Mercury, Nickel, Selenium, Tin and Zinc. Food Surveillance Information Sheet No. 191*. Ministry of Agriculture, Fisheries and Food, London.
25. Anderson, R.A., Bryden, N.A. & Polansky, M.M. (1992) Dietary chromium intake: freely chosen diets, institutional diets, and individual foods. *Biological Trace Elements Research*, **32**, 117–21.
26. Merz, W., Toepfer, E.W., Roginski, E.E. & Palansky, M.M. (1974) Present knowledge of the role of chromium. *Federation Proceedings*, **33**, 2275–80.
27. Petersen, A. & Mortensen, G.K. (1994) Trace elements in shellfish on the Danish market. *Food Additives and Contaminants*, **11**, 365–73.
28. Murthy, G.K., Rhea, U. & Peeler, J.R. (1971) Chromium in the diet of schoolchildren. *Environmental Science and Technology*, **5**, 436–42.
29. National Research Council (1980) *Recommended Dietary Allowances*, 9th edn. National Academy of Sciences, Washington, DC.
30. Hunt, C.D. & Stoecker, B.J. (1996) Deliberations and evaluations of the approaches, endpoints and paradigms for boron, chromium and fluoride dietary recommendations. *Journal of Nutrition*, **126**, 2441S–51S.
31. Nielsen, F.H. (1994) Chromium. In: *Modern Nutrition in Health and Disease*, Vol. 1, 8th edn. (eds. M.E. Shils, J. Olsen & M. Shike), pp. 264–8. Lea & Febiger, Philadelphia, Pa.
32. Seaborn, C.D. & Stoecker, B.J. (1992) Effects of ascorbic acid depletion and chromium status on retention and urinary excretion of ^{51}chromium. *Nutrition Research*, **12**, 1229–34.
33. Hill, C.H. (1976) Mineral interrelationships. In: *Trace Elements in Human Health and Disease* (ed. A. Prasad), pp. 201–10. Academic Press, New York.
34. Anderson, R.A. (1998) Effects of chromium on body composition and weight loss. *Nutrition Reviews*, **56**, 266–70.
35. Hopkins, L.L. & Schwarz, K. (1964) Chromium(III) binding to serum proteins, specifically siderophilin. *Biochimica Biophysica Acta*, **90**, 484–91.
36. Gibson, R.A. (1990) *Principles of Nutritional Assessment*. Oxford University Press, Oxford, UK.
37. Holm, R.H., Kennepohl, P. & Solomon, E.I. (1996) Structural and functional aspects of metal sites in biology. *Chemical Reviews*, **96**, 2239–314.
38. Mertz, W. (1969) Chromium occurrence and function in biological systems. *Physiological Reviews*, **49**, 163–239.
39. Jeejeebhoy, K.N., Chu, R.C., Marliss, E.B., Greenberg, G.R. & Bruce-Robertson, A. (1977) Chromium deficiency, glucose intolerance, and neuropathy reversed by chromium supplementation, in a patient receiving long-term total parenteral nutrition. *American Journal of Clinical Nutrition*, **30**, 531–8.
40. Mertz, W., Toepfer, E.W., Roginski, E.E. & Polansky, M.M. (1974) Present knowledge of the role of chromium. *Federation Proceedings*, **33**, 2275–80.
41. Vincent, J.B. (1999) Mechanisms of chromium action: low-molecular-weight chromium-binding substance. *Journal of the American College of Nutrition*, **18**, 6–12.

42. Offenbacher, E.G. & Pi-Sunyer, F.X. (1980) Beneficial effects of chromium-rich yeast on glucose tolerance and blood lipids in elderly subjects. *Diabetes*, **29**, 919–25.
43. Anderson, R.A. (1995) Chromium, glucose tolerance, diabetes and lipid metabolism. *Journal of the Advancement of Medicine*, **8**, 37–50.
44. Kobla, H.V. & Volpe, S.L. (2000) Chromium, exercise, and body composition. *Critical Reviews in Food Science and Nutrition*, **40**, 291–308.
45. Kaufman, D.B., Nicola, W. & McIntosh, R. (1970) Acute potassium dichromate poisoning. *American Journal of the Diseases of Childhood*, **119**, 374–6.
46. Hathcock, J.N. (1997) Vitamins and minerals: efficacy and safety. *American Journal of Clinical Nutrition*, **66**, 427–37.
47. Fox, G.N. & Sabovic, Z. (1998) Chromium picolinate supplementation for diabetes mellitus. *Journal of Family Practice*, **46**, 83–6.
48. Veillon, C. (1986) Trace element analysis of biological samples. Problems and precautions. *Analytical Chemistry*, **58**, 851A–66A.
49. Cary, E.E. & Rutzke, E.E. (1983) Atomic absorption spectrometric determination of chromium in plants. *Journal of Agricultural and Food Chemistry*, **22**, 1037–42.
50. Tinggi, U., Reilly, C. & Patterson, C. (1997) Determination of manganese and chromium in foods by atomic absorption spectrometry after wet digestion. *Food Chemistry*, **60**, 123–8.
51. Gambelli, L., Belloni, P., Ingrao, G., Pizzoferrato, L. & Santorini, G.P. (1999) Minerals and trace elements in some Italian dairy products. *Journal of Food Composition and Analysis*, **12**, 27–35.
52. Nielsen, F.H. (1994) Ultratrace elements. In: *Modern Nutrition in Health and Disease* (eds. M.E. Shils, J.A. Olsen & M. Shike), pp. 275–7. Lea & Febiger, Philadelphia.
53. Pennington, J.A.T. & Schoen, S.A. (1996) Total Diet Study: estimated dietary intakes of nutritional elements, 1982–91. *International Journal of Vitamin and Nutrition Research*, **66**, 350–62.
54. Wenlock, R.W., Buss, D.H. & Dixon, E.J. (1979) Trace nutrients 2: manganese in British foods. *British Journal of Nutrition*, **41**, 253–61.
55. Gregory, J., Foster, K., Tyler, H. & Wiseman, M. (1990) *The Dietary and Nutritional Survey of British Adults*. The Stationery Office, London.
56. Nielsen, F.H. (1994) Ultratrace elements. In: *Modern Nutrition in Health and Disease* (eds. M.E. Shils, J.A. Olsen & M. Shike), pp. 275–7. Lea & Febiger, Philadelphia.
57. Pennington, J.A.T. & Schoen, S.A. (1996) Total Diet Study: estimated dietary intakes of nutritional elements, 1982–1991. *International Journal of Vitamin and Nutrition Research*, **66**, 350–62.
58. Anonymous (1988) Manganese deficiency in humans: fact or fiction. *Nutrition Reviews*, **46**, 348–52.
59. Freeland-Graves, J.H., Bales, C.W. & Behmardi, F. (1987) Manganese requirements of humans. In: *Nutritional Bioavailability of Manganese* (ed. C. Kies), pp. 90–104. American Chemical Society, Washington, DC.
60. Mertz, W. (1987) Use and misuse of balance studies. *Journal of Nutrition*, **117**, 1811–13.
61. Davidson, L., Cederblad, A. & Lonnderdal, B. (1989) Manganese retention in man: a method for estimating manganese absorption in man. *American Journal of Clinical Nutrition*, **49**, 170–9.
62. Testolini, G., Ciapellano, S., Alberto, A., Paracchini, L. & Jotti, A. (1993) Intestinal absorption of manganese – an in vitro study. *Annals of Nutrition and Metabolism*, **37**, 289–94.
63. Nielsen, F.H. (1994) Ultratrace elements. In: *Modern Nutrition in Health and Disease* (eds. M.E. Shils, J.A. Olsen & M. Shike), pp. 275–7. Lea & Febiger, Philadelphia.
64. Nielsen, F.H. (1994) Ultratrace elements. In: *Modern Nutrition in Health and Disease* (eds. M.E. Shils, J.A. Olsen & M. Shike), pp. 275–7. Lea & Febiger, Philadelphia.
65. Underwood, E.J. (1977) *Trace Elements in Human and Animal Nutrition*, 4th ed., pp. 170–81. Academic Press, New York.
66. Cawte, J. & Florence, M. (1987) Environmental sources of manganese on Groote Eylandt, Northern Territory. *Lancet*, **i**, 62.

67. Bell, A. (1988) Tracing down a mysterious illness. *Ecos*, **57**, 3–8.
68. Cawte, J. Hams, G. & Kilburn, C. (1987) Mechanisms in a neurological ethnic complex in Northern Australia. *Lancet*, **i**, 61.
69. Cawte, J. & Kilburn, C. (1987) *Proceedings of Conference, Darwin, Northern Territory, June 1987*. Department of Health, Darwin, Northern Territory, Australia.
70. Department of Health (1991) *Dietary Reference Values for Food Energy and Nutrients in the United Kingdom*. pp. 181–2. HMSO, London.
71. Yaman, M. (1997) Determination of manganese in vegetables by atomic absorption spectrometry with enrichment using activated carbon. *Chemica Analityczna*, **42**, 79–86.
72. Beard, J.L., Dawson, B.S. & Piñero, D.J. (1996) Iron metabolism: a comprehensive review. *Nutrition Reviews*, **54**, 296–317.
73. Looker, A.C., Dallman, P.R., Carroll, M.D., Gunter, E.W. & Johnson, C.L. (1997) Prevalence of iron deficiency in the United States. *Journal of the American Medical Association*, **227**, 973–6.
74. Coultate, T.P. (1985) *Food: the Chemistry of its Components*, 282–5. The Royal Society of Chemistry, London.
75. Gregory, J., Foster, K., Tyler, H. & Wiseman, M. (1990) *The Dietary and Nutritional Survey of British Adults*. The Stationery Office, London.
76. Pennington, J.A.T. & Schoen, S.A. (1996) Total Diet Study: estimated dietary intakes of nutritional elements, 1982–1991. *International Journal of Vitamin and Nutrition Research*, **66**, 350–62.
77. Crosby, N.T. (1977) Determination of heavy metals in food. *Proceedings of the Institute of Food Science and Technology*, **10**, 65–70.
78. Biffoli, R., Chiti, F., Mocchi, M. & Pinzauti, S. (1980) Contamination of canned foods with metals. *Rivista Società Italiano Scienza Alimentazione*, **27**, 11–16.
79. Monson, E.R. (1971) The need for iron fortification. *Journal of Nutrition Education*, **2**, 152–5.
80. Brady, M.C. (1996) Addition of nutrients: current practice in the UK. *British Food Journal*, **98**, 12–19.
81. Statutory Instrument (1995) No 3202. *Bread and Flour Regulations*. HMSO, London.
82. Statutory Instrument (1995) No 77. *The Infant Formula and Follow-on Formula Regulations*. HMSO, London.
83. Fairweather-Tait, S.J. (1997) Iron fortification in the UK: how necessary and effective is it? In: *Trace Elements in Man and Animals – 9: Proceedings of the Ninth International Symposium on Trace Elements in Man and Animals* (eds. P.W.P. Fischer, M.R. L'Abbé, K.A. Cockell & R.S. Gibson), pp. 392–6. National Research Council Research Press, Ottawa, Canada.
84. Hurrell, R.F. (1985) Non elemental sources. In: *Iron Fortification of Foods* (eds. F.M. Clydesdale & K.L. Wiemer), pp. 39–53. Academic Press, London.
85. Powell, I.W., Halliday, J.W. & Bassett, M.C. (1982) The case against iron supplementation. In: *The Biochemistry and Physiology of Iron* (eds. P. Saltmann & J. Hegenauer), pp. 811–43. Elsevier, New York.
86. Department of Health and Social Security (1981) *Nutritional Aspects of Bread and Flour: Report of the Panel on Bread, Flour and Other Cereal Products*. Committee on Medical Aspects of Food Policy. HMSO, London.
87. Roe, M.A. & Fairweather-Tait, J. (1999) High bioavailability of reduced iron added to UK flour. *Lancet*, **353**, 1938–9.
88. Hurrell, R.F. (1977) Preventing iron deficiency through food fortification. *Nutrition Reviews*, **55**, 210–22.
89. Roe, M.A. & Fairweather-Tait, J. (1999) High bioavailability of reduced iron added to UK flour. *Lancet*, **353**, 1938–9.
90. Crosby, W.H. (1977) Current concepts in nutrition: who needs iron? *New England Journal of Medicine*, **297**, pp. 225–68.
91. Arens, U. (1996) Iron nutrition in health and disease. *British Nutrition Foundation Nutrition Bulletin*, **21**, 71–3.

92. Reilly, C. (1996) Too much of a good thing? The problem of trace element fortification of foods. *Trends in Food Science and Technology*, 7, 139–42.
93. Boirgiato, E.V.M. & Martinez, F.E. (1998) Iron nutritional status is improved in Brazilian preterm infants fed food cooked in iron pots. *Journal of Nutrition*, **128**, 855–9.
94. Mandishona, E.M., Moyo, V.M., Gordeuk, V.R. *et al.* (1999) A traditional beverage prevents iron deficiency in African women of childbearing age. *European Journal of Clinical Nutrition*, **53**, 722–5.
95. Walker, A.R.P. & Segal, I. (1999) Iron overload in Sub-Saharan Africa: to what extent is it a public health problem? *British Journal of Nutrition*, **81**, 427–34.
96. Hallberg, L. (1983) Prevention of iron deficiency in industrialised countries. In: *Groupes à Risque de Carence en Fer dans le Pays Industrialisés. Colloque InserimIsta/CNAM* (eds. H. Dupin & S. Hercberg), pp. 289–302. Editions Inserim, Paris.
97. Zijp, I.M., Korver, O. & Tijburg, L.B.M. (2000) Effect of tea and other dietary factors on iron absorption. *Critical Reviews in Food Science and Nutrition*, **40**, 371–98.
98. Disler, P.B., Lynch, S.R. & Charlton, R.W. (1975) The effects of tea on iron absorption. *Gut*, **16**, 193–200.
99. Layrisse, M., Martines-Torres, C. & Gonzales, M. (1974) Measurement of the total daily iron absorption by the extrinsic tag model. *American Journal of Clinical Nutrition*, **27**, 152–62.
100. Nielsen, F.H. (1994) Ultratrace elements. In: *Modern Nutrition in Health and Disease* (eds. M.E. Shils, J.A. Olsen & M. Shike), pp. 275–7. Lea & Febiger, Philadelphia.
101. Walter, T., Olivares, M., Pizarro, F. & Muñoz, C. (1997) Iron, anaemia and infection. *Nutrition Reviews*, **55**, 111–24.
102. Expert Scientific Working Group (1985) Summary of a report on assessment of the iron nutritional status of the United States population. *American Journal of Clinical Nutrition*, **42**, 1318–30.
103. Brock, J.H. & Mulero, V. (2000) Cellular and molecular aspects of iron and immune function. *Proceedings of the Nutrition Society*, **59**, 537–40.
104. Cook, J.D. (1999) Defining optimal body iron. *Proceedings of the Nutrition Society*, **58**, 489–95.
105. Department of Health (1991) *Dietary Reference Values for Food Energy and Nutrients in the United Kingdom. Report of the Panel on Dietary Reference Values of the Committee on Medical Aspects of Food Policy. Report on Health and Social Subjects*, **41**. HMSO, London.
106. Barr, D.B.G. & Fraser, D.K.B. (1968) Acute iron poisoning in children: role of chelating agents. *British Medical Journal*, **1**, 737–41.
107. Reilly, C. (1998) Cooking pots and iron status. *BNF Nutrition Bulletin*, **23**, 138–40.
108. Conrad, M.E., Uzel, C., Berry, M. & Latour, L. (1994) Ironic catastrophies – one's food another's poison. *American Journal of the Medical Sciences*, **307**, 434–7.
109. Schmutzer, E. (1993) Welcome remarks. *Trace Elements in Man and Animals – TEMA 8* (eds. M. Anke, C.F. Meissner & C.F. Mills), pp. 1–2. Verlag Media Touristik, Gersdorf, Germany.
110. Manousoff, M. (1998) Cobalt's calm before the storm. *American Metal Market*, January 27, 1–3.
http://www.amm.com/ref/hot/COBALT98.htm
111. Gambelli, L., Belloni, P., Ingrao, G., Pizzoferrato, L. & Santorini, G.P. (1999) Minerals and trace elements in some Italian dairy products. *Journal of Food Composition and Analysis*, **12**, 27–35.
112. Schroeder, H.A., Nason, A.P. & Tipton, I.H. (1967) Essential trace elements in man: cobalt. *Journal of Chronic Disease*, **20**, 869–90.
113. Tylecore, R.F. (1976) *A History of Metallurgy*, pp. 148–9. The Metals Society, London.
114. Varo, P. & Koivistonen, P. (1980) Mineral composition of Finnish foods XII. General discussion and nutritional evaluation. *Acta Agricultura Scandanavica*, **S22**, 165–70.
115. Robberecht, H.J., Hendrix, P., Cauwenbergh, R. & Deelstra, H.A. (1994) Daily dietary manganese intake in Belgium, using duplicate portion sampling. *Zeitschrift für Lebensmittel Untersuchung und-Forschung*, **199**, 446–8.

116. Anonymous (1968) Epidemic cardiac failure in beer drinkers. *Nutrition Reviews*, **26**, 173–5.
117. Grinvalsky, H.T. & Fitch, D.M. (1969) A distinctive myocardiopathy occurring in Omaha, Nebraska: pathological aspects. *Annals of the New York Academy of Science*, **156**, 544–65.
118. Toskes, P.P., Smith, G.W. & Conrads, M.E. (1973) Cobalt absorption in sex-linked anaemic mice. *American Journal of Clinical Nutrition*, **26**, 435–7.
119. Wahner-Roedler, D.L., Fairbanks, V.F. & Linman, J.W. (1975) Cobalt excretion tests as an index of iron absorption and diagnostic test for iron deficiency. *Journal of Laboratory and Clinical Medicine*, **85**, 253–8.
120. Werner, E. & Hansen, C.H. (1991) Measurement of intestinal absorption of inorganic and organic cobalt. In: *Trace Elements in Man and Animals* 7 (ed. B. Momĉilović), pp. 11/16–11/17. Imi, University of Zagreb, Zagreb, Croatia.
121. Yamagata, N., Muratan, S. & Morii, T. (1962) The cobalt content of the human body. *Journal of Radiation Research*, **3**, 4–8.
122. Blokhima, R.I. (1970) Cobalt and goitre. In: *Trace Element Metabolism in Animals* (ed. C.F. Mills), pp. 426–32. Livingstone, Edinburgh.
123. Silió, F., Santos, A. & Ribas, B. (1999) The metallic component of the glucose tolerance factor, GTF. Cobalt against chromium. In: *Trace Elements in Man and Animals 10* (eds. A.M. Roussel, R.A. Anderson & A.E. Favier), pp. 540–1. Kluwer, New York.
124. Hillerdahl, G. & Hartung, M. (1983) Cobalt toxicity. *International Archives of Occupational and Environmental Health*, **53**, 89–90.
125. Edel, J., Sabbioni, E., Devos, A. & Rebvecchi, A. (1993) Metabolic aspects of cobalt. In: *Trace Elements in Man and Animals – TEMA 8* (eds. M. Anke, D. Meissner & C.F. Mills), pp. 544–8. Verelag Media Touristik, Gersdorf, Germany.
126. Smith, R.M. (1987) Cobalt. In: *Trace Elements in Human and Animal Nutrition* (ed. W. Mertz), Vol. 1, pp. 143–83. Academic Press, London.
127. Ministry of Agriculture, Fisheries and Food (1998) *Cadmium, Mercury and Other Metals in Food. 53rd Report of Steering Group on Chemical Aspects of Food Surveillance*. The Stationery Office, London.
128. Poupon, J., Gleizes, V., Saillant, G. & Gaillot-Guilley, M. (1999) Determination of serum cobalt at the nanomolar level by direct electrothermal atomic absorption spectrometry. In: *Trace Elements in Man and Animals 10* (eds. A.M. Roussel, R.A. Anderson & A.E. Favier), pp. 1122–3. Kluwer, New York.
129. Andersen, I. & Hegetveit, A.C. (1984) Analysis of cobalt in plasma by electrothermal atomic spectrometry. *Fresenius' Journal of Analytical Chemistry*, **318**, 41–8.
130. Tylecote, R.F. (1976) *A History of Metallurgy*, pp. 148–9. The Metals Society, London.
131. Barceloux, D.G. (1999) Nickel. *Journal of Toxicology – Clinical Toxicology*, **37**, 239–58.
132. Tylecore, R.F. (1976) *A History of Metallurgy*, pp. 148–9. The Metals Society, London.
133. Smart, G.A. & Sherlock, J.C. (1987) Nickel in foods and diets. *Food Additives and Contaminants*, **4**, 61–71.
134. Mitchell, R.L. (1945) Trace elements in soil. *Soil Science*, **60**, 63–75.
135. Smart, G.A. & Sherlock, J.C. (1987) Nickel in foods and diets. *Food Additives and Contaminants*, **4**, 61–71.
136. Ellen, G., Van Den Bosch-Tibbesma & Douma, F.F. (1987) *Zeitschrift für Lebensmittel Unterforschung*, **179**, 145–56.
137. Pennington, J.A.T. & Jones, J.W. (1987) Molybdenum, nickel, cobalt, vanadium and strontium in total diets. *Journal of the American Dietetic Association*, **87**, 1644–50.
138. Koops, J., Klomp, H. & Westerbeek, D. (1982) Spectroscopic determination of nickel and furildioxime with special reference to milk and milk products and to the release of nickel from stainless steel by acidic dairy products and by acid cleaning. *Netherlands Milk and Dairy Journal*, **36**, 333–5.
139. Ysart, G., Miller, O., Crews, H. *et al.* Dietary exposure estimates of 30 elements from the UK Total Diet Study. *Food Additives and Contaminants*, **16**, 391–403.

140. Varo, P. & Koivistonen, P. (1980) Mineral composition of Finnish foods XII. General discussion and nutritional evaluation. *Acta Agricultura Scandinavica*, **S22**, 165–70.
141. Nielsen, F.H. (1984) Fluoride, vanadium, nickel, arsenic and silicon in total parenteral nutrition. *Bulletin of the New York Academy of Medicine*, **60**, 177–95.
142. Nielsen, F.H. (1984) Fluoride, vanadium, nickel, arsenic and silicon in total parenteral nutrition. *Bulletin of the New York Academy of Medicine*, **60**, 177–95.
143. Sunderman, F.W. (1965) Nickel toxicity. *American Journal of Clinical Pathology*, **44**, 182–200.
144. Ysart, G., Miller, O., Crews, H. *et al.* Dietary exposure estimates of 30 elements from the UK Total Diet Study. *Food Additives and Contaminants*, **16**, 391–403.
145. Barceloux, D.G. (1999) Nickel. *Journal of Toxicology – Clinical Toxicology*, **37**, 239–58.
146. Hunt, C.D. & Stoecker, B. (1996) Deliberations and evaluations of the approaches, endpoints and paradigms for boron, chromium and fluoride dietary recommendations. *Journal of Nutrition*, **126**, 22441S–51S.
147. Stokinger, H.E. (1963) *Nickel in Industrial Hygiene and Toxicology*. Interscience, New York.
148. Sunderman, F.W. (1965) Nickel toxicity. *American Journal of Clinical Pathology*, **44**, 182–200.
149. Hunt, C.D. & Stoecker, B. (1996) Deliberations and evaluations of the approaches, endpoints and paradigms for boron, chromium and fluoride dietary recommendations. *Journal of Nutrition*, **126**, 22441S–51S.
150. Christensen, J.M., Kristiansen, J., Nielsen, N.H., Menne, T. & Byrialsen, K. (1999) Nickel concentrations in serum and urine of patients with nickel eczema. *Toxicology Letters*, **108**, 185–9.
151. Jorhem, L., & Sundstrom, B. (1993) Levels of lead, cadmium, zinc, copper, nickel, chromium, manganese and cobalt in foods on the Swedish market, 1983–90. *Journal of Food Composition and Analysis*, **6**, 223–41.
152. Versieck, K. & Cornelis, R. (1989) *Trace Elements in Human Plasma or Serum*, **68**. CRC Press. Boca Raton, Florida.
153. Gonzalez, M., Gallego, M. & Valcarcel, M. (1999) Determination of nickel, chromium and cobalt in wheat flour using slurry sampling electrothermal atomic absorption spectrometry. *Talanta*, **48**, 1051–60.
154. National Academy of Sciences (1977) *Copper*. National Academy of Sciences, Washington, DC.
155. Nielsen, F.H. (1984) Fluoride, vanadium, nickel, arsenic and silicon in total parenteral nutrition. *Bulletin of the New York Academy of Medicine*, **60**, 177–95.
156. Pennington, J.A.T. & Young, B. (1990) Iron, zinc, copper, manganese, selenium and iodine in foods from the United States Total Diet Study. *Journal of Food Composition and Analysis*, **3**, 166–84.
157. Lombardo-Boccia, G., Aguzzi, A., Cappelloni, M. & Di Lullo, G. (2000) Content of some trace elements and minerals in the Italian total-diet. *Journal of Food Composition and Analysis*, **13**, 525–7.
158. Cuadrado, C., Kumpulainen, J., Carbajal, A. & Moreiras, O. (2000) Cereals contribution to the total dietary intake of heavy metals in Madrid, Spain. *Journal of Food Composition and Analysis*, **13**, 495–503.
159. Booth, C.K., Reilly, C. & Farmakalidis, E. (1996) Mineral composition of Australian ready-to-eat breakfast cereals. *Journal of Food Composition and Analysis*, **9**, 135–47.
160. Reilly, C. (1985) The dietary significance of adventitious iron, zinc, copper and lead in domestically-prepared food. *Food Additives and Contaminants*, **2**, 209–15.
161. Boulay, N. & Edwards, M. (2000) Copper in the urban water cycle. *Critical Reviews in Environmental Science and Technology*, **30**, 297–326.
162. Smart, G.A., Sherlock, J.C. & Norman, J.A. (1987) Dietary intakes of lead and other metals: a study of young children from an urban population in the UK. *Food Additives and Contaminants*, **5**, 85–93.
163. Spitalny, K.C., Brondum, J., Vogt, R.L., Sargent, H.E. & Kappel, S. (1984) Drinking-water-induced copper intoxication in a Vermont family. *Pediatrics*, **74**, 1103–6.

164. Reilly, C. (1985) The dietary significance of adventitous iron, zinc, copper and lead in domestically-prepared food. *Food Additives and Contaminants*, **2**, 209–15.
165. Richardson, K. & George, B. (2000) Acute copper poisoning. *Sunlander* (Queensland, Australia), June, p. 3.
166. Salmon, M.A. & Wright, T. (1971) Chronic copper poisoning presenting as pink disease. *Archives of Diseases of Childhood*, **46**, 108–10.
167. Pandit, A. & Bhave, S.A. (1996) Present interpretation of the role of copper in the Indian childhood cirrhosis. *American Journal of Clinical Nutrition*, **63**, 830S–5S.
168. Müller, T., Feichtinger, H., Berger, H. & Müller, W. (1996) Endemic Tyrolean infantile cirrhosis: an ecogenetic disorder. *Lancet*, **347**, 877–80.
169. Tanner, M.S. (1998) Role of copper in Indian childhood cirrhosis. *American Journal of Clinical Nutrition*, **67**, 1074S–81S.
170. Murthy, G.K., Rhea, U. & Peeler, J.R. (1971) Chromium in the diet of schoolchildren. *Environmental Science and Technology*, **5**, 436–42.
171. Varo, P. & Koivistoinen, P. (1980) Mineral composition of Finnish foods XII. General discussion and nutritional evaluation. *Acta Agricultura Scandanavica*, **S22**, 165–70.
172. Shimbo, S., Hayase, A., Murakami, M. *et al.* (1996) Use of a food composition database to estimate daily dietary intake of nutrient or trace elements in Japan, with reference to its limitations. *Food Additives and Contaminants*, **13**, 775–86.
173. Bro, S., Sandström, B. & Heydorn, K. (1990) Intake of essential and toxic trace elements in a random sample of Danish men as determined by the duplicate portion sampling technique. *Journal of Trace Elements and Electrolytes in Health and Disease*, **4**, 147–55.
174. Van Dokkum, W. (1995) The intake of selected minerals and trace elements in European countries. *Nutrition Research Reviews*, **8**, 271–302.
175. Pennington, J.A.T. (1996) Intakes of minerals from diets and foods: is there a need for concern? *Journal of Nutrition*, **126**, 2304S–8S.
176. Hurrell, R.F. (1985) Non elemental sources. In: *Iron Fortification of Foods* (eds. F.M. Clydesdale & K.L. Wiemer), pp. 39–53. Academic Press, London.
177. EC Scientific Committee for Food (1992) *Reference Nutrient Intakes for the European Community*. European Community, Brussels.
178. WHO (1992) *Trace Elements in Human Nutrition*. World Health Organization, Geneva.
179. Institute of Medicine (2001) *Dietary Reference Intakes for vitamin A, vitamin K, Arsenic, Boron, Chromium, Copper, Iodine, Iron, Molybdenum, Nickel, Silicon, Vanadium and Zinc*. Food and Nutrition Board. National Academy Press, Washington DC.
180. Pennington, J.A.T. (1996) Intakes of minerals from diets and foods: is there a need for concern? *Journal of Nutrition*, **126**, 2304S–8S.
181. Failla, M.L. (1999) Considerations for determining 'optimal nutrition' for copper, zinc, manganese and molybdenum. *Proceedings of the Nutrition Society*, **58**, 497–505.
182. Strain, J.J. (2000) Defining optimal copper status in humans: concepts and problems. In: *Trace Elements in Man and Animals 10* (eds. A.M. Roussel, R.A. Anderson & A.E. Favier), pp. 923–8. Kluwer/Plenum, New York.
183. Turnland, J.R., Keyes, W.R., Anderson, H.I. & Acord, L.L. (1989) Copper absorption and retention in young men at three levels of dietary copper by use of the stable isotope ^{65}Cu. *American Journal of Clinical Nutrition*, **49**, 870–8.
184. Knowles, S.O., Grace, N.D., Rounce, J.R., Litherland, A., West, D.M. & Lee, J. (2000) Dietary Mo as an antagonist to Cu absorption. In: *Trace Elements in Man and Animals 10* (eds. A.M. Roussel, R.A. Anderson & A.E. Favier), pp. 717–21, Kluwer/Plenum, New York.
185. Boyne, R. & Arthur, J.R. (1986) Effects of molybdenum or iron induced copper deficiency on the viability and function of neutrophils from cattle. *Research in Veterinary Science*, **41**, 417–19.
186. Harris, Z.L., Klomp, L.W.J. & Gitlin, J.D. (1998) Aceruloplasminemia: an inherited neurodegenerative disease with impairment of iron homeostasis. *American Journal of Clinical Nutrition*, **67**, 972S–7S.
187. Seaborn, C.D. & Stoecker, B.J. (1992) Effects of ascorbic acid depletion and chromium

status on retention and urinary excretion of ^{51}chromium. *Nutrition Research*, **12**, 1229–34.
188. Danks, D.M. (1988) Copper deficiency in humans. *Annual Review of Nutrition*, **8**, 235–57.
189. Castillo-Duran, C. & Uauy, R. (1988) Copper deficiency impairs growth of infants recovering from malnutrition. *American Journal of Clinical Nutrition*, **47**, 710–14.
190. Klevay, L.M. (1984) The role of copper, zinc and other chemical elements in ischemic heart disease. In: *Metabolism of Trace Elements in Man* (eds. O.M. Rennert & W-Y. Chan), Vol. 1, pp. 129–57. CRC Press, Boca Raton, Florida.
191. Karpel, J.T. & Peden, W.H. (1972) Copper deficiency in long-term parenteral nutrition. *Journal of Pediatrics*, **80**, 32–6.
192. Uauy, R., Olivares, M. & Gonzalez, M. (1998) Essentiality of copper in humans. *American Journal of Clinical Nutrition*, **67**, 952S–9S.
193. Percival, S.S. (1995) Neutropenia caused by copper deficiency: possible mechanisms of action. *Nutrition Reviews*, **53**, 59–66.
194. Gitlin, J.D. (2000) The copper transporting ATPases in human disease. In: *Trace Elements in Man and Animals 10* (eds. A.M. Roussel, R.A. Anderson & A.E. Favier), pp. 9–13. Kluwer/Plenum, New York.
195. Yardick, M.K., Kenney, M.A. & Winterfield, E.A. (1989) Iron, copper and zinc status: response to supplementation with zinc or zinc and iron in adult females. *American Journal of Clinical Nutrition*, **49**, 145–50.
196. Yu, S. & Beynen, A.C. (1995) High tin intake reduces copper status in rats through inhibition of copper absorption. *British Journal of Nutrition*, **73**, 863–9.
197. Food and Agriculture Organization/WHO (1971) Evaluation of food additives. *WHO Technical Report Series No. 462*. World Health Organization, Geneva.
198. Evans, W.H., Dellar, D. & Lucas, B.E. (1980) Observations on the total determination of copper, iron, manganese, and zinc in foodstuffs by flame atomic-absorption spectrophotometry. *Analyst*, **105**, 529–41.
199. Jorhem, L. & Sundström, B. (1993) Levels of lead, cadmium, zinc, copper, nickel, chromium, manganese, and cobalt in foods on the Swedish market, 1983–1990. *Journal of Food Composition and Analysis*, **6**, 223–41.
200. Pokorn, D., Stibilj, V., Gregorič, B., Dermelj, M. & Štupar, J. (1998) Elemental composition (Ca, Mg, Mn, Cu, Cr, Zn, Se and I) of daily diet samples from some old people's homes in Slovenia. *Journal of Food Composition and Analysis*, **11**, 47–53.
201. Taháh, J.E., Sánchez, J.M., Granadillo, V.A., Cubillán, H.S. & Romero, R.A. (1995) Concentration of total Al, Cr, Cu, Fe, Hg, Na, Pb and Zn in commercial canned seafood determined by atomic absorption spectrometric means after mineralization by microwave heating. *Journal of Agriculture and Food Chemistry*, **43**, 910–15.
202. Coughlan, M.P. (1983) The role of molybdenum in human biology. *Journal of Inherited Metabolic Diseases*, **6**, S70–7.
203. Rajagopalan, K. (1987) Molybdenum – an essential trace element. *Nutrition Reviews*, **45**, 321–8.
204. Pennington, J.A.T. & Jones, J.W. (1987) Molybdenum, nickel, cobalt, vanadium and strontium in total diets. *Journal of the American Dietetics Association*, **87**, 1644–50.
205. Coughlan, M.P. (1983) The role of molybdenum in human biology. *Journal of Inherited Metabolic Diseases*, **6**, S7–77.
206. Berrow, M.L. & Webber, J. (1972) Heavy metals in sewage sludge. *Journal of the Science of Food and Agriculture*, **33**, 169–73.
207. Pennington, J.A.T. & Jones, J.W. (1987) Molybdenum, nickel, cobalt, vanadium and strontium in total diets. *Journal of the American Dietetics Association*, **87**, 1644–50.
208. Mills, C.E. & Davis, G.K. (1986) Molybdenum. In: *Trace Elements in Human and Animal Nutrition* (ed. W. Mertz), pp. 429–61. Academic Press, New York.
209. Varo, P. & Koivistonen, P. (1980) Mineral composition of Finnish foods XII. General discussion and nutritional evaluation. *Acta Agricultura Scandinavica*, **S22**, 165–70.
210. Coughlan, M.P. (1983) The role of molybdenum in human biology. *Journal of Inherited Metabolic Diseases*, **6**, S7–77.

211. Chiang, G., Swenseid, M.E. & Turnlund, J. (1989) Studies of biochemical markers indicating molybdenum status in humans. *Federation of American Societies of Experimental Biology*, **3**, Abstract No. 4922.
212. Poole, D.B.R. (1982) Bovine copper deficiency in Ireland – the clinical disease. *Irish Veterinary Journal*, **36**, 169–73.
213. Department of Health (1991) *Dietary Reference Values for Food Energy and Nutrients in the United Kingdom. Report of the Panel on Dietary Reference Values of the Committee on Medical Aspects of Food Policy, Report on Health and Social Subjects*, **41**. HMSO, London.
214. Yarovaya, G.A. (1964) Molybdenum in the diet in Armenia. *Proceedings of the Sixth International Biochemical Congress, New York. Abstracts*, **6**, 440.
215. Momčilovič, B. (2000) Acute human molybdenum toxicity from a dietary supplement – a new member of the 'Lucor Metallicum' family. In: *Trace Elements in Man and Animals 10* (eds. A.M. Roussel, R.A. Anderson & A.E. Favier), pp. 699–70. Kluwer/Plenum, New York.
216. Wu, S., Feng, X., Wittmeier, A. & Xu, J. (1997) Microwave digestion of food and plant reference materials for trace multi-element analysis by ICP-MS with an ultrasonic nebulizer/membrane desolvator. In: *Trace Elements in Man and Animals – 9* (eds. P.W.F. Fischer, M.R. L'Abbé, K.A. Cockell & R.S. Gibson), pp. 257–9. National Research Council Research Press, Ottawa, Canada.
217. Hunt, C.D. & Shuler, T.R. (1989) Open-vessel, wet-ash, low-temperature digestion of biological materials for Inductively Coupled Argon Plasma Spectrometry (ICAP) analysis of boron and other elements. *Journal of Micronutrient Analysis*, **6**, 161–74.

Chapter 9
The other transition metals and zinc

The only other metals of the three transition series that are of some practical significance as possible contaminants of food are titanium and vanadium of the first series, silver of the second, and tungsten of the third. Zinc, as has been noted earlier, is not, strictly speaking, a transition metal. However, it fits in chemically with the metals of the first series, completing the line from scandium to copper in the periodic table. It shares common food sources with several of the biologically active transition metals, and interacts functionally with them in cells of the body. It is also not infrequently found as a contaminant of food in the company of other transition metals. For these reasons, as well as for convenience, zinc will be treated here as an addendum to the transition metals.

9.1 Titanium

It is a surprise for many people to learn that titanium, a metal to which they seldom, if ever, give any thought, is the eighth most common element in the Earth's crust. Food scientists and nutritionists are no less ignorant about the metal, for titanium does not normally have a place in their professional textbooks. However, as with several other ultratrace components of the diet which are becoming of increasing commercial importance, it may be necessary to pay more attention to this element in the future. In fact, titanium is one of the important 'strategic' metals and is used extensively in advanced engineering and space-age technology. It will be some time, however, before the large gap in our knowledge about the nutritional role and metabolic significance of titanium is filled.

9.1.1 *Properties, production and uses of titanium*

Titanium is element number 22 in the periodic table, with an atomic weight of 47.9. It is a hard, light (density 4.5) metal, with a melting point of 1675°C. It is strongly resistant to corrosion, but when heated can ignite and burn in air. It is also the only element that can burn in nitrogen. Like other transition metals, titanium has several oxidation states, but normally exists in compounds in the +4 state.

In spite of its abundance in the environment, titanium metal is difficult to produce in its pure state, because of its high melting point. Major sources are mineral sands which contain the two important ores, ilmenite (ferrous titanate, $FeTiO_3$) and rutile (TiO_2). Since it combines with both oxygen and nitrogen at high temperatures, titanium cannot be extracted from its ores by straightforward reduction. One method used to avoid this problem is to heat the ore with carbon in a stream of chlorine to produce titanium tetrachloride. The tetrachloride, which is a liquid, is condensed, purified by fractional distillation and reduced to the metal with molten magnesium under an atmosphere of argon.

Because of its very high tensile strength titanium metal and its alloys are used extensively in aircraft and spacecraft production, in military equipment, including guided weapons, and in naval vessels. A growing application is in the manufacture of human and veterinary prostheses. A minor, though not unimportant, use is in the production of blenders and homogeniser blades used by food analysts. Because of their high resistance to acids and salt solutions, titanium and its alloys are used in the manufacture of equipment for the food processing industry[1].

Certain titanium compounds are commercially important. Titanium dioxide is used as a white pigment in paints, especially for outdoor use, and now largely replaces white lead, formerly the basis of most white paints. It is also used in the food industry as an additive for whitening flour, confection and non-dairy milk product substitutes[2]. It has a number of applications in the pharmaceutical and cosmetic industries. Titanium chloride, which is a liquid which fumes in moist air, is used in smokescreens and in skywriting. It is also used as a catalyst in the plastics industry for polymerisation of olefins. Various titanium esters are used for waterproofing fabrics and titanium sulphate is an excellent textile mordant. Barium titanate, which is piezoelectric, is used in the electronics industry as a transducer for the interconversion of sound and electricity.

9.1.2 *Titanium in food and beverages*

Very little is known about levels of titanium in foods and beverages. Because of its wide-scale distribution in the environment, the metal is almost certainly a contaminant of everything we consume, but in low concentrations, mainly as external contamination, since it is poorly absorbed by plants from soil[3]. Levels of about 2.0 mg/kg have been reported in cereals, vegetables, dairy products and some other foods. However, these reported concentrations may be suspect, since techniques for the accurate determination of titanium at low levels in biological tissues have not been well established and reliable certified reference materials (CRMs) are not available[4].

Some data have been published on levels of dietary intake of titanium in a number of countries. Daily intakes have been calculated in the US as between 3 and 600 μg for adults living in a non-industrially-polluted area. Intakes may be higher in those exposed to emissions from oil- and coal-fired power stations[5]. A mean daily intake of 271–305 μg has been estimated for Japanese adults[6].

9.1.3 *Absorption and metabolism of titanium*

Few studies appear to have been carried out on the uptake and behaviour of titanium in the human body. Very little of the element appears to be absorbed from digested food, and relatively high levels reported in lung tissue are believed to be due to dust inhalation[7]. Excretion of what is absorbed from the GI tract appears to be rapid. Titanium has no known physiological role. It appears to be non-toxic, at least in the amounts normally consumed in food. There is some evidence that titanium from non-food sources can be taken up by body tissues, but whether this has any significance for health is doubtful. A study of titanium in soft tissues in contact with titanium plates used in craniofacial surgery found that some appears to enter tissues, probably as metallic particles released during insertion of the plates, and that time-dependent leakage is negligible[8].

9.1.4 *Analysis of foodstuffs for titanium*

Little has been reported in the literature on analytical procedures for the determination of titanium in foods and other biological materials. More than a decade ago, it was commented by two experts on trace element analysis that

> *the element titanium in human serum has hardly been searched for. That interest may, however, grow in order to learn the impact on the human body of the titanium material in modern prostheses and in many drugs and cosmetic preparations*[9].

The interest has, indeed, grown, but much still remains to be done. The use of several different instrumental analytical methods has been reported, including GFAAS, with background correction, following acid digestion of samples[10]. A recent study found that ICP-OES was a more convenient and reliable method than AAS for titanium determination in human tissues[11]. Other workers have used NAA to determine the metal in blood[12].

9.2 Vanadium

Interest in the biological effects of vanadium has been growing substantially since claims were made in the 1970s that it was an essential trace element for humans[13]. In recent years vanadium has been the subject of several books and conference proceedings, and though its essentiality remains to be proved beyond doubt, it is becoming widely accepted that the metal has important and unique roles to play in human metabolism[14]. At the same time there is also concern about the increasing level of environmental pollution caused by emission of vanadium from coal- and oil-burning power stations and other industrial plants. It was as an industrial toxin that the metal was known long before its possible role in human metabolism was postulated[15].

9.2.1 *Chemical and physical properties of vanadium*

Vanadium is element number 23 in the periodic table, with an atomic number of 50.9. It is a relatively light metal (density 6.1), steel-grey in colour, tough and corrosion-resistant. Its chemistry is complex[16]. Vanadium has a number of oxidation states, from -1 to $+5$, though the most common valence states are +3, +4 and +5. Tetravalent vanadyl (VO^{2+}) and pentavalent vanadate (VO_4^{3-}) are the predominant forms in biological systems. Vanadium is capable of undergoing oxidation–reduction exchanges and this suggests that some of its known biological activities involve cellular redox activity[17].

9.2.2 *Production and uses of vanadium*

Vanadium occurs widely in the environment, though not in metallic form. In most soils it is present at levels of about 0.1 mg/kg. It has several different ores, all of which are mixed. Carnotite, for instance, also contains uranium. Other ores include patronite, roscoelite and vanadinite. It is mined commercially in the US, as well as in Peru and a few other countries. Much of the commercially used vanadium is a by-product of uranium extraction. An important source is crude oil and fly ash from power stations, especially where oil is used as fuel.

The principal use of vanadium is in alloys with iron and other metals. The presence of even small amounts of vanadium (less than 1%) adds strength, toughness and heat resistance to steel and it also makes cast iron shock-resistant and ductile. The vanadium is usually added in the form of ferrovanadium, a vanadium–iron alloy produced from mixed vanadium–iron ore. Vanadium steels are used to make high-speed machinery and tools as well as springs. Pure vanadium metal and its chemical compounds have numerous industrial applications. They are used as catalysts in the production of ammonia, and of various polymers and rubber. Vanadium compounds, especially the pentoxide, are used in ceramics, glass and paint, as well as in photography.

9.2.3 *Vanadium in food and beverages*

Little information is available on normal levels of vanadium in individual foods and diets. Levels in most foods appear to be low, between 1 and 30 μg/kg[18]. Levels reported in the US in different food groups (expressed as means in μg/kg, with ranges in parentheses) are as follows: cereals/cereal products: 23 (0–150); meat/fish: 10 (0–120); vegetables: 6 (0–72); milk: 1 (0–6)[19]. Among the richest food sources are reported to be skimmed milk, lobster and some other seafoods, vegetable oils, certain vegetables, and some cereals[20]. Fruit, meat, butter and cheese are poor sources[21].

In spite of its low concentrations in most foods, the diet is the major source of vanadium for the general population. Estimated daily intakes range from about 10 μg to 2 mg. Adult daily intakes have been reported to be 13 μg in the UK[22], 10–60 μg in the US[23], and 230 μg in Japan[24]. Beer drinking has been shown to contribute significantly to intake. A German study found that intake by men, at 33–35 μg, was about twice that of women, at 14 $\pm$ 11 μg, and was due to 28 μg provided by beer[25].

A major source of vanadium intake for some individuals is dietary supplements. Various vanadium compounds, such as vanadyl sulphate, are commonly used to enhance weight training by athletes and others. It is believed by many that vanadium can increase muscle mass and generally enhance athletic ability. Though the evidence for this is not strong, some athletes consume 60 mg of vanadyl sulphate/day. A US study has found that some supplements may contain up to 13 mg of vanadium/kg and can contribute up to 25 μg out of a total daily intake of 33 μg of the metal[26].

9.2.4 *Absorption and metabolism of vanadium*

High air levels of vanadium occur in certain occupational situations, such as during boiler cleaning, which disturbs vanadium dust. Soluble vanadium compounds such as vanadium oxides, present in such dusts, are readily absorbed through the lungs, in contrast to vanadium in food, which is poorly absorbed with as little as 2% entering the blood. Daily absorption by an adult is between 5 and 10 μg.[27] Vanadium is found in extremely low levels in biological tissues, including blood. Excretion of accumulated vanadium is rapid and occurs through the kidneys.

There is growing evidence that vanadium has an essential role in animals, possibly including humans[28]. In animals experimental vanadium deficiency has been found to be associated with stunted growth, impaired reproduction, and altered red blood cell formation and iron metabolism, as well as changes in blood lipid levels and hard tissue formation. It is also believed that vanadate and other vanadium compounds increase glucose transport and improve glucose metabolism[29]. It has been found that vanadium can alter the activity of a number of enzymes, including Na-K-ATPases involved in muscle contraction, as well as serine/threonine kinases located in growth factors, oncogenes, phosphatases and receptors for insulin[30]. Whether all these effects occur in humans as well as in experimental animals has not been established.

A vanadium deficiency has not been shown to occur in humans and it has been claimed that many of the deficiency signs reported for vanadium are questionable. However, it is widely believed, though not necessarily on solid grounds, that vanadium compounds can enhance athletic performance by increasing lean body mass and improving glucose metabolism[31].

Vanadium in food has, apparently, a low level of toxicity[32]. Entry into the body by other routes, however, can cause serious consequences. The toxicity of vanadium compounds usually increases with oxidation state, with those containing V(v) the most toxic. Most of the toxic effects result from local irritation of the eyes and upper respiratory tract, though systemic toxicty also occurs, with a wide range of effects, including neurotoxic and hepatoxic damage.

9.2.5 *Analysis of foodstuffs for vanadium*

Problems may be encountered in the determination of vanadium in food and beverages because of low concentration levels. However, these can be overcome by taking normal precautions and, if necessary, use of a pre-concentration procedure, involving solvent extraction or ion-exchange[33]. GFAAS has been used successfully to determine vanadium in foods, as well as in cigarettes[34]. A spectrophotometric method, based on the vanadium(V)-catalysed oxidation of 1,8-diaminonaphthalene (DAN) by potassium bromate, in which absorbance is monitored over time, has been reported to be simple to operate and to give reliable results with food and hair samples, without pre-separation[35].

9.3 Silver

Silver is one of the most ancient of the metals used by mankind. It has a long history of association with food and beverages. Today it is still used for culinary purposes, though in most households it seldom appears on the table except on special occasions when the silver cutlery is taken out. It is only in formal banquets and in top hotels and restaurants that silverware continues to serve as it once did.

9.3.1 *Chemical and physical properties of silver*

Silver is element number 47 in the periodic table, with an atomic weight of 107.87. It is a white, lustrous, soft and malleable metal, capable of taking a high degree of polish. Its melting point is 960°C. It has the highest known electrical and thermal conductivities of any metal, properties which account for its use in the electrical industries as well as in the traditional kitchen. Silver is relatively unreactive chemically, but it blackens in the presence of sulphur compounds as silver sulphide is formed. Several oxidation states are known, though only Ag(I) is of practical consequence. Argentous compounds of some importance are Ag(I) nitrate, acetate and halides.

9.3.2 *Production and uses of silver*

Since Roman times, and earlier, silver has been recovered as a by-product of lead extraction by what is known as the 'cupellation' (oxidation in a blast of air in a heated 'cupel' or flat dish made of porous refractory material) of silver-containing lead ores. It is also extracted from other mixed ores, for example of copper. Small amounts are recovered, also, from the silver ore argentite. Large-scale production of the metal takes place in China, Peru, Russia, the US and Mexico. Apart from its use in jewellery and coins, silver is used in tableware, both in its own right and as silver plate on vessels made of copper and other metals. Silver forms a number of useful alloys and solders with copper, cadmium and lead. Its components are widely used in the photographic industry, as a germicidal agent in water treatment and for other related purposes. Silver metal and solders are also found in computers and other electronic and electric equipment. Various silver compounds of silver are used in the pharmaceutical and cosmetic industries.

9.3.3 *Silver in food and beverages*

Few studies on silver in foods and diets have been reported in the literature. The metal is known to occur naturally at low levels in most plant and animal tissues. Reported levels range from undetectable to a few μg/kg in fruits and vegetables and little more in other types of food[36]. Intake is probably about 20 μg/day for adults not exposed to environmental silver contamination[37]. Where silver or silver-plated utensils are used for preparation and storage of food, intake may be increased. The use of silver compounds to treat domestic water supplies may also make a contribution to intake. There is some evidence that certain types of vegetables, such as *Brassica* species, can accumulate higher than usual levels of silver, especially if grown in soils contaminated by industrial emissions, but this is unlikely to cause excessive intake[38].

9.3.4 *Absorption and metabolism of silver*

Little is known about the level of absorption of silver from food. Animal studies suggest that it is about 10%. Little of the metal appears to be retained in body tissues. When excessive intake occurs, accumulation appears to occur in liver and skin. Most of the absorbed silver is rapidly excreted, mainly in bile with faeces, with only a small amount in urine. Silver is not known to perform any essential function in humans. It has been shown to interact metabolically with both copper and selenium. In experimental animals silver has been reported to accentuate signs of copper deficiency and to induce a selenium-like deficiency state[39]. The addition of both copper and selenium to the diet decreases the toxic effects of silver in turkeys[40].

There is no evidence that silver in normal diets has a toxic effect on humans. However, the silver compounds in drinking water can promote necrotic liver degeneration in rats deficient in vitamin E[41]. The toxicity of silver is related to the solubility and availability of the different silver compounds. In soil, sewage sludge and sediments in which silver sulphide predominates, toxicity is very low, even when total silver concentrations are high. Though silver thiosulphate is highly soluble, its toxicity is also low because of complexing. Highly soluble silver nitrate is one of the most toxic silver compounds[42].

Prolonged ingestion of silver-containing pharmaceutical preparations can bring about the development of a distinctive blue-grey discoloration of the skin, eyes and mucous membranes. This condition, known as *argyria*, is seen sometimes in silversmiths and others who are industrially exposed to the metal[43]. Consumption of some types of 'anti-smoking lozenges' which contain silver salts has been reported to bring about similar effects.

9.3.5 *Analysis of foodstuffs for silver*

Problems likely to be encountered in analysis of silver are often due to the low levels of the metal in foods and other biological samples. The use of closed microwave digestion using acid mixtures is advisable for sample preparation, since silver is readily reduced to the metallic state and volatilises during dry ashing. ETAAS, with background correction, will normally be suitable for endpoint determination. When concentrations are very low, chelate extraction with sulphur-containing ligands, such as dithizone, is recommended.

9.4 Tungsten

Tungsten is the only metal to have two officially recognised names. In 1951 IUPAC, the international body responsible for preserving order in the area of chemical nomenclature, decreed that the element was to be known, officially, as both tungsten and wolfram. The name tungsten, derived from the Swedish *tung*, heavy, and *sten*, stone, was largely used in the English-speaking world, while wolfram, possibly based on the German *wolf*, wolf, and *ram*, the Middle High German for dirt, appeared most frequently in German scientific texts. In spite of the fact that the symbol W is used for the metal, the name tungsten will be used in the present study.

9.4.1 *Chemical and physical properties of tungsten*

Tungsten is element number 74 in the periodic table, with an atomic weight of 183.84. It is a heavy, steel-grey to tin-white, ductile and malleable metal, with a density of 19.3. It has the highest melting point (3410°C) and, at temperatures over 1650°C the highest tensile strength, of all metals. Though oxidised in air at high temperatures, it has excellent corrosion resistance and is only slightly attacked by most acids. Tungsten is a transition metal of the third series and, like other metals of the group, has several oxidation states from +2 to +6, but the most stable derivatives are hexavalent. It is not a very reactive element. Some of its better-known compounds are the trioxide, WO_3, and tungstic acid, the hydrated trioxide, H_2WO_4.

9.4.2 *Production and uses of tungsten*

Tungsten does not occur as the free metal in nature and its compounds are rather rare. Its main ores are wolframite, $FeWO_4$, and scheelite, $CaWO_4$. Major producers of the metal are the US, China, Bolivia and, potentially, Russia. The metal is extracted by converting the ores to oxide by roasting in air and then reducing with hydrogen.

Tungsten has many important industrial uses. It is mainly used in the production of steels for high-speed toolmaking and other purposes where great strength and heat resistance are essential. Tungsten is combined with carbon in the alloy known as *carbaloy*. This is one of the hardest materials known. It can hold a cutting edge under very severe conditions. It is used as a substitute for diamonds on drill and machining surfaces. Tools such as chisels and punches made of tungsten have a close uniform texture and can be ground to a fine edge. Filaments made from tungsten are widely used in electric light bulbs and for wiring in electric furnaces. There are several uses of tungsten compounds, such as for pigments. Magnesium tungstate is used as a phosphor in some fluorescent tubes.

9.4.3 *Tungsten in foods and diets*

Little is known about levels of tungsten in foods. They generally appear to be very low, in spite of wide-scale industrial use of the metal. Daily intakes in Sweden have been estimated to be 8–13 μg, with levels in domestic water supplies between 0.03 and 0.1 μg/litre[44]. A UK study found that daily intake was less than 1 μg[45].

9.4.4 *Absorption and metabolism of tungsten*

Absorption of tungsten from food in the GI tract appears to be very limited. In experimental animals 97% of the ingested metal was eliminated from the body within 72 hours. Excretion was equally divided between urine and faeces. There may be some retention in bones[46]. Dietary tungsten appears to be non-toxic to experimental animals and, presumably, to humans. There is no evidence that it is an essential nutrient for any animal. There may be some interaction between tungsten and molybdenum in the body. Animals fed a tungsten-containing diet were found to have reduced activities of the two molybdenum-containing enzymes, xanthine oxidase and sulphite oxidase.

9.4.5 *Analysis of foodstuffs for tungsten*

The low levels of tungsten in foods and the absence of well-established analytical procedures for the metal in biological samples present problems for the analyst. Inorganic samples of the metal and its compounds can be analysed readily by GFAAS.

9.5 Zinc

While zinc was known to be an essential nutrient for plants and certain animals from early in the twentieth century, zinc deficiency in humans only began to be recognised in the 1960s, when zinc-responsive dwarfism was detected in children in Egypt[47]. Though there was some uncertainty about the possibility of more general zinc deficiency among humans, because of the element's ubiquity in the environment, subsequent studies have confirmed the essential role played by zinc in human metabolism[48]. Today zinc is known to be a key nutrient of worldwide significance, and has joined iodine and iron among trace elements whose deficiency problems urgently need to be addressed[49].

9.5. *Chemical and physical properties of zinc*

Zinc is element number 30 of the periodic table, with an atomic weight of 65.37. Its density is 7.14, which makes it one of the 'heavy metals'. It is a bluish-white, lustrous metal, but tarnishes in air to a blue-grey colour as a coat of zinc carbonate, $Zn_2(OH)_2CO_3$, develops. This forms a tough adhering layer and is the reason why zinc is used extensively, in galvanising, on the surface of other less resistant metals. Zinc has a low melting point of 420°C and is ductile and malleable when heated to 100°C.

Zinc is a chemically reactive element. It combines readily with non-oxidising acids, releasing hydrogen and forming salts. It also dissolves in strong acids to form zincate ions (ZnO_2^{2-}). It reacts with oxygen, especially at high temperatures, to form zinc oxide, ZnO. It also reacts directly with halogens, sulphur and other non-metals. In spite of its completely filled *3d* electron shell, zinc shares with its neighbouring transition metals an ability to form strong complexes with organic ligands. Zinc can form alloys with other metals. One of these is brass, made of copper and zinc and other metals. It may also be added to the copper–tin alloy, bronze.

9.5.2 *Production and uses of zinc*

Zinc ores are widely distributed and are commercially exploited in several European countries as well as on a larger scale in the US, Russia, southern Africa and Australia. The principal ores are sulphides, such as sphalerite or zinc blende (natural ZnS), calamine ($ZnCO_3$) and zincite (ZnO). These ores usually contain other metals, including lead, cadmium and arsenic.

To extract the metal, the ores are crushed and may be concentrated by a flotation process, before being roasted in air to produce the oxide. This is then reduced by heating with coke and the metal is distilled off. The zinc produced in this way is normally contaminated with cadmium and other metals. It is purified by careful redistillation. Zinc is very volatile and processes of smelting produce large emissions, which, unless adequate precautions are taken, can cause serious environmental pollution. In addition to primary production, recycling from scrap accounts for a significant part of the more than 7.5 million tonnes of zinc used each year[50].

The principal use of zinc for many centuries was in the production of brass. This is really a family of alloys composed of 60–82% copper and 18–40% zinc. Brass has been made in China and elsewhere in Asia for some 2000 years. It was used traditionally for the manufacture of cooking equipment and food and beverage containers, a use that has declined considerably in the West, but still continues on a large scale in other parts of the world. A modern large-scale use of zinc is in architectural brasses. These contain at least 30% of the metal and are commonly referred to as 'bronzes', mainly because of their similar uses, colours and weathering characteristics. These zinc-bronzes are used structurally in modern buildings as well as in decorative hardware, plumbing fixtures and architectural details and surface finishes.

A major use of zinc, which accounts for about half the world production, is to protect iron and other metals from corrosion[51]. In the process called *galvanising*, a coat of zinc is applied to other metals by various methods, such as hot dipping or dusting (sherardising). The zinc layer protects the underlying metal in two ways: by forming a tough skin of basic bicarbonate, and by acting as a 'sacrificial metal'. As it corrodes, the zinc releases electrons that flow to the other metal making it more negative and thus reducing its corrosion rate. The electrical circuit, from the zinc anode to the iron or steel cathode, is completed when the electrons return through water or another conducting medium to the zinc. Sacrificial zinc anodes are used to protect ships' hulls, offshore oil platforms, pipelines and other structures exposed to seawater.

Zinc metal itself is used extensively in the construction of buildings, for roofing and rainwater systems, as well as for cladding. It is also found in many manufactured items, such as transistors, circuit boards, photocopiers, dry cell batteries and many others. Zinc compounds have many applications in the chemical and pharmaceutical industries, such as in the manufacture of paints and pigments and of rubber and plastics. They are also found in cosmetics, medicines, nutritional supplements and a variety of household consumables. Its wide use in the home and elsewhere in the community accounts for a very high level of zinc contamination in household dust, as well as in water and air. A study in Hong Kong reported levels of more than 2 mg zinc/

g in dust collected on window ledges in houses[52]. The consequences of such contamination are well known to the food analyst[53].

9.5.3 *Zinc in food and beverages*

In Western societies upwards of 70% of zinc consumed is provided by animal products, especially meat[54]. Liver and other organ meats are particularly rich in the element, as are most seafoods. Another good source is oysters, which can contain as much as 1000 mg/kg. Other good sources are seeds, nuts and wholegrain cereals. Levels of zinc in foods consumed in the UK are shown in Table 9.1. These intakes are comparable to those in the US[55], Australia[56] and several European countries[57]. The average zinc intake by UK adults is about 9–12 mg/day. The zinc content of a typical adult US diet is between 10 and 15 mg/day. Similar levels are reported from many European countries[58]. In many Asian countries zinc intakes are low, especially in the diet of children, because of the absence of appreciable amounts of animal products in the customary diet[59].

Table 9.1 Zinc in UK foods.

Food group	Mean concentration (mg/kg fresh weight)
Bread	9.0
Miscellaneous cereals	8.6
Carcass meat	51
Offal	43
Meat products	25
Poultry	16
Fish	9.1
Oils and fats	0.6
Eggs	11
Sugar and preserves	3.8
Green vegetables	3.4
Potatoes	4.5
Other vegetables	2.6
Canned vegetables	3.9
Fresh fruit	0.8
Fruit products	0.7
Beverages	0.3
Milk	3.5
Dairy products	14
Nuts	31

Adapted from Ysart, G., Miller, P., Crews, H. *et al.* (1999) Dietary exposure estimates of 30 elements from the UK Total Diet Study. *Food Additives and Contaminants*, **16**, 391–403.

9.5.4 *Absorption and metabolism of zinc*

Absorption from the diet, which occurs in the small intestine, is affected by a number of factors. It ranges from less than 10 to more than 90%, with an average of 20–30%[60]. Various other components of the diet can affect uptake. Competition for absorption occurs between zinc and other elements, especially copper, iron and cadmium. Phytate, fibre and calcium can limit uptake, whereas animal protein enhances it. A diet rich in wholemeal bread, which contains these three antagonists, has been shown to cause deficiency of the element[61].

An adult human contains between 1.5 and 2.5 grams of zinc, almost as much as iron and more than 200 times more than copper, the third most abundant trace element in the body. Most of the zinc absorbed from the intestine is found intracellularly, primarily in muscle, bone, liver and other organs[62]. High concentrations are found in the prostate gland in men, and in the eye. Zinc in plasma is mainly loosely bound to albumin and is also transported bound to transferrin. In the liver it is bound to the low molecular weight, metal-binding protein, metallothionein.

Most of the body's zinc reserves turn over slowly and are not readily available for metabolism. Only about 10% makes up a readily available pool of exchangeable zinc, which is used to maintain various zinc-dependent metabolic functions. The body's zinc content is regulated by homeostatic mechanisms, mainly through control of absorption of exogenous zinc from the gut and by regulation of excretion of endogenous zinc in pancreatic and other gastrointestinal secretions[63]. Since only about 0.2% of the body's total zinc is in plasma, a small change in uptake by, or release from, muscle or other tissues can have profound effects on levels in plasma. Plasma zinc concentrations can be affected by stress, surgery, physical exercise, infection and several other factors[64]. Consequently, plasma levels do not give a reliable measure of total body zinc stores under all circumstances.

Zinc is rapidly excreted via the faeces. These contain both unabsorbed zinc as well as some endogenous zinc excreted in bile. Only a small amount appears in urine, with about the same quantity in sweat. This loss may be greatly increased when excessive sweating takes place[65].

9.5.5 *Biological roles of zinc*

Zinc is an essential component of more than 200 different enzymes, of which as many as 50 play important metabolic roles in animals. It occurs in all six classes of enzymes. In addition, the metal provides structural integrity in many proteins. Zinc ligands help maintain the structure of cell membranes and of some ion channels. 'Zinc finger protein' is involved in processes of transcription factors that link with the double helix of DNA to initiate gene expression[66]. The expression of certain genes is known to be regulated by the quantity of zinc absorbed from the diet. It is believed that zinc has a role in regulation of cell growth and differentiation.

Its multiple roles are indicated by what occurs when zinc is deficient in the diet of young animals. There is a rapid development of clinical symptoms, which include growth retardation, poor and erratic consumption of food, loss of appetite, alopecia, and development of scaly keratinous lesions on the skin. Resistance to infection is reduced through failure of the body's immune response mechanism. Wound healing is slowed. Similar symptoms are seen in humans whose dietary intake of zinc fails to meet the body's requirements. An extreme case is the rare inherited disease known as *acrodermatitis enteropathica* in which a genetic defect results in inability to absorb an adequate amount of zinc from food. It occurs in early childhood and is characterised by severe bullous-pustular dermatitis of the hands and feet and the oral, anal and genital areas, and total hair loss. The disease, which is readily treated with supplements, was always fatal until the discovery of its relation to zinc deficiency[67]. Similar symptoms have been observed in patients on long-term TPN whose zinc supply was inadequate.

During the 1960s zinc deficiency of epidemic proportions was observed in young people, mainly males, in Iran, Egypt and other parts of the Middle East. Those affected belonged to poor communities, whose diet consisted mainly of unleavened wheat bread. Though the cereal contained zinc, it was also rich in phytate and fibre which made the zinc unavailable for absorption, resulting in severe zinc deficiency. Those affected showed signs of poor growth, stunting, anaemia, hypogonadism and delayed sexual maturity. Zinc supplementation was followed by dramatic changes, with rapid development of puberty and accelerated growth[68].

There is considerable evidence that many people, especially in poorer and developing countries, such as Bangladesh[69], suffer from marginal zinc deficiency. Though its effects are less dramatic than those reported in Iran in the 1970s, marginal deficiency can also have serious health consequences, especially for children. These include depressed immunity, impaired taste and smell, night blindness, impaired memory and decreased spermatogenesis in men[70]. Because of the serious implications of the problem, considerable efforts are currently being made to improve zinc nutrition, especially in poorer countries, by provision of supplements as well as by other means, including fortification of staple foods[71].

9.5.6 *Zinc requirements and dietary reference values*

Considerable problems have been encountered in trying to establish dietary zinc requirements. This is largely due to the difficulty of assessing zinc status and optimal zinc nutriture in humans. Plasma zinc levels do not give a true measure of the body's available zinc, since this can be affected by many other factors besides diet, such as stress and illness. Other biomarkers of zinc status, such as activity of zinc-dependent enzymes, are not sufficiently specific to be more than of supportive value[72]. The use of estimates of habitual intake in populations without evidence of zinc deficiency can be of some use, but lacks precision. In practice, a factorial approach, which is based on the quantity of absorbed zinc required to match endogenous excretion of the element, has been used to estimate requirements in several countries, including the UK and the US. Its use, however, has not produced uniform recommendations. Thus, while the UK RNI for adult males and females is 9.0 mg/day, the 1989 US RDA was 15 mg/day for adult males and 12 mg/day for females up to the age of 50 years.

9.5.7 *Toxic effects of zinc in food and beverages*

Zinc salts are intestinal irritants and can cause nausea, vomiting and abdominal pain, which usually requires an intake of about 250 mg or more of a zinc salt in adults[73]. This has been known to occur when water which has been stored in galvanised containers has been consumed. Prolonged exposure to high intakes of zinc are believed to result in copper deficiency and subsequent anaemia. It may also interfere with iron metabolism. Because of evidence that similar effects may be produced by ingestion of high doses of zinc supplements, an intake of more than 15 mg/day in this way is not recommended[74].

9.5.8 *Analysis of foodstuffs for zinc*

Zinc is readily determined in food by AAS. Depending on levels in samples, flame AAS or graphite furnace AAS may be used, with good results. Because of the fact that zinc is a pervasive and universally distributed component of air and dust, precautions have to be taken to prevent contamination. Both FAAS, with deuterium background correction, and GFAAS with pyrolytic-graphite coated tubes have been used with success to investigate levels of zinc and other metals in dietary supplements[75]. The method of choice for zinc in several large-scale multielement investigations is ICP-MS. The UK TDS uses this method on samples which have been homogenised and digested with nitric acid in plastic pressure vessels, using microwave heating[76].

References

1. Feliciani, R., Marcoaldi, R., Gramiccioni, L., Caiazza, S. & Giamberardini, S. (1999) Titanium in the alimentary field. *Industrie Alimentari*, **38**, 13–15.
2. Phillips, L.G. & Barbano, D.M. (1997) The influence of fat substitutes based on protein and titanium dioxide on the sensory properties of low fat milks. *Journal of Dairy Science*, **80**, 2726–31.
3. Mitchell, R.L. (1957) Mineral composition of grasses and clovers grown on different soils. *Research* (London), **10**, 357–62.
4. Iyengar, V. & Wottiez, J. (1988) Trace elements in human clinical specimens: evaluation of literature data to identify reference values. *Clinical Chemistry*, **34**, 474–81.
5. Nielsen, F.H. (1986) Other elements. In: *Trace Elements in Human and Animal Nutrition*. 5th edn. (ed. W. Mertz), Vol. 2, pp. 415–62. Academic Press, New York.
6. Shimbo, S., Hayase, A., Murakami, M. *et al.* (1996) Use of a food composition database to estimate daily dietary intakes of nutrient or trace elements in Japan, with reference to its limitation. *Food Additives and Contaminants*, **13**, 775–86.
7. Tipton, I.H. & Cook, M.J. (1963) Titanium in human tissues. *Health Physics*, **9**, 103–45.
8. Poupon, J., Méningaud, J.P., Chenevier, C., Galliot-Guillery, M. & Bertrand, J.C. (2000) Determination of total and soluble titanium in soft tissues covering titanium microplates in stomatology. In: *Trace Elements in Man and Animals 10* (eds. A.M. Roussel, R.A. Anderson & A.E. Favier), pp. 230–1. Kluwer/Plenum, New York.
9. Versieck, J. & Cornelis, R. (1989) *Trace Elements in Human Plasma or Serum*, p. 182. CRC Press, Boca Raton, Florida.
10. Van Loon, J.C. (1987) The determination of titanium and vanadium in whole blood. *Trace Elements in Medicine*, **4**, 28–34.
11. Poupon, J., Méningaud, J.P., Chenevier, C., Galliot-Guillery, M. & Bertrand, J.C. (2000) Determination of total and soluble titanium in soft tissues covering titanium microplates in stomatology. In: *Trace Elements in Man and Animals 10* (eds. A.M. Roussel, R.A. Anderson & A.E. Favier), pp. 230–1. Kluwer/Plenum, New York.
12. Cantone, M.C., Molho, N. & Pirola, L. (1985) Cadmium and titanium in human serum determined by proton nuclear activation. *Journal of Radiology and Nuclear Chemistry*, **91**, 197–203.
13. Nriagu, J.O. (1998) *Vanadium in the Environment*, Part 2, *Health Effects*. Wiley, New York.
14. McNeill, J.H. (2000) Insulin enhancing effects of vanadium. In: *Trace Elements in Man and Animals 10* (eds. A.M. Roussel, R.A. Anderson & A.E. Favier), pp. 491–6. Kluwer/Plenum, New York.
15. Leonard, A. & Gerber, G.B. (1998) Mutagenicity, carcinogenicity and teratogenicity of vanadium. In: *Vanadium in the Environment* (ed. J.O. Nriagu), Part 2, pp. 39–53. Wiley, New York.
16. Barceloux, D.G. (1999) Vanadium. *Journal of Toxicology–Clinical Toxicology*, **37**, 265–78.
17. Tracey, A.S. & Crans, D.C. (eds.) (1998) *Vanadium Compounds: Chemistry, Biochemistry and Therapeutic Applications. ACS Symposium Series*, **711**. American Chemical Society, Washington, DC.
18. Myron, D.R., Givand, S.H. & Nielsen, F.H. (1977) Vanadium content of selected foods as determined by flameless atomic absorption spectroscopy. *Journal of Agriculture and Food Chemistry*, **25**, 297–300.
19. Pennington, J.A.T. & Jones, J.W. (1987) Molybdenum, nickel, cobalt, vanadium and strontium in total diets. *Journal of the American Dietetic Association*, **87**, 1644–50.

20. Schroeder, H.A. & Nason, A.P. (1971) Trace element analysis in clinical chemistry. *Clinical Chemistry*, **17**, 461–74.
21. Badmaev, V., Prakash, S. & Majeed, M. (1999) Vanadium: a review of its potential role in the fight against diabetes. *Journal of Alternative and Complementary Medicine*, **5**, 273–91.
22. Evans, W.H., Read, J.I. & Caughlin, D. (1985) Quantification of results for estimating elemental dietary intakes of lithium, rubidium, strontium, molybdenum, vanadium and silver. *Analyst*, **110**, 873–86.
23. Clarkson, P.M. & Rawson, E.S. (1999) Nutritional supplements to increase muscle mass. *Critical Reviews in Food Science and Nutrition*, **39**, 317–28.
24. Tsalev, D.L. (1984) *Atomic Absorption Spectrometry in Occupational and Environmental Health Practice*, p. 273. CRC Press, Boca Raton, Florida.
25. Anke, M., Illing-Gunther, H., Gürtler, H. *et al.* (2000) Vanadium – an essential element for animals and humans? In: *Trace Elements in Man and Animals 10* (eds. A.M. Roussel, R.A. Anderson & A.E. Favier), pp. 221–5. Kluwer, New York.
26. Hight, S.C., Anderson, D.L., Cunningham, W.C., Capar, S.G., Lamont, W.H. & Sinex, S.A. (1993) Analysis of dietary supplements for nutritional, toxic and other elements. *Journal of Food Composition and Analysis*, **6**, 121–39.
27. Myron, D.R., Zimmerman, T.J. & Shuler, T.R. (1978) Intake of nickel and vanadium by humans: a survey of selected diets. *American Journal of Clinical Nutrition*, **31**, 527–31.
28. Nielsen, F.H. & Uthus, E.O. (1990) The essentiality and metabolism of vanadium. In: *Vanadium in Biological Systems* (ed. N.D. Chasteen), pp. 51–62. Kluwer, Dordrecht.
29. Cam, M.C., Rodrigues, B. & McNeill, J.H. (1999) Distinct glucose lowering and beta cell protective effects of vanadium and food restriction in streptozocin-diabetes. *European Journal of Endocrinology*, **141**, 546–54.
30. Berner, Y.N., Shuler, T.R. & Nielsen, F.H. (1989) Selected ultratrace elements in total parenteral nutrition solutions. *American Journal of Clinical Nutrition*, **50**, 1079–83.
31. Evans, W.H., Read, J.I. & Caughlin, D. (1985) Quantification of results for estimating elemental dietary intakes of lithium, rubidium, strontium, molybdenum, vanadium and silver. *Analyst*, **110**, 873–86.
32. Thompson, K.H., Battell, M. & McNeill, J.H. (1998) Toxicology of vanadium in mammals. In: *Vanadium in the Environment* (ed. K.H. Thompson), pp. 21–38. Wiley, New York.
33. Lugowski, S., Smith, D.C. & Van Loon, J.C. (1987) The determination of titanium and vanadium in whole blood. *Trace Elements in Medicine*, **4**, 28–34.
34. Adachi, A., Asai, K., Koyama, Y., Matsumoto, Y. & Okano, T. (1998) Determination of vanadium in cigarettes by atomic absorption spectrometry. *Analytical Letters*, **31**, 1769–76.
35. Gao, J.Z., Zhang, X., Yang, W., Zhao, B.W., Hou, J.G. & Kang, J.W. (2000) Kinetic-spectrophotometric determination of trace amounts of vanadium. *Talanta*, **51**, 447–53.
36. Nielsen, F.H. (1986) Other elements. In: *Trace Elements in Human and Animal Nutrition*, 5th edn. (ed. W. Mertz), Vol. 2, pp. 415–62. Academic Press, New York.
37. Harvey, S.C. (1980) Silver. In: *The Pharmacological Basis of Therapeutics* (eds. L.S. Goodman & A. Gillman), pp. 967–9. Macmillan, New York.
38. Headlee, A.J.W. & Hunter, R.G. (1953) Uptake of metals by vegetables in the vicinity of coal-fired power stations. *Industrial Engineering Chemistry*, **45**, 548–51.
39. Underwood, E.J. (1977) *Trace Elements in Human and Animal Nutrition*, 4th edn., pp. 444–5. Academic Press, New York.
40. Jensen, S.L. (1974) Trace elements in the diet of turkeys. *Poultry Science*, **53**, 57–64.
41. Nielsen, F.H. (1986) Other elements. In: *Trace Elements in Human and Animal Nutrition*, 5th edn. (ed. W.Mertz), Vol. 2, pp. 415–62. Academic Press, New York.
42. Ratte, H.T. (1999) Bioaccumulation and toxicity of silver compounds: a review. *Environmental Toxicology and Chemistry*, **18**, 89–108.
43. Hill, W.B. & Pillsbury, D.M. (1939) *Argyria. The Pharmacology of Silver*. Williams, Baltimore, Md.
44. Wester, P.O. (1974) Tungsten in Swedish diets. *Atherosclerosis*, **20**, 207–15.

45. Hamilton, E.I. & Minski, M.J. (1972) Tungsten in the UK diet. *Science of the Total Environment*, **1**, 375–94.
46. Nielsen, F.H. (1986) Other elements. In: *Trace Elements in Human and Animal Nutrition*, 5th edn. (ed. W.Mertz), Vol. 2, pp. 415–62. Academic Press, New York.
47. Prasad, A.S. (1990) Discovery of human zinc deficiency and marginal deficiency of zinc. In: *Trace Elements in Clinical Medicine* (ed. H. Tomita), pp. 3–11. Springer, Tokyo.
48. Brown, K.H., Wuehler, S.E. & Peerson, J.M. (2001) The importance of zinc in human nutrition and estimation of the global prevalence of zinc deficiency. *Food and Nutrition Bulletin*, **22**, 113–25.
49. Ranum, P. (2001) Zinc enrichment of cereal staples. *Food and Nutrition Bulletin*, **22**, 169–72.
50. http:/www.zincworld.org/zgd_org/pg2000/pg08.htm
51. Anonymous (1988) Lead and zinc in the market place, *Metals Review*, No. 16, pp. 1–2. Australian Lead and Zinc Development Association, Melbourne.
52. Tong, S.T.Y. & Lam, K.C. (2000) Home sweet home? A case study of household dust contamination in Hong Kong. *Science of the Total Environment*, **256**, 115–23.
53. Kosta, L. (1982) Contamination as a limiting parameter in trace analysis. *Talanta*, **29**, 985–92.
54. Welsh, S.O. & Marston, R.M. (1982) Zinc levels in the US food supply: 1909–1980. *Food Technology*, **36**, 70–6.
55. Pennington, J.A.T. & Young, B. (1990) Iron, zinc, copper, manganese, selenium, and iodine in foods from the United States Total Diet Study. *Journal of Food Composition and Analysis*, **3**, 166–84.
56. English, R. (1981) The Market Basket Survey: zinc content of Australian diets. *Journal of Food and Nutrition*, **38**, 63–5.
57. Van Dokkum, W. (1995) The intake of selected minerals and trace elements in European countries. *Nutrition Research Reviews*, **8**, 271–302.
58. Van Dokkum, W. (1995) The intake of selected minerals and trace elements in European countries. *Nutrition Research Reviews*, **8**, 271–302.
59. Osendarp, S.J.M., Van Raaij, J.M.A., Darmstadt, G.L., Baqui, A.H., Hautvast, J.G.A.G. & Fuchs, G.J. (2001) Zinc supplementation during pregnancy and effects on growth and morbidity in low birthweight infants: a randomised placebo controlled trial. *Lancet*, **357**, 1080–5.
60. Forbes, R.M. & Erdman, J.W.Q. (1983) Bioavailability of trace mineral elements. *Annual Reviews of Nutrition*, **3**, 213–31.
61. Prasad, A.S. (1983) Human zinc deficiency. In: *Biological Aspects of Metals and Metal-Related Diseases* (ed. A.B. Sakar), pp. 107–19. Raven Press, New York.
62. Jackson, M.J. (1989) Physiology of zinc: general aspects. In: *Zinc in Human Biology* (ed. C.F. Mills), pp. 1–14. Springer, London.
63. King, J.C., Shames, D.M. & Woodhouse, L.R. (2000) Zinc homeostasis in humans. *Journal of Nutrition*, **130**, 1360S–6S.
64. Brown, K.H., Wuehler, S.E. & Peerson, J.M. (2001) The importance of zinc in human nutrition and contamination of the global prevalence of zinc deficiency. *Food and Nutrition Bulletin*, **22**, 113–25.
65. Jackson, M.J. (1983) Physiology of zinc: general aspects. In: *Zinc in Human Biology* (ed. C.F. Mills), pp. 1–14. Springer, London.
66. Berg, J.M. & Shi, Y. (1996) The galvanising of biology: a growing appreciation of the role of zinc. *Science*, **271**, 1081–5.
67. Moynahan, E.J. (1974) Acrodermatitis enterohepatica; a lethal inherited human zinc disorder. *Lancet*, **ii**, 399–400.
68. Prasad, A.S., Miale, A., Farid, Z., Sandstead, H.H. & Shubert, A.R. (1963) Zinc metabolism in patients with the syndrome of iron deficiency anaemia, hepatosplenomegaly, dwarfism and hypogonadism. *Journal of Laboratory and Clinical Medicine*, **61**, 537–49.
69. Osendarp, S.J.M., Van Raaij, J.M.A., Darmstadt, G.L., Baqui, A.H., Hautvast, J.G.A.G. & Fuchs, G.J. (2001) Zinc supplementation during pregnancy and effects on growth and

morbidity in low birthweight infants: a randomised placebo controlled trial. *Lancet*, **357**, 1080–5.
70. Walsh, C.T., Stanstead, H.H., Prasad, A.S., Newberne, P.M. & Frakere, P.J. (1994) Zinc health effects and research priorities for the 1990s. *Environmental Health Perspectives*, **102**, 5–46.
71. Gibson, R.S. & Ferguson, E.L. (1998) Nutrition interventions to combat zinc deficiencies in developing countries. *Nutrition Research Reviews*, **11**, 115–31.
72. Hambridge, K.M. & Krebs, N.F. (1995) Assessment of zinc status in man. *Indian Journal of Pediatrics*, **62**, 169–80.
73. Failla, M.L. (1999) Considerations for determining optimal nutrition for copper, zinc, manganese and molybdenum. *Proceedings of the Nutrition Society*, **58**, 497–505.
74. Department of Health (1991) *Dietary Reference Values for Food Energy and Nutrients in the United Kingdom*. HMSO, London.
75. Hight, S.C., Anderson, D.L., Cunningham, W.C., Capar, S.G., Lamont, W.H. & Sinex, S.A. (1993) Analysis of dietary supplements for nutritional, toxic and other elements. *Journal of Food Composition and Analysis*, **6**, 121–39.
76. Ysart, G., Miller, P., Crews, H. *et al.* (1999) Dietary exposure estimates of 30 elements from the UK Total Diet Study. *Food Additives and Contaminants*, **16**, 391–403.

Chapter 10
The metalloids: arsenic, antimony, selenium, tellurium and boron

The metalloids, arsenic, antimony, selenium, tellurium and boron, chemically and physically lying between metals and non-metals, have considerable interest for both the toxicologist and the food scientist. All of them are to a greater or lesser extent toxic and some appear as contaminants in the food regulations of several countries.

Arsenic has been traditionally associated with homicide and the forensic scientist. No one can take the presence of this element in the diet lightly. Yet arsenic is almost universally distributed in foods and drinks, and daily intake, even in the absence of pollution, can be as high as 0.5 mg for some. It may even be that this poisonous element is essential for human life.

Antimony, though also ancient and long used as a medicine and in cosmetics, is not believed to play a nutritional role in humans. Along with its compounds, such as the hydride stibine, it is a well-known industrial hazard and its toxicity to animal life has been reported.

Selenium was for a long time better known for its toxicity than for any other quality. Farmers whose cattle died from 'blind staggers' after eating certain selenium-accumulating plants had no idea that the toxin in the fodder was, in fact, an essential nutrient for man and animals. When Berzelius first isolated selenium in 1817 he initially thought that what he had obtained from the residues of the roasted copper pyrites was the element tellurium discovered by his friend and fellow scientist M.H. Klaproth a few years earlier. However, though the properties of the substance he had isolated were very like those of the other element, he soon realised that they were not identical. As he later wrote; 'for this reason I gave it the name selenium from the Greek word *selene*, which signifies the moon, while *tellus* is the name of our own planet'[1]. In spite of these similarities in chemical properties, tellurium shares none of the important metabolic functions of selenium.

Artisans in ancient cultures relied on borax, and other compounds of boron, to make their glass and ceramics, as do glassmakers and potters today. Boric acid has long been used in medicine and in cosmetics, and there is increasing evidence pointing towards a role for the element in human nutrition. However, its toxicity, both chronic and acute, is also recognised.

10.1 Arsenic

Until relatively recently arsenic contamination of food was considered by many to be a rare occurrence, usually the result of accidental addition to a manufactured product, as in the infamous tragedy of 1900 when arsenic-contaminated beer poisoned more than 6000 drinkers in the north of England[2]. Apart from such cases, arsenic was little known outside the books of the murder mysteries writers. That is no longer the case. Today arsenic is recognised as one of the most widespread and serious of the inorganic contaminants of the diet in several countries. According to the WHO, there are more than 20 countries in which there is serious arsenic contamination of groundwater[3]. In West Bengal in India and in neighbouring Bangladesh, millions of people are at risk of arsenic poisoning from this source. It has been estimated that, within a few years, one in ten adult deaths in that part of the world could be from cancers triggered by the element[4].

10.1.1 Chemical and physical properties of arsenic

Arsenic is one of the Group 15/V elements along with nitrogen, phosphorus, antimony and bismuth. Its atomic number is 33 and the density of its stable form is 5.7. Elemental arsenic exists in several allotropic forms. The stable form is a silver-grey, brittle crystalline solid that tarnishes rapidly in air, and at high temperature burns to form a white cloud of arsenic trioxide. Yellow crystalline and black amorphous forms are also known.

The chemistry of arsenic is similar to that of phosphorus in many respects. It has oxidation states of −3, 0, +3 and +5. It combines readily with many other elements, for example with hydrogen to form arsine, AsH_3, an extremely poisonous gas, and with oxygen to form a trioxide (As_2O_3, or As_4O_6). The trioxide is also known as 'white arsenic' and is very poisonous. It was probably this substance that the ancient Greeks, as well as the medieval alchemists, who were not able to prepare the pure element itself, knew as arsenic. Other important compounds are arsenic trichloride and the various arsenates (compounds of arsenic acid, H_3AsO_4), such as lead arsenate and copper aceto-arsenate. Arsenic also forms organic compounds such as dimethylarsinic acid $(CH_3)_2AsO(OH)$ and arsenobetaine ($CH_3As^+CH_2COO^-$). These are of considerable interest since they occur naturally in certain foods, especially of marine origin, and are far less toxic than inorganic arsenic compounds.

10.1.2 Production and uses of arsenic

Arsenic is distributed widely in the Earth's crust. It occurs in soils, waters, both marine and fresh, and in almost all plants and animal tissues. Its principal native ores are arsenopyrite, FeAsS, realgar, As_2S_2, and orpiment, As_2S_3. Arsenopyrite, which is also known as mispickel, is one of the more important sources of the element. Arsenic is widely distributed in other ores in association with various metals, and in fact the bulk of commercial arsenic is obtained as a by-product of the extraction of copper, lead and other metals from their ores. The arsenic is recovered from the dust deposited from the flue gases produced during smelting. The flue dust contains up to 97% arsenic trioxide (with about 3% of oxides of other metals, the most important of which is antimony).

Arsenic has many industrial applications. It forms alloys with other metals which it makes strong, hard and corrosion-resistant. A major use is in the chemical industry, in pharmaceuticals, agricultural chemicals, preservatives and pigments. It is such uses that at one time helped confer on arsenic the dubious honour of being accounted second only to lead as a toxicant in the farm and household[5].

Agricultural chemicals and related arsenic-containing substances have, until recently, been widely used in herbicides, fungicides, wood preservatives, insecticides, rodenticides and sheep dips. Though this use is now restricted, and even totally banned in some countries, arsenic is a persistent contaminant. Even many years after they ceased to be used, some streams and locations on farms remain toxic because they were once exposed to arsenicals. The most important of the farm chemicals of this type are lead arsenate, copper arsenate, copper aceto-arsenate (Paris green), and sodium arsenate (Wolman salts), all formerly well known to farmers. A deadly arsenic compound, is 2-chlorovinyldichloroarsine, $ClCH{=}CHA_sHCl_2$, also known as lewisite. This was one of the poison gases used in enormous quantities, with horrendous consequences, by both sides during World War I.

Other uses of arsenic and its compounds are in the making of glass and enamels, in textile printing, as preservatives in tanning and taxidermy, and in pyrotechnics. Arsenic is added to lead in the manufacture of shot. Its presence slows down the solidification of the lead and allows it to flow more slowly and thus assure perfectly spherical pellets. The arsenic also helps harden the lead. Arsenic is added to germanium in the production of semiconductors for transistors and integrated circuits.

Arsenic compounds were once widely used in medicine. A traditional home remedy was Fowler's solution, a preparation containing 1% arsenic trioxide which was used for treatment of skin problems and even taken internally for the prevention of leukaemia and anaemia[6]. Organic arsenicals, such as salvarsan, were used for treating syphilis. Today arsenic compounds continue to be used in certain herbal remedies, for example, in Ayurvedic medicine[7].

10.1.3 *Arsenic in food and beverages*

Arsenic is to be found in most foods. Levels rarely exceed 1 mg/kg except in seafoods, including fish, crustaceans and seaweeds. Indeed, arsenic intake is usually positively related to the quantity of seafoods in the diet[8]. Levels of total arsenic found in UK food in the 1994 TDS are shown in Table 10.1 The investigators observed that though fish consumption was low, with an average intake per household per day of 0.018 kg, compared, for example, to 0.133 kg of potatoes, the fish group contributed 89% of the total national arsenic intake compared to 8% contributed by potatoes. Bread and cereals, though also eaten in greater quantities (0.120 kg/day) than fish, contributed a respectable 27% of the total arsenic, still less than the fish group[9]. A Canadian study found that 18.1% of intake came from cereals and cereal products, 14.9% from starchy vegetables, and 32.1% from the meat/fish group[10]. It is important to note that all these intake figures are for total arsenic and that most of this is organic arsenic, not the more toxic inorganic form.

10.1.3.1 Arsenic in mushrooms

Certain types of mushroom investigated in a Swiss study have been reported to contain surprisingly high levels of arsenic[11]. While levels in cultivated commercial mushrooms were 0.2–0.5 mg/kg dry weight, which is of the same order of magnitude as levels in most fruits and vegetables, some edible wild mushrooms, including those of the genus *Agaricus*, which are traditionally consumed in Britain, contained more than 10 mg/kg dry weight. One species, *Laccaria amethystina*, described as of 'minor culinary importance', contained up to 200 mg/kg dry weight. However, as the authors of the report point out, moderate consumption of most edible mushrooms will contribute little to the arsenic intake of the average consumer. Even eating *L. amethystina* may be harmless, since its arsenic is present mostly in the organic form, as methylated arsinic acids, which have a low toxicity.

Table 10.1 Total arsenic in UK foods.

Food Group	Mean concentration (mg/kg fresh weight)
Bread	0.008
Miscellaneous cereals	0.01
Carcass meat	0.004
Offal	0.004
Meat products	0.004
Poultry	0.003
Fish	4.3
Oils and fats	0.007
Eggs	0.002
Sugar and preserves	0.008
Green vegetables	0.003
Potatoes	0.005
Other vegetables	0.007
Canned vegetables	0.003
Fresh fruit	0.002
Fruit products	0.003
Beverages	0.002
Milk	0.002
Dairy products	0.003
Nuts	0.009

Adapted from Ysart, G., Miller, P., Crews, H. *et al.* (1999) Dietary exposure estimates of 30 elements from the UK Total Diet Study. *Food Additives and Contaminants*, **16**, 391–403.

10.1.3.2 Arsenic in fish: organic and inorganic forms

As has been seen above, fish and other seafoods can accumulate high levels of arsenic and contribute significantly to daily intake of the element. Up to 26 mg/kg have been found in crab meat, and 40 mg/kg in shrimps, and a massive 170 mg/kg in prawns in one UK study[12]. Free-swimming fish do not generally contain as much arsenic as do shellfish. A UK study found a range of 1.0–6.0 mg/kg in haddock and < 0.5–2.4 mg/kg in herring taken from the open sea, with 0.2–34 mg/kg in flounder and 0.5–24 mg/kg in sole from a polluted estuary[13]. Australian marine fish contain up to 4.4 mg/kg, with levels in 21% of those tested above the legally permitted level of 1.0 mg/kg[14].

It is appropriate to mention that the food regulations in force in Australia at the time of this study, while setting a maximum permitted level for arsenic in fish, crustaceans and molluscs, made it clear that this applied to inorganic arsenic only[15]. In the revised Food Standards Code for Contaminants and Restricted Substances (Standard 1.4.1), issued by the joint Australia New Zealand Food Authority (ANZFA) in 2000, a similar approach is taken, with the addition of 'seaweed (edible kelp)' to the seafood group. In the case of cereals, however, a limit is set (of 1 mg/kg) for total arsenic[16]. In contrast to the Australian/New Zealand Regulations, the UK *Arsenic in Food Regulations* simply note that the permitted level for arsenic in unspecified food is a maximum of 1 mg/kg; and in certain specified foods between 0.1 and 10 mg/kg. No distinction is drawn in the regulations between organic and inorganic forms of the element. A footnote to the regulations states that 'the limit of 1 ppm arsenic does not apply to: ... fish, edible seaweed, or any product containing fish or edible seaweed, where arsenic is naturally present at levels above 1 ppm in that fish or seaweed'[17].

10.1.3.3 Arsenic in water

Arsenic is detectable in almost all potable waters. The usual levels of concentration range from 0 to 200 μg/litre[18], with a mean of 0.5 μg/litre. Since water is a significant source of dietary arsenic, and, moreover, the element in water is mainly in the more toxic inorganic form, most countries set limits for arsenic in drinking water. In the US, for instance, this was 50 μg/litre in 1975 but has since been reduced to 10 μg/litre, as a result of concerns about the long-term effects of high intakes in water[19]. Particular sources, such as spas and hot springs, often have higher concentrations of arsenic than are found in domestic drinking water. This is recognised in some countries in regulations relating to bottled natural mineral waters. In the UK, for instance, the allowance is a maximum of 50 μg/litre[20].

Natural groundwater, including wells and springs, sometimes contains exceptionally high levels of arsenic. What has been described as 'regional endemic chronic arsenicism' or 'endemic arseniasis', has been attributed to domestic use of such naturally contaminated waters[21]. Levels of 1–4 mg/litre of inorganic arsenic, as the trioxide, have been detected in domestic water in volcanic regions of several South American countries, including Argentina and Chile[22]. In Taiwan similarly contaminated water has been associated with chronic arsenic poisoning[23].

Far more widespread contamination of domestic water by arsenic, with serious consequences for human health, has in recent years been reported from several regions in South-east Asia. In such countries as Bangladesh, and in Bengal in India, deep tube wells supply drinking water for as much as 90% of the population. These wells were introduced as a reliable source of clean groundwater to replace normally used and often disease-contaminated surface water. Unfortunately, many of these deep wells tapped formerly unavailable sources of naturally occurring arsenic. As a result, drinking water used by many millions of people has become contaminated. The magnitude of the tragedy is shown by findings that, in Bangladesh alone, water in 40% of wells is unfit for human consumption. Levels of 100 μg of arsenic/litre are common, with many samples above 1000 μg/litre.

10.1.3.4 Arsenic from industrial pollution

Industrial pollution and accidental contamination can result in higher than normal levels of arsenic in food and beverages. Environmental contamination around a coal-burning power station in the Czech Republic was responsible for considerable increases in arsenic levels in water and locally grown crops. More than one tonne of arsenic was emitted each day from the plant's smokestacks. Drinking water was found to contain 70 μg arsenic/litre and surface water 210 μg arsenic/litre[25]. Chronic arsenic poisoning, in animals as well as humans, has been reported in the UK in the vicinity of coal-fired brick kilns[26]. An investigation of the effects of past mining activities in Cornwall found evidence of increased arsenic levels in locally grown crops and in hair of local residents[27]. Emissions from metal industries, with resulting environmental consequences, have been reported from the US, Sweden and other countries[28].

10.1.3.5 Accidental contamination of foods by arsenic

Accidental contamination of food with arsenic has resulted in a number of serious cases of poisoning. An early and fully recorded case, which resulted in the setting up of a Royal Commission, occurred in the UK in 1900 when arsenic-contaminated beer poisoned more than 6000 drinkers in Lancashire and Staffordshire, 70 of them fatally. The cause of the outbreak was the use of arsenical pyrites to make the sulphuric acid that was used in the beer fermentation[29]. An even more serious accident occurred in Japan 55 years later when more than 12 000 infants fed on arsenic-contaminated formula were poisoned, killing 120. The arsenic was traced to the sodium phosphate used to stabilise the formula. This had been produced as a by-product during refining of aluminium from bauxite which contained considerable amounts of arsenic[30].

The former wide-scale use of arsenicals as pesticides in vineyards has resulted in some cases of arsenic poisoning of wine drinkers. Increased dietary intake has also occurred from residual arsenic in horticultural chemicals on vegetables and fruit. Though there has been a significant decrease in the use of arsenicals in agriculture in recent years, they are still used extensively on certain non-food crops, including cotton. Cotton seed by-products, such as cooking oil, are used as human foods and there can be some danger that they may be a pathway for arsenic into the diet[31].

10.1.4 Dietary intakes of arsenic

Typical estimates of total arsenic intakes for Europe and North America range from 10 to 30 μg/day. WHO/FAO have set a PTWI of 15 μg/kg body weight[32]. This applies only to inorganic arsenic and is equivalent to an intake of 130 μg each day for a 60 kg adult. Since most of the arsenic ingested in the diet is in the organic form there seems to be little cause for toxicological concern about normal intakes of the general population in those countries surveyed, even when, as some of the following data show, arsenic intakes are apparently high compared to the WHO limits.

Mean adult intakes in the UK are 120 μg, with a 97.5th percentile of 420 μg. These levels, based on the 1997 TDS, are similar to those found in previous surveys[33]. US intakes are lower than UK levels, with mean intakes of 51 μg and 38 μg/day found in the 1986 and the 1991 TDS[34]. Levels of intake closer to the UK results have been reported from other countries, such as New Zealand (adult intake 150 μg/day) and Spain (Basque region) with a mean daily total arsenic exposure of 286 μg[35]. In all these studies, fish consumption was found to contribute most of the arsenic in the diet.

10.1.5 Absorption and metabolism of arsenic

All arsenic compounds, including the inorganic tri- and penta-valent forms as well as its organic compounds, are easily absorbed from the GI tract. Once absorbed arsenic is rapidly distributed to lungs, liver, kidney and spleen and subsequently redistributed to the skin, nails and hair where it is tightly bound to keratin. After about 24 hours the organs begin to release their arsenic, which is excreted through the kidneys. Skin and hair, however, hold onto it, and levels may increase over several days[36].

The ability of hair to retain absorbed arsenic is the basis for the use of this tissue to investigate intakes of the toxin. Levels in hair from 5 to 700 mg/kg arsenic have been found in cases of acute arsenic poisoning[37]. It has been claimed that the death of the exiled French Emperor Napoleon on St Helena in 1826 was caused by chronic arsenic poisoning. In support of this theory is the finding that a lock of his hair, removed within a day or two of his death, contained 10.38 mg/kg of arsenic, more than 12 times the normal level in hair[38]. Whether this finding, which was made nearly 150 years after his death, justifies the claim that Napoleon was murdered by his enemies by putting the toxin into his wine over a long period of time, is impossible to say.

Metabolism of arsenic in mammals varies between species. In humans ingested inorganic arsenic is reduced and methylated in the liver to generate much less toxic organic compounds, such as monomethylarsonic and dimethylarsinic acids, which are then excreted in urine. Inorganic arsenic species are large ligands and their toxicity is a result of their tendency to attach themselves to sulphydryl groups of proteins and thus block their action. Arsenite, in particular, binds preferentially to the enzyme pyruvate dehydrogenase, causing inhibition of cell energy production and raised pyruvate concentrations. Arsenate oxyanions, because their stereochemical structure resembles sulphate and phosphate, can enter mitochondria and compete with phosphate as a substrate, to form high-energy compounds which are unstable and uncouple oxidative phosphorylation[39].

In contrast to the inorganic compounds, organic arsenicals are non-toxic. Arsenobetaine, which is the main arsenic compound in fish and shellfish, where it occurs along with trimethylarsenine and other species, is stable. When ingested it is rapidly absorbed into the blood and is excreted, unchanged, in urine. Inorganic arsenic can cause both acute and chronic poisoning. Arsenic trioxide is a common cause of the former, with a fatal dose of between 70 and 180 mg. The pentavalent form is less toxic than the trivalent form.

Ingestion of inorganic arsenic can cause vomiting and haemorrhagic diarrhoea within minutes or hours, if taken with food. Haematuria and acute renal failure may follow. The principal pharmacological effects are dilation and increased permeability of capillaries, especially in the intestines. Chronic poisoning from long-term, low-dose intakes, causes decreased appetite and weight loss. Other symptoms include abdominal pain, facial oedema, respiratory difficulty and obstructive jaundice. Later, features of neuritis appear, with weakness, trembling and other effects[40]. Transverse white 'Mees' lines are seen on the nails. Longer-term effects include peripheral neuritis, conjunctivitis and skin problems. These are often seen in people living in areas where the groundwater is contaminated with arsenic. Within a few years, they develop hyperkeratosis, pigmentation and skin ulcers. Incidences of cancers of skin, lung and gastrointestinal tract are significantly increased, as extensive studies in Bangladesh and India have shown[41].

Arsenic is known to interact with other toxic elements in the animal body. It has been shown to counteract the toxicity of an excess of selenium. For this reason it is sometimes added to the feed of poultry and cattle in areas of naturally high selenium. The reverse can also occur, when selenium counteracts the toxic effect of arsenic. Sodium selenite, injected into a pregnant animal simultaneously with sodium arsenite, which is teratogenic, has been shown to prevent malformations of the fetus. Similar interactions between arsenic and cadmium have been demonstrated. It is possible that common modes of action of the different elements, especially through interactions with protein —SH groups, account for some of these observations[42].

It has been argued that, in spite of its evident toxicity, arsenic is an essential nutrient for humans[43]. Arsenic deprivation in farm animals has been associated with depressed growth and abnormal reproduction[44]. There is some evidence that arsenic may play an essential metabolic role, possibly as an activator of certain enzymes[45]. It could also act as a substitute for phosphate. However, whether this justifies the inclusion of arsenic among the established essential trace elements is still far from certain[46].

10.1.6 Analysis of foodstuffs for arsenic

A traditional and sensitive test for arsenic is Marsh's test. This semi-quantitative test depends upon the formation of the hydride when hydrogen gas is bubbled through a solution containing arsenic or its compounds. When the arsine is passed through a narrow heated glass tube, it is decomposed leaving a deposit of metallic arsenic. The test was well known to the forensic scientist as well as to Sherlock Holmes' fans in former years, but has now been replaced by more sensitive instrumental procedures.

Arsine is still the basis of one of the widely used methods for the determination of arsenic in food. HGAAS is recommended for nanogram quantities of the element[47]. A recent development is the flow-injection adaptation (FI-HGAAS), which provides a simple and reliable method for both total and inorganic arsenic[48].

Because of the importance of determining the species of arsenic compounds in foods, several methods have been developed for this purpose. Separation of arsenic compounds by high-performance liquid chromatography (HPLC) allowed determination of different arsenic species using on-line decomposition in a UV reactor prior to HGAAS and atomic fluorescence spectrometry (AFS)[49]. In large-scale studies such as the UK TDS, hydride generation coupled to ICP-MS is used to determine total arsenic in most foods[50].

10.2 Antimony

Antimony came to prominence in the news media in recent years when a possible link between it and sudden infant death syndrome (SIDS), or cot death, was reported. The suspected link was through a fungus, *Scopulariopsis brevicaulis*, found in some PVC mattresses. The fungus was believed to act on fire-retardant chemicals, including antimony, present in the mattresses. This resulted in the generation of toxic antimony trihydride or stibine, which could be responsible for cot deaths. In support of this hypothesis was the finding of higher than normal antimony levels in blood of SIDS victims[51]. However, further investigations showed that the fungus was not always present and that antimony was not present in flame retardants used on cot mattresses.

10.2.1 *Chemical and physical properties of antimony*

Antimony, which has been given the chemical symbol Sb from its ancient classical name of *stibium*, has an atomic weight of 121.8, and is number 51 in the periodic table of elements. It is a fairly heavy metal with a density of 6.7. Though closely related to arsenic, its physical properties are more metallic. Its crystalline form is lustrous silvery grey in colour and is very brittle and flaky. It melts at 631°C and vaporises at 1750°C. It is a poor conductor of heat and electricity. Oxidation states are +3 and +5. Antimony is not affected by air at room temperature, but when heated burns brilliantly. Among its most important compounds are the tri- and pent–oxides, the trichloride, and the tri- and penta-sulphides. The trihydride, stibine, SbH_3, is a gas.

10.2.2 *Production and uses of antimony*

The main ore of antimony is stibnite, Sb_2S_3, which is mined in China and in South and Central America. Antimony also occurs along with nickel in ullmanite, NiSbS. To extract the metal, the ores are heated with scrap iron, which removes the sulphur as FeS. Antimony is also obtained as a by-product of the refining of other metals. Antimony is used in the metallurgy industry, especially in alloys. When it is added to other metals, such as lead, it hardens them. It is employed for this purpose in, for example, the manufacture of battery plates and in type metal, as well as in solders, ammunition and electric cable coverings.

Antimony is widely used in the chemical industry to produce fireproofing materials, as well as paint and lacquer, rubber, glazes and pigments for ceramic and glass manufacture. Antimony compounds have long been used medicinally for the treatment of parasitic infestations. One of these, which is still widely used in the tropics, is antimony potassium tartrate, $2K(SbO)C_4H_4O_6.H_2O$, commonly known as 'tartar emetic'. This and other antimonials are used to treat schistosomiasis (bilharzia) and certain types of leishmaniasis (kala-azar). In ancient times the metal was valued as a cosmetic, especially as an eyeshadow.

10.2.3 *Metabolism and biological effects of antimony*

Not a great deal is known about the uptake of antimony from food or its behaviour in the human body. From animal experiments it would appear that about 15% of the ingested metalloid is absorbed in the GI tract. The absorbed antimony seems to be concentrated in organs, including the liver, kidneys and skin. Excretion is rapid initially, but there may be a long-term component which is retained in the body. There is no evidence that antimony is an essential nutrient in humans[52].

Physical contact with antimony and its compounds, as well as with its fumes and dust, can cause dermatitis, conjunctivitis and nasal septum ulceration[53]. The hydride, stibine, is very dangerous. Exposure to it causes headache, nausea and vomiting, and can be fatal. Exposure to other antimonials affects the skin and can cause myocardial infarction. Though most of the known antimony poisoning incidents are due to industrial exposure, a few caused by ingestion have been reported. Most of these have been due to contamination of soft drinks stored in enamelled containers. Symptoms include colic, nausea and weakness, with slow or irregular respiration[54].

10.2.4 *Antimony in food and beverages*

Levels of antimony in foods are generally very low, in many cases below detection limits. They range from < 0.001 to 0.004 mg/kg fresh weight, as shown in Table 10.2. A study of infant foods in the UK detected antimony in most samples but generally at very low levels, with a mean of 0.004, and a range of < 0.0001–0.36 mg/kg[55]. Little information is available on antimony levels in foods in other countries. Intakes of 0.25–1.25 mg/day have been reported for children in the US[56]. Mean adult intake in the UK is estimated to be 0.003 mg/day[57], and in Japan 80 $\pm$ 68 μg/day[58]. Levels of 0.1–0.2 μg/litre have been reported in some samples of both river and seawater[59]. High levels of antimony can occur in beverages in contact with enamelled surfaces. Low-pH fruit drinks have been shown to be capable of dissolving as much as 100 mg/litre of antimony from enamel[60]. The US EPA limit for antimony in drinking water is 0.1 mg/litre.

10.2.5 *Analysis of foodstuffs for antimony*

Because of low concentrations in foods and beverages, it may be necessary to preconcentrate antimony by solvent extraction before analysis. Solid samples may be digested either by dry ashing, using magnesium nitrate, or wet acid digestion. To avoid loss by volatilisation, temperatures must be kept low in open digestion. Use of closed microwave digestion is recommended to overcome such problems. GFAAS, with appropriate background correction, can be used for determination of even very low levels of antimony. The presence of chloride can cause problems and it may have to be removed by back-extraction into alkaline ammonium tartrate. HGAAS is

Table 10.2 Antimony in UK foods.

Food group	Mean concentration (mg/kg fresh weight)
Bread	0.001
Miscellaneous cereals	0.004
Carcass meat	0.002
Offal	0.001
Meat products	0.004
Poultry	0.001
Fish	0.003
Oils and fats	0.002
Eggs	0.001
Sugar and preserves	0.002
Green vegetables	<0.001
Potatoes	0.001
Other vegetables	0.001
Canned vegetables	0.002
Fresh fruit	0.001
Fruit products	0.001
Beverages	<0.001
Milk	<0.001
Dairy products	0.001
Nuts	0.001

Adapted from Ysart, G., Miller, P., Crews, H. *et al.* (1999) Dietary exposure estimates of 30 elements from the UK Total Diet Study. *Food Additives and Contaminants*, **16**, 391–403.

effective for low concentrations of antimony. There may, however, be problems with stability of the hydride in the presence of some other elements[61].

10.3 Selenium

Though selenium itself was only discovered by the Swedish chemist Jöns Jakob Berzelius in 1817, its toxicity to animals had brought the element to the attention of humans many centuries earlier. In about 1295 the Venetian traveller, Marco Polo, travelling in what was then called Cathay, described what we now know as selenosis in horses[62]. They had eaten certain plants, native to the part of China in which he was travelling, which seem to have had the ability to accumulate selenium in their leaves[63]. The same type of toxicity has been recognised in many investigations in other parts of the world. It occurs over a wide area of semi-arid regions of the mid-US, where it can poison horses and cattle and was given the name 'alkali disease' by stockmen. The earliest reports were of poisoning of cavalry horses fed on herbage growing in what became known as 'poison areas' and 'bad lands'. It has been suggested that one of the plants, *Astragalus racemosus* , also known as milk vetch or 'locoweed', was responsible for the failure of the relief cavalry to reach General Custer at the Battle of Little Big Horn[64].

Symptoms of selenosis in animals include lameness, hair loss, hoof cracking, blindness and paralysis, accounting for its other names of 'change hoof disease' in horses, and 'blind staggers' in cattle. The plants responsible can accumulate extraordinarily high levels of selenium: in *A. racemosus* as much as 15 g/kg (dry weight), and in the Australian accumulator *Neptunia amplexicaulis* 4 g/kg, more than enough to kill an animal[65].

Selenium poisoning of farm stock has been reported in many parts of the world besides China, the US and Australia, including Mexico, Venezuela, Colombia, Canada, Ireland and Israel. Selenosis is still recognised as a danger for farm animals under certain soil conditions. It can also be a danger to humans who consume foodstuffs grown on selenium-rich soils[66].

It was a considerable surprise, therefore, when this highly toxic substance turned out to be an essential trace element, for farm animals as well as humans. Indeed, as was found in the country where Marco Polo first came across its toxic trail, there are some regions of the world where the element is in such short supply that endemic selenium deficiency occurs, with serious consequences for health. China is in the unhappy situation of having within its borders three endemic diseases, all selenium-related, one of excess (human selenosis) and the others of deficiency (Keshan and Kashin-Beck)[67].

10.3.1 Chemical and physical properties of selenium

Selenium occurs in Group16/VI of the periodic table, along with sulphur. The two elements, with tellurium and polonium, make up the sulphur family. Selenium has an atomic weight of 78.96 and is number 34 in the table. Its density is 4.9. It boils at 684°C and volatilises easily on heating. It occurs in a number of allotropic forms, one of which is metallic or grey selenium. This crystalline form contains parallel 'zigzag' chains of atoms and is the stable form at room temperature. Another form is monoclinic or red selenium which is an amorphous powder, analogous to yellow 'flowers of sulphur'.

Selenium has unique electrical properties that make it of exceptional value industrially. Its electrical conductivity, which is low in the dark, is increased several hundredfold in light, which also generates a small electrical current in the selenium. It is, in addition, a semiconductor, possessing asymmetrical conductivity, which allows it to transmit a current more easily in one direction than in another. Selenium is chemically close to sulphur and forms similar types of compounds. It reacts with metals to form ionic compounds containing the selenide ion, Se^{2-}. It also forms covalent compounds with most other elements. Normal oxidation states are −2 (e.g. Na_2Se, sodium selenide), 0 (Se, elemental selenium), +4 (e.g. Na_2SeO_3, sodium selenite) and +6 (e.g. Na_2SeO_4, sodium selenate).

Other inorganic compounds of particular interest include hydrogen selenide, H_2Se, which can be formed by the direct action of hydrogen on selenium, but the reaction is sluggish. It is speeded up by the addition of sodium borohydride, as is done in the determination of the element by HGAAS. The hydride is a flammable, highly toxic gas, with an offensive odour. Selenium dioxide, SeO_2, is produced when the element is burned in air. It is a strong oxidising agent, which dissolves in water to form selenous acid, H_2SeO_3. It is used in Kjeldahl digestions to catalyse the oxidation of nitrogen compounds. Selenium trioxide corresponds to sulphur trioxide, and dissolves in water to form selenic acid, H_2SeO_4. It is a strong oxidant and reacts with many inorganic and organic substances. The two series of compounds formed by these acids, selenites and selenates, are of importance in the pharmaceutical and agricultural industries as dietary supplements, and in fertilisers.

The organic compounds of selenium are of considerable interest and some of them play important roles in cell biochemistry and nutrition. Organoselenium compounds are similar to organosulphur compounds, but are not identical in chemical and biochemical properties. Nearly all of the selenium compounds have a notoriously unpleasant and pervasive odour, a property that may formerly have restricted investigations by some chemists[68]. Today they are among the most widely investigated of the organometallic compounds. Of most interest are the selenoamino-carboxylic acids, selenium-containing peptides and selenium derivatives of the nucleic acids. Particularly important are selenomethionine, the Se analogue of the *S*-amino acid methionine, and selenocysteine. The latter is of crucial importance in human metabolism and is known as the '21st amino acid', the only non-S amino acid to have its own specific codon to direct its incorporation into a protein in ribosome-mediated synthesis[69].

These different organic and inorganic forms of selenium are of considerable significance in relation to biological function and toxicity of the element. There is evidence that its nutritional role depends on whether the selenium occurs in the organic or inorganic form in the diet[70]. Food normally contains only organoselenium compounds, such as selenocysteine and selenomethionine. Its inorganic compounds, such as sodium selenite and selenate, are normally consumed only in supplements.

10.3.2 Production and uses of selenium

Selenium is widely, though unevenly, distributed in the lithosphere. Soils and rocks generally contain about 50–200 μg/kg, but in some places, depending on geological and other factors, it can be at considerably greater or lesser concentrations. It occurs at high concentrations particularly in sedimentary rocks, such as limestone, coal and shale, as well as in volcanic deposits where it is isomorphous with sulphur. Selenium is concentrated in soils in some dry regions, such as the US Midwest. There it is found in alkaline soil as selenates. Elsewhere, in acid soils, it is present as selenides and to some extent as elemental selenium.

Selenium occurs in a number of ores, though none of these is commercially worked as a primary source. It occurs in ores along with other metals, usually as the selenide, for example in clausthalite, PbSe. When an ore such as iron pyrites, which contains selenium, is roasted, the selenium is deposited in the flue dust, from which it can be extracted. It is also obtained as a residue of copper pyrites oxidation in the lead chamber used in the manufacture of sulphuric acid. Most selenium is obtained today from the slimes left when copper is refined electrolytically. Refinery slimes can contain up to 50% selenium, along with other valuable metals, including gold, germanium and tellurium. Most of the world's supply of selenium comes as a by-product of metal refining in Canada, the US and Belgium, with smaller amounts from Zambia and a few other countries. World production is about 2300 tonnes per year.

Use of selenium has been growing in recent years. It has many industrial applications, and is also used extensively in agriculture and in the pharmaceutical industry. The electrical and electronic industries take about one-third of total production. It is used on photoreceptive drums of plain paper copiers (e.g. xerox machines), though this has been decreasing as selenium is replaced by other photo-sensitive materials. Other related uses are in laser printers, rectifiers, photovoltaic cells and X-ray machines. Selenium is also extensively used both to decolorise and to colour glass. The addition of cadmium sulphoselenide to the glass mix makes one of the most brilliant reds known to glass-makers. Other selenium compounds are used to produce a variety of other colours, as well as the bronze or smoky glass used in many modern buildings.

In metallurgy, addition of small amounts of selenium to alloys improves the machinability of wrought products and castings. It enchances corrosion resistance in chromium and other alloys. Other industrial uses of selenium include pigments in plastics, a hardener for grids of lead–acid batteries, as a catalyst in chemical reactions and in rubber curing.

Selenium is used extensively as an additive in animal feeds and as a dietary supplement. Soil deficiencies are corrected by adding selenium compounds to fertilisers and top dressings. Potasssium ammonium sulphoselenide, which is a strong pesticide, was the first systemic insecticide to be marketed. It was once widely used on vegetables, but is now restricted to non-food plants because of its toxic properties. Considerable amounts of selenium are used in the pharmaceutical industry, particularly as dietary supplements. A surprisingly large amount is used as an anti-dandruff preparation. A stabilised buffered solution of selenium sulphide is marketed as a shampoo under the trade name Selsun.

10.3.3 Selenium in food and beverages

Levels of selenium in plant foods, and in animals that feed on them, reflect levels in soils on which they grow. Soil selenium concentrations are subject to considerable regional variations, and consequently levels in different foods can show a wide range. This has important consequences for dietary intakes. Thus, in the US where much of the grain-producing regions have selenium-rich soils, the average intake of the element is 62–216 μg/day, while in parts of China, where the soil is depleted, intakes are 3–22 μg/day. In the UK, where, as in most other European countries, soil levels are relatively low, average intake is about 40 μg/day. Levels of selenium in the different food groups consumed in the UK, as found in the 1997 TDS, are shown in Table 10.3[71]. These are similar to those reported for other European countries, such as Germany[72]. They are higher than levels in New Zealand[73], but lower than those reported for the US[74].

10.3.3.1 Variability in selenium levels in foods

It should be pointed out that the figures reported for selenium levels in foods in the UK and other countries represent average concentrations. However, the coefficient of variation (CV) for selenium concentrations in individual foods can be considerable. In the US TDS which reported the concentration of selenium in some 234 different foods, the CV ranged from 19 to 47% with an average of 32%, higher than for other metals[75]. The high CV in animal tissues was attributed to variable amounts of selenium consumed, and in plant material to the high variability of the soil content. Variations in selenium levels in foods in other countries, in some cases of up to a hundredfold, have been attributed to similar causes[76].

An example of the level of variability in selenium levels in a particular foodstuff is seen in the case of milk, which ranges from 10 to 260 μg/kg in different parts of the US[77]. However, there can even be significant differences between levels in milk produced on different farms in the same district. An Australian study found that raw milk from one farm contained 38.5 μg/litre compared to 21.0 μg/litre in milk collected on the same day on a neighbouring farm. Overall, concentrations in processed milk collected over two years across Australia ranged from 38.34 ± 6.01 to 15.87 ± 4.49 μg/litre[78]. The range of selenium in milk in the UK has been reported to be 7–43 μg/litre[79], and in a high soil selenium region of the US (South Dakota) 32–138 μg/litre[80]. In contrast, milk products in a region of low soil selenium in China contained 2–10 μg/litre[81]. It is also interesting to note that though wheat grown in North America is generally a good source of selenium, the concentration range is 6–66 μg/

Table 10.3 Selenium in UK foods.

Food group	Mean concentration (μg/kg fresh weight)
Bread	44
Miscellaneous cereals	39
Carcass meat	115
Offal	492
Meat products	130
Poultry	185
Fish	360
Oils and fats	30
Eggs	194
Sugar and preserves	9
Green vegetables	8
Potatoes	3
Other vegetables	22
Canned vegetables	14
Fresh fruit	1
Fruit products	0.7
Beverages	0.4
Milk	14
Dairy products	32
Nuts	251

Adapted from British Nutrition Foundation (2001) *Briefing Paper: Selenium and Health*. BNF, London.

kg[82]. This may be compared with the reported range of 2–53 μg/kg in the UK, where dietary selenium intake is moderate to low[83].

An important conclusion from a consideration of the wide variations in selenium in foodstuffs is that, in general, food composition tables cannot be used to provide an accurate assessment of selenium intakes. This has implications for official investigations which annually determine the range of selenium, and other nutrients, in a variety of foods. The contribution of each group to average intake is assessed, using population average consumption data, but, as can be seen from the above considerations, estimations of intakes of selenium can only be relatively crude.

10.3.3.2 Good sources of dietary selenium

The richest sources of selenium are organ meat, such as liver (0.05–1.33 mg/kg), muscle meat (0.06–0.42 mg/kg) and fish (0.05–0.54 mg/kg). Though cereals contain only 0.01–0.31 mg/kg, cereal products make a major contribution to intake because of the relatively large amount of them consumed in most diets. Another good source of the element is nuts. Vegetables, fruit and dairy products are poor sources of selenium.

10.3.3.3 Brazil nuts

Brazil nuts have been reported to be the richest natural source of dietary selenium. Levels of up to 53 μg/g have been found in some sold in the UK[84]. In the US, nuts purchased in supermarkets averaged 36 $\pm$ 50 μg selenium/g, with the extraordinarily high level of 512 μg/g in an individual nut. However, it has to be pointed out that high levels of selenium in Brazil nuts on sale in the UK and other countries cannot be guaranteed[85]. The content is highly variable and depends on how effective the rain forest tree *Bertholletia excelsa*, from which the nuts are harvested, takes up the element from the soil. This is determined by the maturity of the root system and the tree variety, as well as by selenium levels and chemical form in the soil, pH and other factors.

Brazil nuts are gathered over an enormous area of the Amazon basin, in Brazil, Bolivia and neighbouring countries in tropical South America. Not all soil types and conditions across the growing areas of *B. excelsa* are the same. In some, such as the Manaus to Belem area, stretching for nearly 1000 miles across the lower reaches of the Amazon, soil levels are high and readily available for absorption. Nuts from this region contain between 1.25 and 512.0 μg/g, with an average of 36.0 ± 50.0 μ/g. In contrast nuts from the Acre–Rondonia region, on the upper Amazon, where soil levels are low, contain an average of 3.06 ± 4.01 μg/g, with a range of 0.03–3.17 μg/g[86]. It is probable that differences in place of origin account for the range of selenium levels found in a recent survey of those sold in the UK, which was 0.085–6.86 μg/g[87].

Brazil nut kernels average about three grams in weight. If these were to contain selenium even at about 50 μg/g, which is approximately the level found in some nuts[88], consumption of three nuts would result in ingestion of about 450 μg of the element. This is the amount of selenium that the UK Department of Health considers to be the safe maximum intake of selenium for an adult[89]. Half a nut would do the same for a 10 kg infant.

10.3.4 Dietary intakes of selenium

As might be expected, since selenium concentrations in the diet generally depend on levels of the element in the soils in the area in which foods are produced, and soil selenium can show considerable differences across the globe, estimates of per capita intakes vary widely between countries. This is shown in Table 10.4, in which intakes range from a low of 3 μg/day in parts of China, where human disease has been correlated with selenium deficiency, to a high of nearly 7 mg/day in another part of the same country where human selenosis occurs. In a country like New Zealand that has a history of selenium-deficiency diseases in farm animals, human intakes are almost two to ten times higher than in selenium-deficient areas of China. In contrast, residents of the US normally have an intake twice as high, or more, than that of New Zealanders[90].

In the diets of most countries, the main sources of selenium are cereals, meat and fish, with only a small contribution from dairy products and vegetables. A US survey found that five foods: beef, white bread, pork, chicken and eggs, contributed about 50% of the total selenium in a typical diet[91]. In the UK diet, bread, cereals, fish, poultry and meat are the main contributors. Percentage intakes from different foods were: bread and cereals 22, meat and meat products 32, fish 13, dairy products and eggs 22, vegetables 6 and other foods 5, and, from nuts 1.3%[92].

The dominance of cereal-based foods as core sources of dietary selenium means that intakes can be affected by a variety of unexpected circumstances. These include the success or failure of domestic harvests, international grain prices, and national agricultural and trade policies, that affect the importation of grain from the world market[93]. Europe, for example, which relied in the past on imported North American wheat to provide high-quality flour for bread making, has, with the development of new bakery technology, been able to rely more and more on home-grown wheat.

In contrast to the Canadian and US wheat, which mostly comes from areas where the soil is relatively selenium-rich, European soils usually contain low levels of the

Table 10.4 Estimated selenium intakes of adults in different countries.

Country	Intake (mean and/or range, μg/day)
Australia	57–87
Bangladesh	63–122
Belgium	30
Canada	98–224
China:	
Enshi Province	3200–6690
Keshan area	3–11
Finland:	
Pre-1984	25–60
Post-1984	67–110
France	47
Germany	47
Greece	110–220
Italy	49
Japan	133
Mexico	9.8–223.0
New Zealand	5–102
Spain	60
Sweden	38
Turkey	32
UK:	
1974	60
1997	39
USA	98 (60–220)
Venezuela	80–500

Adapted from Reilly, C. (1996) *Selenium in Food and Health*. Blackie/Chapman & Hall, London.

element. In consequence selenium levels today in bread and other bakery products sold in Europe are lower than they were in immediate post World War II years. This change has significant implications for dietary selenium intake in Europe. In the UK, for instance, there has been a substantial drop in selenium intake, from an average in 1978 of 60 μg/day, to 43 μg in 1994 and 39 μg in 1997[94]. Corresponding to this decline has been a fall in blood levels of selenium in British residents over the past twenty years[95]. Similar falls in intake and status have been reported elsewhere in Europe[96].

There is concern among some nutritionists about the possible adverse health effects of low selenium intakes. In some countries steps have been taken to protect the population against possible adverse effects of this situation. In Finland the law requires that selenium be added to all fertilisers and, as a result, the selenium status of the population has been more than doubled in recent years[97]. In New Zealand the law permits, but does not require, farmers to use selenium-enriched top dressings on grazing land, to combat deficiency in farm animals. Self-medication with selenium supplements is widely practised, and is actively promoted by the pharmaceutical industry and the media in many countries[98]. Increasingly, selenium-enriched food products are becoming available to meet the desire of consumers to increase levels in their diets. Several types of 'functional foods', ranging from fortified table salt to enriched sports drinks, are sold in some countries[99]. In addition to natural accumulators, such as Brazil nuts, other plant foods which are encouraged to accumulate selenium by cultivation on selenium-enriched soil, such as some of the genus *Brassicae*, as well as garlic and other members of the *Allium* family, are widely promoted, especially in health food stores[100].

10.3.5 Dietary requirements and recommended intakes for selenium

Selenium has been added relatively recently to the dietary recommendations in many countries, as evidence establishing its important role in human health has become accepted. Until 1989 the only official guideline for selenium levels in diets was the US ESADDI of 50–200 μg/day, when the Food and Nutrition Board introduced the first US adult RDA of 70 μg/day for males and 55 μg/day for females. In 2000 the RDA was replaced by a Dietary Reference Intake of 55 μg/day for all adults[101]. In the UK the Department of Health has set an RNI of 75 μg/day and 60 μg/day for men and women respectively[102]. This is higher than the US DRI for men and also exceeds the WHO recommendation of 40 μg/day for men and 30 μg/day for women[103].

It is believed by some health experts that these recommendations are insufficient to meet human needs and should be increased. There is considerable debate as to whether dietary recommendations should apply only to the prevention of deficiency diseases, or generally to the promotion of growth, maintenance of good health and the reduction of risk of other diseases. In the case of selenium in particular, it has been argued that intake recommendations should be increased to reflect possible beneficial effects for prevention of chronic diseases, such as cancer and heart disease[104].

10.3.6 Uptakes from food and metabolism of selenium

Selenium is present in foods mainly as the amino acids selenomethionine (mainly in cereals) and selenocysteine (in animal products). In some plants it also occurs as the selenate. In the leaves of cabbage, beets and in garlic, up to 50% may be present in the inorganic form[105]. Various other selenium compounds are found, generally in small amounts, in some plants; for example, methyl derivatives, responsible for the distinctive odour of garlic[106], and selenocystathionine, which is a toxic factor in the South American nut, Coco de Mono[107]. Selenium is also consumed in nutritional supplements in the form of selenite, selenate, selenomethionine and other selenium compounds, as well as in enriched yeast preparations.

Absorption of selenium in the GI tract is efficient, with uptake of around 80%. Selenomethionine appears to be actively absorbed, sharing a transport mechanism with the amino acid methionine. Selenocysteine may share a common active transport mechanism with basic amino acids. Selenate is absorbed by a sodium-mediated carrier transport mechanism shared with sulphur, while selenite uses passive diffusion[108]. A number of dietary factors, in addition to the chemical form, can affect the absorption of selenium from food. Enhancers of bioavailability include protein, vitamin E and vitamin A. Inhibitors include arsenic, mercury, sulphur, guar gum and vitamin C.

Retention of absorbed selenium appears to vary according to its chemical form. Thus selenomethionine is absorbed and retained more efficiently than selenate or selenite, but is not as efficient at maintaining selenium status. The reason for this appears to be that though selenomethionine is retained in muscle and other tissue proteins to a greater extent than selenocysteine or the inorganic forms, this retention is non-specific and is as a replacement for methionine. As a consequence, unlike the other forms, the selenium in selenomethionine is not immediately available for synthesis of functional selenoproteins[109].

Excretion of selenium is primarily in urine, in the form of the trimethylselenonium ion, $(CH_3)_3\,Se^{+}$[110]. A small amount of selenium is exhaled through the skin and lungs as dimethyl selenide, CH_3—Se—CH_3, and dimethyl diselenide, CH_3—Se—Se—CH_3. These are the compounds that account for the strong 'garlic' smell on the breath of people who have consumed excessive amounts of selenium.

10.3.7 Biological roles of selenium

The primary fate of all selenium absorbed by the body, whatever form is ingested, is to be incorporated into selenoprotein enzymes which perform a number of essential functions in the body. After absorption, selenium is transported from the gut and reduced to selenide within red blood cells, liver and other tissues. It is then deposited in various organs where it is incorporated into specific selenoproteins as selenocysteine. This incorporation is a unique and quite extraordinary activity, the discovery of which has been hailed as one of the most remarkable achievements of modern molecular biology.

Synthesis of selenoproteins is a complex process which has not yet been fully worked out[111]. However, what is clear is that genes for the process contain an in-frame UGA codon which directs the cotranslational insertion of selenocysteine into protein. This is the first time that a non-standard amino acid has been shown to act in this manner. As a result, selenocysteine is recognised, in terms of ribosome-mediated protein synthesis, as the 21st amino acid[112]. Some 30–35 selenoproteins have been detected and about half of these have been characterised in mammalian cells[113]. Those which have been identified in human tissues are listed in Table 10.5. All of these enzymes play an essential role in human metabolism.

The glutathione peroxidases function as intracellular antioxidants that protect cell systems against free radicals. It was the recognition of its role in the enzyme glutathione peroxidase (GSHPx) in 1973 that helped to identify selenium as an essential trace element[114]. Since then four distinct glutathione peroxidases have been identified, as listed in Table 10.5. These GSHPx enzymes protect the body against free-radical damage from lipid and phospholipid peroxides and hydrogen peroxide. The iodothyronine deiodinases are involved in thyroid hormone metabolism, particularly in the conversion of the inactive form of the hormone thyroxine, T4, into the active T3, tri-iodothyronine form. The three thioredoxin reductases are involved in DNA synthesis and thus in cell growth and division.

Less is known about some of the other functional selenoproteins. Selenoprotein P may have an antioxidant role in extracellular fluids. It may protect endothelial cells against peroxidative damage during inflammation, particularly from peroxynitrite, a reactive nitrogen species formed by the reaction between nitric oxide and superoxide under inflammatory conditions[115]. Seleoprotein W may also have a role in antioxidant defences, as well as in muscle metabolism. Seleoprotein synthetase is an enzyme required for the incorporation of selenocysteine into selenoproteins. Sperm capsule selenoprotein protects sperm cells from oxidative damage and may be necessary for the structural integrity of mature spermatazoa[116].

As is only to be expected, deficiency of a trace element such as selenium, with so many roles in cellular metabolism, will have significant health effects. Selenium deficiency is associated with several diseases of major economic importance in farm animals. In humans chronic low intake of dietary selenium is responsible for Keshan

Table 10.5 Selenoprotein-containing enzymes in humans.

Glutathione peroxidases	classical or intracellular GSHPx1 gastrointestinal GSHPx3 plasma GSHPx3 phospholipid hydroperoxide GSHPx4
Sperm capsule selenoprotein	(form of GSHPx4)
Selenoprotein P	
Iodothyronine 5′-deiodinases	Type 1 (IDI-1) Type 2 (IDI-2) Type 3 (IDI-3)
Thioredoxin reductases	TR1, TR2, TR3
Selenoprotein W	
Selenoprotein synthetase (SPS2)	

disease, a sometimes fatal cardiomyopathy which occurs especially in children and young women, as well as for Kashin-Beck disease, a chronic osteoarthropathy, which also affects mainly children. These diseases are found in parts of China and other areas of central Asia where soil levels of selenium are very low.

Several other selenium-responsive conditions occur in humans, including cardiomyopathies and muscular problems in patients on total parenteral nutrition (TPN). Normal function of the thyroid gland is also dependent on an adequate supply of the element[117]. There is evidence that selenium deficiency can also cause a range of problems, including defective immune response[118], and increased susceptibility to various forms of cancer and to coronary artery disease. A connection between selenium deficiency and increased susceptibility to infectious disease has been postulated on the basis of evidence that low selenium levels in the host can trigger increased virulence in a viral pathogen[119]. A decline in selenium status in human immunodeficiency virus (HIV) has been shown to occur. There is increasing evidence that selenium supplementation can have a beneficial effect on HIV and AIDS patients[120].

10.3.7.1 Protective role of selenium against toxic metals

Ingested selenium has been shown to be able to counteract the toxic effects of heavy metals in food. It has been used as an antidote against cadmium and mercury poisoning. There is evidence that the simultaneous presence of high levels of selenium along with high levels of methylmercury can protect consumers against mercury toxicity. Inhabitants of Greenland, and other communities that rely on fish and other marine foods as the major component of their diet, can ingest large amounts of methylmercury[121]. However, as was found in a study of an Inuit fishing community, though blood mercury levels often exceed 200 μg/litre, a level at which symptoms of methylmercury poisoning normally appear, there was no sign of mercury intoxication. This was attributed to the presence of high levels of selenium in the fish and seal meat the people consumed[122].

10.3.8 Selenium toxicity

Selenium toxicity is well documented in farm animals. Human selenosis occurs in the central Chinese district of Enshi County, where the condition has been extensively studied. Intakes of up to nearly 7000 µg/day were recorded in one particular village where selenosis was widespread[123]. There is also some evidence of diet-related selenosis occuring in other seleniferous areas of the world[124]. Sometimes it is caused by excessive intake of dietary supplements[125].

Symptoms of chronic selenium poisoning include vomiting, diarrhoea, hair and nail loss and lesions of the skin and the nervous system. There is some debate about the levels of intake which bring about such symptoms. Residents of some high soil selenium areas appear to have none of them, though they consume as much as 700 µg/day. According to the Environmental Protection Agency in the US, a daily intake of 5 µg/kg body weight (350 µg for a 70 kg adult) is not toxic[126]. WHO recommends an upper safe limit of 400 µg/day. In the UK the recommended maximum safe daily intake from all sources for adults is 6 µg/kg body weight or 450 µg for an adult male.

Some investigators believe that the WHO upper safe limit, and presumably also the slightly higher UK maximum safe limit, are too conservative[127]. It is argued that this view is supported by Chinese findings of no adverse effects among individuals with blood concentrations as great as 1000 µg/litre (about ten times the normal range) and intakes of 835 µg selenium/day[128].

In the case of selenium supplements, because of concern at the possibility of overconsumption, a maximum upper safe level of 200 µg/day has been proposed by the European Federation of Health Food Manufacturers. A similar level of 200 µg is recommended as a maximum daily dose for long-term supplementation, with 700 µg/day for short-term use[129]. A recent large-scale human intervention trial in the US found no dermatological changes or other signs of selenosis with long-term supplementation of up to ten years, using a dose of 200 µg/day[130].

10.3.9 Analysis of foodstuffs for selenium

Until relatively recently, the determination of selenium in foodstuffs was a challenge to analysts, because of the element's often very low levels in samples and the ease with which it can be lost by volatilisation during preparation. Even today, with all the advantages of modern analytical equipment and techniques available, there can, as has been commented, 'still exist some problems with the analytical methodology, without, however, assuming enormous proportions'[131].

Careful attention to sample pretreatment is essential to prevent losses. Dry ashing should be avoided. An acid digest, using nitric acid alone or a nitric–sulphuric acid mixture, carried out on a temperature-controlled heating block or in conical flasks on a plate heater, has been found appropriate, though care must be paid to temperature setting and maintenance[132]. Digestion times can be shortened, and danger of volatilisation reduced, by using a closed digestion system with microwave heating[133].

Though spectrofluorimetry, using 2,3-diaminonaphthalene (DAN), has been shown to be an accurate and sensitive analytical procedure, and was widely used for selenium until relatively recently, it is now largely replaced by less cumbersome instrumental techniques[134,135]. HGAAS is now the method of choice in many of the studies reported[136]. Combining the method with flow injection (FI) to introduce standards and test solutions in a 10% HCl matrix stream provides, it is claimed, several advantages such as lower reagent consumption and higher tolerance to interfering elements and, in addition, permits automation of the procedures[137].

Other techniques and procedures for the determination of selenium in foods used today include instrumental neutron activation analysis (INAA), inductively coupled plasma atomic emission spectrometry (ICP-AES), inductively coupled plasma mass spectrometry (ICP-MS), isotope dilution mass spectrometry (IDMS) and X-ray fluorescence (XRF), as well as less elaborate techniques such as polarography, gas chromatography and GFAAS. The method used to determine selenium in foods collected in the UK TDS is hydride generation–inductively coupled plasma–mass spectrometry (HG-ICP-MS).

There is increasing interest in determining different species of selenium compounds in foods[138]. Methods for doing so are being developed and several procedures have been reported. Some of these have recently been evaluated in a review published by the Commission on Microchemical Techniques and Trace Analysis of the International Union of Pure and Applied Chemistry (IUPAC)[139]. Among them were methods based on radioimmunological assay (RIA) for selenoproteins, which were found to offer very low detection limits, but required that the proteins be first isolated in amounts sufficient for antibody production. More promising were another group of methods, using hyphenated techniques, based on the coupling of electrophoretic or chromatographic separation with an atomic spectrometric or other selenium-specific technique. These hybrid systems include polyacrylamide gel electrophoresis (SDS-PAGE)-INAA, and high-performance liquid chromatography (HPLC)-ICP-MS. The IUPAC report noted that though progress has been made in developing such procedures, much still remains to be done. A major challenge is to refine the methods to improve sensitivity and reduce vulnerability to matrix interference. Now that the importance of knowing not only the total amount of selenium present in a food or other biological sample, but also which and how much of the various species are present, has been established, undoubtedly it will not be long before this challenge is overcome[140].

10.4 Tellurium

Tellurium is a rare element, with a natural abundance of a few mg/kg, and its importance as a food contaminant is minor. It is, however, included in this section to round off the story of selenium, with which it is closely related, in chemistry and in origins. In addition, tellurium is also one of those metals at present of little concern to toxicologists but which, because of their growing industrial uses, could become of greater significance as potential contaminants of the human diet.

10.4.1 Chemical and physical properties of tellurium

Tellurium, Te, is number 52 in the periodic table of the elements, with an atomic weight of 127.6. It is a silvery-white brittle metalloid and exists in several allotropic forms. It is a member of the sulphur family, with very similar chemical properties to sulphur and selenium. It combines with oxygen to form a variety of oxides, including the dioxide, TeO_2, and trioxide, TeO_3. These in turn give rise to the two weak acids, tellurous, H_2TeO_3, and telluric, H_6TeO_6. Tellurium forms two series of tellurides, examples of which are sodium telluride, Na_2Te, and sodium hydrogen telluride, NaHTe. Hydrogen telluride, H_2Te, is formed by the action of weak acid on tellurides.

10.4.2 Production and uses of tellurium

Tellurium has been found in nature as the free element, but most frequently occurs in ores combined with gold, lead or silver as tellurides. There are several different ores, but none is used for production of the metalloid. All commercially used tellurium is recovered from the slimes from electrolytic processing of non-ferrous metals, particularly copper. World production of tellurium is estimated to be about 450 tonnes each year, with production mainly confined to Canada, Japan, Peru and the US[141].

Tellurium has a number of industrial uses, largely in metallurgy. It is used to improve the properties of steels and other metal alloys for special purposes. About a quarter of world production is used as chemicals and catalysts in, for example, hydrogenation and halogenation reactions. Tellurium compounds are used as curing agents in rubber manufacture, in explosives and the manufacture of glass and plastics. A growing and increasingly important use is in electronics as a semiconductor in thermoelectric and photoelectric devices. A mercury–cadmium telluride alloy is used as a sensing material for thermal-imaging devices as well as in medical diagnosis[142].

10.4.3 Tellurium in foods and diets

Little information is available on tellurium levels in most foods. Some of what is available, especially in older reports, is of doubtful reliability, due to inadequate analytical techniques which inflated results. A more recent study found that tellurium concentrations, in the limited number of foods and beverages examined, was of the order of 1 μg/kg or below. This is about two orders of magnitude lower than previous recorded values[143]. Levels of tellurium in a variety of foods and beverages are given in Table 10.6, which is based on this study.

From the limited data available, it is possible to estimate that daily intake of tellurium is between 1 and 10 μg/day. This is considerably lower than the intake of 100 μg/day formerly reported[144]. It has been suggested that differences between intake in different regions are due either to regional variations in the tellurium content of plants, or to contact of some of the foods analysed with metal containers that had been contaminated by tellurium[145].

10.4.4 Uptakes and metabolism of tellurium

About 10–20% of ingested tellurium appears to be absorbed by the body[146]. This is followed by rapid excretion, mainly in urine[147]. Some tellurium may be exhaled from the lungs after being metabolised into dimethyl telluride. This has a strong and persistent garlic-like odour that has been detected on the breath of volunteers who consumed 100 μg of sodium tellurite[148].

There is no evidence that tellurium is an essential trace element or performs any specific metabolic function in humans. The toxicity of orally administered tellurium has been shown to be low in experimental animals. Rats fed 375–1500 μg of elemental tellurium per gram of diet for 21 days developed a garlic-like odour on their breath, but showed no pathological changes. When the tellurium was fed as TeO_2, they suffered from temporary paralysis of hind limbs, hair loss and other symptoms. Soluble salts were toxic at concentrations of 25–50 μg/g of diet[149].

Table 10.6 Tellurium levels in some foods and beverages.

Food/Beverage	Te content (μg/kg, wet wt)
Water (tap)	< 0.2
Red wine	0.5
Milk	< 4
Meat (beef)	< 5–< 10
Fish (imported)	< 2
White flour	< 5
Wholemeal flour	< 5
Corn	0.22
Garlic	0.9–< 5
Mushrooms	1.0
Onions	0.8
Potatoes	3.2
Tomatoes	< 0.7
Lettuce	< 10
Apples	< 0.7–3.2
Curry powder	< 10
Cheese (parmesan)	13
Herbs (mixed, dried)	12
Paprika powder	< 10

Adapted from Kron, T., Hansen, C.H. & Werner, E. (1991) Tellurium ingestion with foodstuffs. *Journal of Food Composition and Analysis*, **4**, 196–205.

Note: most of the concentrations are below detection limits, indicated by <. The different detection limits were the result of different amounts of sample prepared for the AAS analytical technique used.

10.4.5 *Analysis of foodstuffs for tellurium*

Problems have been encountered by analysts in determining tellurium in food because of its generally very low concentrations. There are other difficulties, including analyte loss, preconcentration problems, background and matrix effects[150]. Some of the earlier determinations which were carried out using NAA appear to have given doubtfully high results. HGAAS has been used with some success, though problems were experienced with foods containing fats and oils, which had the effect of reducing the tellurium signal.

10.5 Boron

In the official UK report on dietary reference values published in 1991, the element boron was covered in a total of just twelve lines of text. The final sentence reads: 'the role of B in humans is unknown and the essentiality of B remains to be demonstrated. There do not appear to be any published studies relating to the question of B toxicity'[151]. That situation has changed dramatically over the past decade. There is now increasingly strong evidence that boron is indeed an essential trace element for humans and there is no shortage of published studies which deal with the question, not just of boron's toxicity, but also of its unique metabolic role.

10.5.1 Chemical and physical properties of boron

Boron, B, is number 5 in the periodic table and has an atomic weight of 10.8. It is a metalloid with predominantly non-metallic properties and exists in two forms, as an amorphous dark brown powder and as yellow crystals. Crystalline boron is transparent and nearly as hard as diamonds. Its electrical resistance is remarkable in that at room temperature it is about two million times greater than at red heat. Chemical properties of boron are similar to those of silicon. It forms compounds with oxygen, hydrogen, halogens, nitrogen, phosphorus and carbon. Boron burns in oxygen to form boric oxide, B_2O_3. This dissolves in water to form orthoboric acid, H_3BO_3. If dry orthoboric acid is heated it is converted first into metaboric acid, HBO_2, then into tetraboric acid, $H_2B_4O_7$, and finally into boric anhydride, B_2O_3.

Among the most important compounds of boron are sodium tetraborate decahydrate, $Na_2B_4O_7.10H_2O$, commonly known as borax. Sodium perborate, $NaBO_2.3H_2O.H_2O_2$, is made by treating borax with sodium hydroxide and hydrogen peroxide. On hydrolysis the perborate liberates hydrogen peroxide. Boron nitride, BN, is a crystalline substance analagous to graphite and is sometimes called 'inorganic graphite'. Boron carbide is an extremely hard substance and can be used as a substitute for diamonds in cutting and abrading tools. Boron forms a number of organic compounds. It has a tendency to form macromolecules with organic molecules, such as carbohydrates, as well as with ATP, vitamin C, steroids and other compounds of biological importance[152].

10.5.2 Production and uses of boron

Boron is widely distributed in the lithosphere, but usually at low concentrations. It occurs in concentrated form as orthoboric acid in volcanic regions and as borax and colemanite, $Ca_2B_6O_{11}.5H_2O$, in dry lakes of the world's deserts. Though borax was known since ancient times and was used in medicine, glass-making and alchemy, the element was first produced in a semi-pure form in 1808 by Davy in England and by Gay-Lussac in France. The element is obtained commercially by reducing the oxide with magnesium and leaching with hydrochloric acid.

Boron has a considerable number of industrial applications[153]. It is used as a deoxidiser and degasifier in metallurgy and to improve hardenability and facilitate annealing of steel and other alloys. Boric acid and perborates are used as brazing fluxes. It is also widely used in the production of enamels, as well as in the manufacture of special glass, such as that used to make laboratory glassware, as well as in fibreglass and ceramics. The ability of boron to absorb neutrons makes it ideal for use in steel alloys for making shielding materials and control rods in nuclear reactors. Boron fibres, which have a very high tensile strength, are added to plastics to make a material that is stronger, though far lighter, than steel.

Borates are used in magnets, sandpaper and grinding wheels. In the automobile industry boron compounds are used to make antifreeze, motor oil, brake fluid and power steering fluids. The electrical and electronic industries use borates to make capacitors, transistors, semiconductors and other computer components. Borax and boric acid were once used extensively to preserve and extend the palatability of foods, such as fish, meat and meat products[154]. They also have a long history of use as an antiseptic and in several other pharmaceutical compounds. However, because of toxicological considerations, this use is now discouraged, if not actually prohibited, in some countries. Boron compounds are found in a very wide range of household chemicals, especially in washing powders and cleaning agents, as well as in cosmetic preparations, such as creams, soaps, shampoos and many others. They are also used in wood treatment, insecticides and microbiocides, as well as in water treatment plants and in fertilisers.

10.5.3 Boron in food and beverages

Boron is a normal component of the diet, occurring in all food groups, as shown in Table 10.7. It occurs at highest concentrations in plant foods, especially in vegetables and fruit, with a particularly high level in nuts, but at low concentrations in animal products. The UK data correspond closely to levels in foods in the US[155]. Particularly good sources of boron in the US diet, according to a recently compiled database, are peanuts (17 mg/kg), peanut butter (14.5 mg/kg) and raisins (22 mg/kg)[156]. Boron concentrations in beverages ranged from 6.1 mg/kg in dry table wine, to 1.8 mg/kg in apple juice and milk, 0.29 mg/kg in coffee, 0.13 mg/kg in cola-type soft drinks, 0.12 mg/kg in beer and 0.09 mg/kg in brewed leaf tea. The median level of boron in drinking water in the US has been reported to be 0.031 mg/litre, with an upper range of 2.44 mg/litre[157]. Interestingly, coffee and milk, in spite of their low concentrations, are the top two contributors to boron intake in the US because of the volumes in which they are consumed. For the same reason, several of the other beverages also make a major contribution.

Mean adult intake in the UK has been estimated to be 1.4 mg/day, with an upper range of 2.6 mg/day[158]. The US average daily adult intake has been estimated as 1.5 mg, which is very close to the UK value[159]. However, a recent report, based on a greater number of foods, found that intakes by adults in the US are as low as 1.17 $\pm$

Table 10.7 Boron in UK foods.

Food group	Mean concentration (μg/kg fresh weight)
Bread	0.5
Miscellaneous cereals	0.9
Carcass meat	<0.4
Offal	<0.4
Meat products	0.4
Poultry	<0.4
Fish	0.5
Oils and fats	0.4
Eggs	<0.4
Sugar and preserves	0.8
Green vegetables	2.0
Potatoes	1.4
Other vegetables	1.4
Canned vegetables	1.2
Fresh fruit	3.4
Fruit products	2.4
Beverages	0.4
Milk	<0.4
Dairy products	0.4
Nuts	14

Adapted from Ysart, G., Miller, P., Crews, H. *et al.* (1999) Dietary exposure estimates of 30 elements from the UK Total Diet Study. *Food Additives and Contaminants*, **16**, 391–403.

0.65 for males, and 0.96 ± 0.55 for females[160]. In contrast, intakes by Canadian women have been reported to be 1.33 ± 13 mg/day[161].

Because of higher levels of the element in plants than in animals, vegetarians might be expected to have a higher than average intake, with about 20 mg/day[162]. However, it was found that boron intakes by adult vegetarians in the US are actually 1.47 ± 0.70 for males and 1.29 ± 1.12 μg/day for females. It has been suggested that the reason for these unexpectedly low intakes may be that, although vegetarians consume more boron-rich fruits, nuts and legumes, their diet also includes more low-boron grain products and less total energy than that of non-vegetarians[163].

10.5.4 Absorption and metabolism of boron

Ingested boron is rapidly taken up from the GI tract, with possibly more than 90% absorbed[164]. The boron is rapidly distributed throughout body water. It is not metabolised and most of it is rapidly excreted in urine[165]. A small amount appears to be retained in body tissues, especially bone and spleen[166]. Boron has long been known to be essential for the growth of plants and there is increasing evidence that it is also essential for animals and humans. The WHO Expert Committee on Trace Elements has declared that boron is 'probably essential'[167], and the USDA believes that a provisional RDA could be established for the element[168]. The 1989 edition of the US *Recommended Dietary Allowances* stated that there was substantial evidence that boron was an essential micronutrient in animals. However, more information about actual boron intakes was needed before recommended intake levels could be determined[169]. The same reasoning was followed by authorities in the UK and so far there are no official recommendations on boron intakes in the diet.

There is considerable experimental evidence that boron deprivation affects the composition or function of several components of the body, including the skeleton and brain. It also affects biochemical indices associated with the metabolism of other nutrients, including calcium, copper, nitrogen and cholecalciferol[170]. Effects of low boron intake were enhanced when the animal simultaneously suffered from nutritional stress, for example by deficiencies of vitamin D, calcium, cholecalciferol, magnesium and potassium. Boron appears to affect calcium and magnesium metabolism positively, and may be needed for energy substrate use and membrane function[171]. A number of clinical studies have indicated that boron deficiency can also occur in humans and may be of significance in relation to osteoporosis. There is evidence that supplementation with boron can influence the metabolism of major minerals and bone-related hormones, for example in post-menopausal women[172].

10.5.5 Boron toxicity

Dietary boron has a low level of toxicity, most probably because almost all that is absorbed is rapidly excreted. However, this control can be overwhelmed by high doses, as has been shown in the case of experimental animals and in a few humans who have been accidentally poisoned by pharmaceutical and household products containing boron[173]. Acute toxicity causes nausea, vomiting, dermatitis, lethargy and other symptoms. In extreme cases kidney damage occurs, followed by circulatory collapse and death. The minimum lethal dose for humans is not known, though single doses of 18 to 20 g have been fatal to adults. An ADI of 0.3 mg/kg body weight/day, equivalent to about 20 mg boron/day for a 70 kg man, has been proposed by the US FDA on the basis of a NOAEL (no observable adverse effect level) of 9.6 mg/kg/day.

10.5.6 Analysis of foodstuffs for boron

The accurate determination of boron in biological materials has been described by expert analysts as 'exceptionally difficult'[174]. The reasons for this are not simply because of the very low levels of the element usually encountered in samples. There are also many other problems, including carbide formation and matrix effects, which affect AAS determinations in particular. ICP-MS was used successfully in the UK TDS, though the high limit of detection (0.4 mg/kg) indicates difficulties encountered[175]. A method has been described that is said to circumvent many of the problems, such as volatilisation, mineral residue and low boron concentrations. This uses a very sensitive instrument, an inductively coupled argon plasma spectrometer (ICAP), after an open-vessel, low-temperature wet digestion[176].

References

1. Berzelius, J.J. (1818) Lettre de M. Berzelius à M. Bertollet sur deux métaux nouveaux. *Annales de Chimie et de Physique, Série 2*, 7, 199–202.
2. Ministry of Agriculture, Fisheries and Food (1982) *Survey of Arsenic in Food*. HMSO, London.
3. Biswas, B.K., Dhar, R.K., Samanta, G. *et al.* (1998) Detailed study report of Samata, one of the arsenic affected villages of Jessore district, Bangladesh. *Current Science*, **74**, 134–45.
4. Pearce, F. (1998) Arsenic in water. *Guardian* (London) 19 February, Online 1–4.
5. Buck, W.B. (1978) Toxicity of inorganic and aliphatic organic mercurials. In: *Toxicity of Heavy Metals in the Environment* (ed. F.W. Oehme), pp. 357–69. Dekker, New York.
6. Fierz, U. (1965) Skin-cancer and arsenic-containing pharmaceuticals. *Dermatology*, **131**, 41–58.
7. Kew, J., Morris, C., Aihie, A., Fysh, R., Jones, S. & Brooks, D. (1993) Arsenic and mercury intoxication due to Indian ethnic remedies. *British Medical Journal*, **306**, 506–7.
8. Department of Health (1991) *Dietary Reference Values for Food Energy and Nutrients in the United Kingdom*, p. 191. HMSO, London.
9. Ysart, G., Miller, P., Crews, H. *et al.* (1999) Dietary exposure estimates of 30 elements from the UK Total Diet Study. *Food Additives and Contaminants*, **16**, 391–403.
10. Dabeka, R.W., McKenzie, A.D. & Lacroix, G.M.A. (1987) Dietary intakes of lead, cadmium, arsenic and fluoride by Canadian adults: a 24-hour duplicate diet study. *Food Additives and Contaminants*, **4**, 89–102.
11. Stijve, T. & Bourqui, B. (1991) Arsenic in edible mushrooms. *Deutsche Lebensmittel-Rundschau*, **87**, 307–10.
12. Ministry of Agriculture, Fisheries and Food (1982) *Survey of Arsenic in Food*. HMSO, London.
13. Ministry of Agriculture, Fisheries and Food (1982) *Survey of Arsenic in Food*. HMSO, London.
14. Bebbington, G.N., Mackay, N.J. & Chvojka, R. (1977) Heavy metals, selenium and arsenic in nine species of commercial fish. *Australian Journal of Marine and Freshwater Research*, **28**, 277–86.
15. National Health and Medical Research Council (1986) *Model Food Legislation. A12. Metals and Contaminants in Food*. Australian Government Publishing Services, Canberra, ACT.
16. Australia New Zealand Food Authority (2000) *Standard 1.1.4. Contaminants and Restricted Substances* http://www.anzfa.gov.au/draftfoodstandards/Chapter1/Part1.4/1.4.1.htm
17. Jukes, D.J. (1997) *Food Legislation of the UK. A Concise Guide*, pp. 137, 149–51. Butterworth/Heinemann, Oxford.

18. Bowen, H.J.M. (1966) *Trace Elements in Biochemistry*. Academic Press, London.
19. United States Environmental Protection Agency (1996) *Arsenic in Domestic Waters*. http://www.epa.gov/OGWDW/ars/arsenic.html
20. Flowerdew, D.W. (1990) *A Guide to the Food Regulations in the United Kingdom*, p. 156. British Food Manufacturing Industries Research Association, Leatherhead, Surrey, UK.
21. Borgono, J.M. & Greiber, R. (1972) *Trace Substances in Environmental Health* (ed. D. Hemphill), Vol. 5, pp. 13–24. University of Missouri Press, Columbia, Mo.
22. Queirolo, F., Stegen, S., Restovic, M. *et al.* (2000) Total arsenic, lead and cadmium levels in vegeteables cultivated at the Andean villages of northern Chile. *Science of the Total Environment*, **255**, 75–84.
23. Shibata, A., Ohneseit, P.F., Tsai, Y.C., Spruck, C.H., Nichols, P.W. & Chang, H.S. (1994) Mutational spectrum in the p53 gene in bladder tumours from the endemic area of blackfoot disease in Taiwan. *Carcinogenesis*, **15**, 1085–7.
24. Dhar, R.K., Biswas, B.K., Samanta, G. *et al.* (1997) Groundwater arsenic calamity in Bangladesh. *Current Science*, **73**, 48–59.
25. Wickstrom, G. (1982) Arsenic emission from the Novaky power station. *Work Environment and Health*, **9**, 2–8.
26. Anonymous (1979) Arsenic poisoning near brick kilns. *Observer* (London), 22 January.
27. Peach, D.F. & Lane, D.W. (1998) A preliminary study of geographic influence on arsenic concentrations in human hair. *Environmental Geochemistry and Health*, **20**, 231–7.
28. Milton, A. & Johnson, A. (1999) Arsenic in the food chains of a revegetated metalliferous mine tailings pond. *Chemosphere*, **39**, 765–79.
29. Ministry of Agriculture, Fisheries and Food (1982) *Survey of Arsenic in Food*. HMSO, London.
30. Tsuchiya, K. (1977) Arsenic contamination of infant formula. *Environmental Health Perspectives*, **19**, 35–42.
31. Woolson, E.A. (1975) Arsenic in cotton seed byproducts. *Journal of Agriculture and Food Chemistry*, **23**, 677–83.
32. WHO (1989) Evaluation of certain food additives and contaminants. 33rd Report of the Joint FAO/WHO Expert Committee on Food Additives. *WHO Technical Report Series, Number 776*. World Health Organization, Geneva.
33. Ministry of Agriculture, Fisheries and Food (1999) *Food Surveillance Information Sheet, Number 191*. http://www.foodstandards.gov.uk/ma...ood/infsheet/1999/no191/191tds.htm
34. Gunderson, E.L. (1995) FDA Total Diet Study, July 1986–April 1991. Dietary intakes of pesticides, selected elements and other chemicals. *Journal of the Association of Official Analytical Chemists International*, **78**, 1353–63.
35. Vannort, R.W., Hannah, M.L. & Pickston, L. (1995) *1990/1991 New Zealand Total Diet Study. Part 2. Contaminant Elements*. New Zealand Health, Wellington, New Zealand.
36. Stoeppler, M. & Vahter, M. (1994) Arsenic. In: *Trace Element Analysis in Biological Specimens* (eds. R.F.M. Herber & M. Steoppler), pp. 291–320. Elsevier, Amsterdam.
37. Lander, H., Hodges, P.R. & Crisp, C.S. (1965) Arsenic levels in hair. *Journal of Forensic Medicine*, **12**, 52–67.
38. Smith, H., Forshufvud, S. & Wassén, A. (1962) Distribution of arsenic in Napoleon's hair. *Nature*, **194**, 725–6.
39. Clarkson, T.W. (1983) Molecular targets of metal toxicity. In: *Chemical Toxicology and Clinical Chemistry of Metals* (eds. S.S. Brown & J. Savory), pp. 211–26. Academic Press, London.
40. Lafontaine, A. (1978) Health effects of arsenic. In: *CEC Trace Metals: Exposure and Health Effects* (ed. E.D. Ferrante), pp. 107–16. Pergamon, Oxford.
41. Chen, C.J., Kuo, T.L. & Wu, M.M. (1988) Arsenic and cancers. *Lancet*, **i**, 414–15.
42. Rhian, M. & Moxon, A.L. (1943) Interaction of arsenic and selenium. *Journal of Pharmacology and Experimental Therapy*, 78, 249–64.

43. Nielsen, F.H. (1990) Other trace elements. In: *Present Knowledge in Nutrition* (ed. M.L. Brown), pp. 294–307. International Life Sciences Institute, Washington, DC.
44. Anke, M., Groppel, B. & Krause, U. (1991) The essentiality of the toxic elements cadmium, arsenic and nickel. In: *Trace Elements in Man and Animals – 7* (ed. B. Momĉilović), pp. 11/6–11/8. Institute for Medical Research and Occupational Health, Zagreb, Croatia.
45. Uthus, E.O., Poellot, R., Brossart, B. & Nielsen, F.H. (1989) Effect of arsenic deprivation on polyamine content in rat liver. *Federation of American Societies of Experimental Biology Journal*, **3**, A1072.
46. Department of Health (1991) *Dietary Reference Values for Food Energy and Nutrients in the United Kingdom*, p. 191. HMSO, London.
47. Tsalev, D.S. (1984) *Atomic Aborption Spectrometry in Occupational and Environmental Health Practice*, Vol. II, p. 15. CRC Press, Boca Raton, Florida.
48. Oygard, J.K., Lundebye, A.K. & Julshamm, K. (1999) Determination of inorganic arsenic in marine food samples by hydrochloric acid distillation and flow-injection hydride-generation atomic absorption spectrometry. *Journal of AOAC International*, **82**, 1217–23.
49. Van Elteren, J.T. & Slejkovec, Z. (1999) Ion-exchange separation of eight arsenic compounds by high-performance liquid chromatography–UV decomposition–hydride generation atomic fluorescence spectrometry and stability tests for food treatment procedures, *Journal of Chromatography*, **91**, 337–48.
50. Ysart, G., Miller, P., Crews, H. *et al.* (1999) Dietary exposure estimates of 30 elements from the UK Total Diet Study. *Food Additives and Contaminants*, **16**, 391–403.
51. Richardson, B.A. (1994) Sudden infant death syndrome: a possible primary cause. *Journal of the Forensic Science Society*, **34**, 199–204.
52. Nielsen, F.H. (1986) Other elements: antimony. In: *Trace Elements in Human and Animal Nutrition* (ed. W. Mertz), Vol. 2, pp. 156–60. Academic Press, New York.
53. White, G.P., Mathias, C.G. & Davin, J.S. (1993) Dermatitis in workers exposed to antimony in a smelting process. *Journal of Occupational Medicine*, **35**, 392–5.
54. Monier-Williams, G.M. (1934) Antimony in enamelled hollow-ware. *Ministry of Health Report on Public Health and Medical Subjects, Number 7*. HMSO, London.
55. Ministry of Agriculture, Fisheries and Food (1999) Metals and other elements in infant foods. *MAFF Food Surveillance Information Sheet, Number 190*. Department of Health, London.
56. Borgono, J.M. & Greiber, R. (1972) *Trace Substances in Environmental Health* (ed. D. Hempill), Vol. 5, pp. 13–24. University of Missouri Press, Columbia, Mo.
57. Department of Health (1991) *Dietary Reference Values for Food Energy and Nutrients in the United Kingdom*, p. 191. HMSO, London.
58. Shimbo, S., Hayase, A., Murakami, M. *et al.* (1996) Use of a food composition database to estimate daily dietary intake of nutrient or trace elements in Japan, with reference to limitation. *Journal of Food Additives and Contaminants*, **13**, 775–86.
59. Soukup, A.V. (1972) Trace elements in water. In: *Proceedings of the Congress of Environmental Chemistry: Human and Animal Health* (ed. A.V. Soukup). Fort Collins, 7–11 August. Colorado University Press, Colorado.
60. Monier-Williams, G.M. (1934) Antimony in enamelled hollow-ware. *Ministry of Health Report on Public Health and Medical Subjects, Number 7*. HMSO, London.
61. MacKay, K.M. (1960) *Hydrogen Compounds of the Metallic Elements*. Spon, London.
62. Polo, M. (1967) *The Travels of Marco Polo*, translated by E.W. Marsden, revised by T. Wright, Chapter XL, pp. 110–11. Everyman's Library, Dent, London.
63. Yang, G., Wang, S., Zhou, R. & Sun, S. (1983) Endemic selenium intoxication of humans in China. *American Journal of Clinical Nutrition*, **37**, 872–81.
64. Jukes, T.H. (1983) Nuggets on the surface: selenium an 'essential poison'. *Journal of Applied Biochemistry*, **5**, 233–34.
65. Knott, S.G. & McCray, C.W.R. (1959) Two naturally-occurring outbreaks of selenosis in Queensland. *Australian Veterinary Journal*, **35**, 161–5.
66. Rosenfeld, I. & Beath, O.A. (1964) *Selenium Geobotany, Biochemistry, Toxicity and Nutrition*. Academic Press, New York.

67. Reilly, C. (1996) *Selenium in Food and Health*, pp. 4–21. Chapman & Hall/Blackie, London.
68. Frost, D.V. (1972) The two faces of selenium – can selenophobia be cured? In: *Critical Reviews of Toxicology* (ed. D. Hemphill), pp. 467–514. CRC Press, Boca Raton, Florida.
69. Stadman, T.C. (1990) Selenium biochemistry. *Annual Review of Biochemistry*, **59**, 111–27.
70. WHO (1987) *Environmental Health Criteria 58: Selenium*. United Nations Environmental Programme, the International Labour Organisation and the World Health Organization. World Health Organization, Geneva.
71. BNF (2001) *Briefing paper: Selenium and Health*, 5. British Nutrition Foundation, London.
72. Combs, G.F. Jr & Combs, S.B. (1986) The biological availability of selenium in foods and feeds. In: *The Role of Selenium in Nutrition*, pp. 127–77. Academic Press, New York.
73. Frost, D.V. (1972) The two faces of selenium – can selenophobia be cured? In: *Critical Reviews of Toxicology* (ed. D. Hemphill), pp. 467–514. CRC Press, Boca Raton, Florida.
74. Pennington, J.A.T. & Young, B. (1990) Iron, zinc, copper, manganese, selenium and iodine in foods from the United States Total Diet Study. *Journal of Food Composition and Analysis*, **3**, 166–84.
75. Molnar, J., MacPherson, A., Barclay, I. & Molnar, P. (1995) Selenium content of convenience and fast foods in Ayrshire, Scotland. *International Journal of Food Science and Nutrition*, **46**, 343–52.
76. Molnar, J., MacPherson, A., Barclay, I. & Molnar, P. (1995) Selenium content of convenience and fast foods in Ayrshire, Scotland. *International Journal of Food Science and Nutrition*, **46**, 343–52.
77. Combs, G.F. (2001) Selenium in global food systems. *British Journal of Nutrition*, **85**, 517–47.
78. Tinggi, U., Patterson, C. & Reilly, C. (2001) Selenium levels in cow's milk from different regions of Australia. *International Journal of Food Sciences and Nutrition*, **52**, 43–51.
79. Barclay, M.N.I., MacPherson, A. & Dixon, J. (1995) Selenium content of a range of UK foods. *Journal of Food Composition and Analysis*, **8**, 307–18.
80. Olson, O.E. & Palmer, I.S. (1984) Selenium in foods purchased or produced in South Dakota. *Journal of Food Science*, **49**, 446–52.
81. Combs, G.F. Jr & Combs, S.B. (1986) The biological availability of selenium in foods and feeds. In: *The Role of Selenium in Nutrition* (eds. G.F. Combs & S.B. Combs), pp. 127–77. Academic Press, New York.
82. United States Department of Agriculture (1999) Nutrient database for standard reference. http://www.nal.usda.gov/fnic/foodcomp/Data/SR13/sr13.html
83. Barclay, M.N.I., MacPherson, A. & Dixon, J. (1995) Selenium content of a range of UK foods. *Journal of Food Composition and Analysis*, **8**, 307–18.
84. Thorn, J., Robertson, J. & Buss, D.H. (1978) Trace nutrients. Selenium in British foods. *British Journal of Nutrition*, **39**, 391–6.
85. Reilly, C. (1999) Brazil nuts – a selenium supplement? *BNF Nutrition Bulletin*, **24**, 177–84.
86. Sector, C.L. & Lisk, D.J. (1989) Variation in the selenium content of individual Brazil nuts. *Journal of Food Safety*, **9**, 279–81.
87. Molnar, J., MacPherson, A. Barclay, I. & Molnar, P. (1995) Selenium content of convenience and fast foods in Ayreshire, Scotland. *International Journal of Food Science and Nutrition*, **46**, 343–52.
88. United States Department of Agriculture (1999) Nutrient database for standard reference. http://www.nal.usda.gov/fnic/foodcomp/Data/SR13/sr13.html
89. Department of Health (1991) *Dietary Reference Values for Food Energy and Nutrients for the United Kingdom*. Report on Health and Social Subjects 41. HMSO, London.
90. Combs, G.F. (2001) Selenium in global food systems. *British Journal of Nutrition*, **85**, 517–47.
91. Schubert, A., Holden, J.M. & Wolf, W.R. (1987) Selenium content of a core group of foods based on a critical evaluation of published analytical data. *Journal of the American Dietetic Association*, **87**, 285–92.

92. MAFF (1999) 1997 Total Diet Study – aluminium, arsenic, cadmium, chromium, copper, lead, mercury, nickel, selenium, tin and zinc *Food Surveillance Sheet, No. 191*. Ministry of Agriculture, Fisheries and Food, London.
93. Combs, G.F. (2001) Selenium in global food systems. *British Journal of Nutrition*, **85**, 517–47.
94. Rayman, M.P. (2000) The importance of selenium to human health. *Lancet*, **356**, 233–41.
95. Rayman, M.P. (1997) Dietary selenium: time to act. *British Medical Journal*, **314**, 387–8.
96. Rayman, M.P. (2000) The importance of selenium to human health. *Lancet*, **356**, 233–41.
97. Varo, P., Alfthan, G., Huttunen, J.K. & Aro, A. (1994) Nationwide selenium supplementation in Finland – effects on diet, blood and tissue levels, and health. In: *Selenium in Biology and Human Health* (ed. R.F. Burk), pp. 198–218. Springer Verlag, New York.
98. Reilly, C. (1996) 'Selenium supplementation – the Finnish experiment', *BNF Nutrition Bulletin*, **21**, 167–73.
99. Reilly, C. (1998) Selenium; a new entrant into the functional food arena. *Trends in Food Science and Technology*, **9**, 114–18.
100. Barclay, M.N.I., MacPherson, A. & Dixon, J. (1995) Selenium content of a range of UK foods. *Journal of Food Composition and Analysis*, **8**, 307–18.
101. Institute of Medicine, Food and Nutrition Board (2000) *Dietary Reference Intakes for Vitamin C, Vitamin E, Selenium and Carotenoid*. National Academy Press, Washington, DC.
102. Department of Health (1991) *Dietary Reference Values for Food Energy and Nutrients in the United Kingdom*, p. 191. HMSO, London.
103. Levander, O. (1997) Selenium requirements as discussed in the 1996 FAO/IAEA/WHO expert consultation on trace elements in human nutrition. *Biomedical and Environmental Sciences*, **10**, 214–19.
104. Levander, O.A. & Whanger, P.D. (1996) Deliberations and evaluations of the approaches, endpoints and paradigms for selenium and iodine dietary recommendations. *Journal of Nutrition*, **126**, 2427S–34S.
105. Agency for Toxic Substances and Disease Registry (1996) *Toxicological Profile for Selenium* (Update). US Department of Health and Human Services, Washington, DC.
106. Block, E. (1998) The organosulfur and organoselenium components of garlic and onion. In: *Phytochemicals – a New Paradigm* (eds. W.R. Bidlack, S.T. Omaye, M.S. Meskin & D. Jahner), pp. 129–41. Technomic Publishing Co., Lancaster, Pa.
107. Kerdel-Vergas, F. (1966) The depilatory and cytotoxic action of 'Coco de Mono' (*Lecythis ollaria*) and its relation to chronic selenosis. *Economic Botany*, **20**, 187–95.
108. Fairweather-Tait, S.J. (1997) Bioavailability of selenium. *European Journal of Clinical Nutrition*, **51**, S20–3.
109. Thomson, C. (1998) Selenium speciation in the human body. *Analyst*, **123**, 827–31.
110. Levander, O.A. (1972) Metabolic interrelationships and adaptations in selenium toxicity. *Annals of the New York Academy of Sciences*, **192**, 181–92.
111. Stadman, T.C. (1990) Selenium biochemistry. *Annual Review of Biochemistry*, **59**, 111–27.
112. BNF (2001) *Briefing paper: Selenium and Health*, 5. British Nutrition Foundation, London.
113. Patching, S.G. & Gardiner, P.H.E. (1999) Recent developments in selenium metabolism and chemical speciation: a review. *Journal of Trace Elements in Medicine and Biology*, **13**, 193–214.
114. Rotruck, J., Pope, A. & Ganther, H. (1973) Selenium: biochemical role as component of glutathione peroxidase. *Science*, **179**, 588–90.
115. Arteel, G.E., Briviba, K. & Sies, H. (1999) Protection against peroxynitrite. *Federation of European Biological Societies Letters*, **445**, 226–30.
116. Ursini, F., Heim, S., Keiss, M. *et al.* (1999) Dual function of the selenoprotein PHGPx during sperm maturation. *Science*, **285**, 1393–6.

117. Arthur, J.R., Beckett, G.J. & Mitchell, J.H. (1999) The interaction between selenium and iodine deficiencies in man and animals. *Nutrition Research Reviews*, **12**, 55–73.
118. Beck, M.A. (1999) Selenium and host defense towards viruses. *Proceedings of the Nutrition Society*, **58**, 707.
119. Beck, M.A. (1999) Selenium and host defense towards viruses. *Proceedings of the Nutrition Society*, **58**, 707.
120. Baum, M.K. & Shor-Posner, G. (1998) Micronutrient status in relationship to mortality in HIV-1 disease. *Nutrition Reviews*, **56**, S135–9.
121. Margolin, S. (1980) Mercury in marine seafood: the scientific medical margin of safety as a guide to the potential risk to public health. *World Review of Nutrition and Diet*, **34**, 182–265.
122. Hansen, J.C., Kromann, N. & Wulf, H.C. (1984) Selenium and its interrelation with mercury in wholeblood and hair in an East Greenland population. *Science of the Total Environment*, **38**, 33–40.
123. Yang, G., Wang, S., Zhou, R. & Sun, S. (1983) Endemic selenium intoxication of humans in China. *American Journal of Clinical Nutrition*, **37**, 872–81.
124. Jaffe, W.G., Ruphael, M.D., Mondragon, M.C. & Cuevas, M.A. (1972) Clinical and biochemical studies on schoolchildren from a seleniferous zone. *Archivos Latinoamericanos de Nutricion*, **22**, 595–611.
125. Civil, I.D.S. & McDonald, M.J.A. (1978) Acute selenium poisoning: case report. *New Zealand Medical Journal*, **87**, 345–6.
126. Baum, M.K. & Shor-Posner, G. (1998) Micronutrient status in relationship to mortality in HIV-1 disease. *Nutrition Reviews*, **56**, S135–9.
127. Barclay, M.N.I. MacPherson, A. & Dixon, J. (1995) Selenium content of a range of UK foods. *Journal of Food Composition and Analysis*, **8**, 307–18.
128. Poirier, K.A. (1994) Summary of the derivation of the reference dose for selenium. In: *Risk Assessment of Essential Elements* (eds W. Mertz, C.O. Abernathy & S.S. Olin), pp. 157–66. International Life Sciences Institute, Washington, DC.
129. Molnar, J., MacPherson, A., Barclay, I. & Molnar, P. (1995) Selenium content of convenience and fast foods in Ayrshire, Scotland. *International Journal of Food Science and Nutrition*, **46**, 343–52.
130. Clark, L.C. & Jacobs, E.T. (1998) Environmental selenium and cancer: risk or protection? *Cancer Epidemiology, Biomarkers and Prevention*, **7**, 847–8.
131. Versieck, J. & Cornelis, R. (1989) *Trace Elements in Human Plasma or Serum*, p. 76. CRC Press, Boca Raton, Fa.
132. Tinggi, U. (1999) Determination of selenium in meat products by hydride generation atomic absorption spectrophotometry. *Journal of the Association of Official Analytical Chemists International*, **82**, 364–7.
133. Oles, P.J. & Graham, W.M. (1991) Microwave acid digestion of various food matrices for nutrient determination by atomic absorption spectrometry. *Journal of the Association of Official Analytical Chemists International*, **74**, 812–14.
134. Olson, O.E. & Palmer, I.S. (1984) Selenium in foods purchased or produced in South Dakota. *Journal of Food Science*, **49**, 446–52.
135. Tinggi, U., Patterson, C.M. & Reilly, C. (1992) Determination of selenium in foodstuffs, using spectrofluorimetry and hydride generation atomic absorption spectrophotometry. *Journal of Food Composition and Analysis*, **5**, 269–80.
136. Federov, P.N., Ryabchuk, G.N. & Zverev, A.V. (1997) Comparison of hydride generation and graphite furnace atomic absorption spectrometry for the determination of arsenic in food. *Spectrochimica Acta*, **B52**, 1517–23.
137. Mindak, W.R. & Dolan, S.P. (1999) Determination of arsenic and selenium in food using a microwave digestion dry ash preparation and flow injection hydride generation atomic absorption spectrometry. *Journal of Food Composition and Analysis*, **12**, 111–22.
138. Crews, H.M., Lewis, D.J., Fairweather-Tait, S.J., Fox, T., Arthur, J.R. & Brown, K.M. (1997) The analyst's viewpoint with special reference to selenium. *Nutrition and Food Science*, **6**, 221–4.
139. Lobinski, R., Edmonds, J.S., Suzuki, K.T. & Uden, P.C. (2000) Species-selective deter-

mination of selenium compounds in biological materials. *Pure and Applied Chemistry*, **72**, 447–61.
140. Crews, H.M. (1998) Speciation of trace elements in foods, with special reference to cadmium and selenium: is it necessary? *Spectrochimica Acta*, **B53**, 213–19.
141. Brown, R.D. (1998) Selenium tellurium supply–demand relationship. *Proceedings of the 6th International Symposium on the Uses of Selenium and Tellurium*, 10–12 May 1998. Scottsdale, Arizona, pp. 13–23. Selenium–Tellurium Development Association, Grimbergen, Belgium.
142. Gross, N. (1997) A revolution in medical imaging. *Business Week*, 21 July, **3536**, 52–3.
143. Kron, T., Hansen, C.H. & Werner, E. (1991) Tellurium ingestion with foodstuffs. *Journal of Food Composition and Analysis*, **4**, 196–205.
144. Schroeder, H.A., Buckman, J. & Balassa, J. (1967) Abnormal trace elements in man: tellurium. *Journal of Chronic Diseases*, **20**, 147–61.
145. Glover, J.R. & Vouk, V. (1979) Tellurium. In: *Handbook on the Toxicology of Metals* (eds. L. Friberg, G.F. Nordbergy & V. Vouk), Elsevier/North Holland, Amsterdam.
146. ICRP (1979) *Limits for Intakes of Radionuclides by Workers*, International Commission on Radiological Protection Publication, **30**. Pergamon, Oxford.
147. Beliles, R.P. (1975) Metals. In: *Toxicology: the Basic Science of Poisons* (ed. L.J. Casarett), p. 493. Macmillan, New York.
148. Demeio, R.H. & Henriques, F.C. (1947) Excretion and distibution in tissues studied with a radioactive isotope. *Journal of Biological Chemistry*, **169**, 609–23.
149. Nielsen, F.H. (1986) Other elements: Sb, Ba, B, Br, Cs, Ge, Rb, Ag, Sr, Sn, Ti, Zr, Be, Bi, Ga, Au, In, Nb, Sc, Te, Tl, W. In: *Trace Elements in Human and Animal Nutrition* (ed. W. Mertz), Vol. 2, pp. 452–3. Academic Press, New York and London.
150. Tsalev, D.S. (1984) *Atomic Absorption Spectrometry in Occupational and Environmental Health Practice*, Vol. II, p. 15. CRC Press, Boca Raton, Florida.
151. Department of Health (1991) *Dietary Reference Values for Food Energy and Nutrients in the United Kingdom*, p. 191. HMSO, London.
152. Nielsen, F.H. (1988) Boron – an overlooked element of potential nutritional importance. *Nutrition Today*, **2**, 4–7.
153. Woods, W.G. (1994) An introduction to boron – history, sources, uses and chemistry. *Environmental Health Perspectives*, **102**, 5–11.
154. Nielsen, F.H. (1994) Ultratrace minerals. In: *Modern Nutrition in Health and Disease*, (eds. M.E. Shils, J.A. Olsen & M. Shike), 8th edn., Vol. 1, pp. 272–4. Lea & Fabiger, Philadelphia, Pa.
155. Nielsen, F.H. (1994) Ultratrace minerals. In: *Modern Nutrition in Health and Disease*, (eds. M.E. Shils, J.A. Olsen & M. Shike), 8th edn., Vol. 1, pp. 272–4. Lea & Fabiger, Philadelphia, Pa.
156. Rainey, C.J., Nyquist, L.A., Christensen, R.E., Strong, P.L., Culver, B.D. & Coughlin, J.R. (1999) Daily boron intake from the American diet. *Journal of the American Dietetics Association*, **99**, 335–40.
157. Murray, F.J. (1995) A human health risk assessment of boron (boric acid and borax) in drinking water. *Regulatory Toxicology and Pharmacology*, **22**, 221–30.
158. Murray, F.J. (1995) A human health risk assessment of boron (boric acid and borax) in drinking water. *Regulatory Toxicology and Pharmacology*, **22**, 221–30.
159. Ysart, G., Miller, P., Crews, H. *et al.* (1999) Dietary exposure estimates of 30 elements from the UK Total Diet Study. *Food Additives and Contaminants*, **16**, 391–403.
160. Clarke, W.B. & Gibson, R. (1988) Lithium, boron and nitrogen in 1-day composites and a mixed-diet standard. *Journal of Food Composition and Analysis*, **1**, 209–20.
161. Clarke, W.B. & Gibson, R. (1988) Lithium, boron and nitrogen in 1-day composites and a mixed-diet standard. *Journal of Food Composition and Analysis*, **1**, 209–20.
162. Coughlin, J.R. (1996) Inorganic borates: chemistry, human exposure, and health and regulatory guidelines. *Journal of Trace Elements in Experimental Medicine*, **9**, 137–51.
163. Draper, A., Lewis, J., Malhotra, N. & Wheeler, E. (1993) The energy and nutrient intakes of different types of vegetarians: a case for supplements? *British Journal of Nutrition*, **69**, 3–19.

164. Murray, F.J. (1995) A human health risk assessment of boron (boric acid and borax) in drinking water. *Regulatory Toxicology and Pharmacology*, **22**, 221–30.
165. Murray, F.J. (1996) Issues in boron risk assessment: pivotal study, uncertainty factors, and ADIs. *Journal of Trace Elements in Experimental Medicine*, **9**, 231–43.
166. Nielsen, F.H. (1987) Other elements. In: *Trace Elements in Human and Animal Nutrition* (ed. W. Mertz), pp. 415–63. Academic Press, San Diego, Ca.
167. WHO (1996) Boron. In: *Trace Elements in Human Nutrition and Health. Report of the WHO Expert Committee on Trace Elements in Human Nutrition*, pp. 175–9. World Health Organization, Geneva.
168. Hunt, C.D. & Stoecker, B.J. (1996) Deliberations and evaluations of the approaches, endpoints and paradigms for boron, chromium and fluoride dietary recommendations. *Journal of Nutrition*, **126**, 2441S–2451S.
169. Hunt, D. (1996) Biochemical effects of physiological amounts of dietary boron. *Journal of Trace Elements in Experimental Medicine*, **9**, 185–213.
170. Nielsen, F.H. (1996) Evidence for the nutritional essentiality of boron. *Journal of Trace Elements in Experimental Medicine*, **9**, 215–29.
171. Hunt, D. (1996) Biochemical effects of physiological amounts of dietary boron. *Journal of Trace Elements in Experimental Medicine*, **9**, 185–213.
172. Beattie, J.H. & Peace, H.S. (1993) The influence of a low-boron diet and boron supplementation on bone, major mineral and sex steroid metabolism in postmenopausal women. *British Journal of Nutrition*, **69**, 871–84.
173. Murray, F.J. (1995) A human health risk assessment of boron (boric acid and borax) in drinking water. *Regulatory Toxicology and Pharmacology*, **22**, 221–30.
174. Hunt, C.D. & Shuler, T.R. (1989) Open-vessel, wet-ash, low-temperature digestion of biological materials for inductively coupled argon plasma spectroscopy (ICAP) analysis of boron and other elements. *Journal of Micronutrient Analysis*, **6**, 161–74.
175. Ysart, G., Miller, P., Crews, H. *et al.* (1999) Dietary exposure estimates of 30 elements from the UK Total Diet Study. *Food Additives and Contaminants*, **16**, 391–403.
176. Ysart, G., Miller, P. Crews, H. *et al.* (1999) Dietary exposure estimates of 30 elements from the UK Total Diet Study. *Food Additives and Contaminants*, **16**, 391–403.

Chapter 11
The new metal contaminants

In 1988 the distinguished Canadian environmentalist Jerome Nriagu, in a paper on the problems posed by industrialisation to the environment, made the following thought-provoking observation:

> *Metals and their compounds are indispensable to the safety and economy of most nations and have been key factors in the liberation of modern civilization from hunger, disease and discomfort. Few, if any, of the metals known to mankind has not found some application in industry, and the number of commercial uses continues to grow with the development of modern science and technology. Inevitably, each industrial process generates wastes which must be discharged into the environment, along with the ever growing list of new metallic compounds*[1].

This chapter will deal with some of those new metallic compounds for which industry continues to find novel applications. It will examine the significance of some of these metals for human health, especially in relation to food contamination. Many were formerly unknown to the majority of people and seldom encountered except by the specialist scientist and metallurgist. In some cases little is known about them from the human health point of view, but an attempt will be made to at least flag potential problems which could arise with increasing use in the future. Since it is not be possible to cover all the metals that are actually used commercially or are still at the stage of experimental application, a selection of those about which most is known will be made. They will be grouped under three different headings

(1) the radioactive metals
(2) the catalytic metals
(3) the electronic metals.

11.1 The radioactive metals

In 1896 the French scientist Henri Becquerel discovered that the element uranium emitted rays which could affect the emulsion on a photographic plate. A decade and a half later Madame Marie Curie and her husband Pierre showed that atoms of uranium and radium undergo spontaneous disintegration to form atoms of other elements and in the process emit penetrating rays. This spontaneous decomposition of atomic nuclei is what we call radioactivity. In 1903, Becquerel and the two Curies shared the Nobel Prize in physics for their discovery.

Several radioactive elements and isotopes of elements occur in nature, and people have been exposed to their penetrating rays, usually at very low levels, ever since the human race first evolved. But during recent decades a new form of radioactivity, artificially produced, has been introduced into the environment. Atomic weapons and nuclear energy production, as well as a variety of applications in medicine and industry, have greatly added to the world's radioactive load.

The penetrating rays that are emitted by radioactive elements are commonly referred to as ionising radiation, because of the effects they produce. The rays are high-velocity particles emitted from the nuclei of atoms. Two principal kinds of particles are produced: α-particles, which consist of two neutrons and two protons (identical to the helium nucleus); and β-particles, which are high-velocity electrons. These emissions may be accompanied by γ-rays. These are electromagnetic rays, like X-rays, and are more penetrating than either α- or β-particles.

When a nucleus loses an α-particle, its atomic mass is decreased by four units and its atomic number by two. The loss of two protons from the nucleus is accompanied by the loss of two electrons, thus maintaining electrical balance. When a β-particle is lost, there is no change in the mass of the nucleus, but there is an increase of one in the net positive charge, so that the atomic number increases by one unit. The emission of γ-rays has no effect either on the atomic number or the atomic weight of an element, since these rays possess neither mass nor charge. All elements with atomic numbers greater than that of bismuth (83) have one or more isotopes, which are naturally radioactive. A few elements of lower atomic number, such as potassium and rubidium, also have naturally occurring radioactive isotopes. Artificially produced radioisotopes of many atoms are also known.

It may be useful to clarify some of the terms used with regard to isotopes and radioactive elements, since they are sometimes confused in non-specialist literature. *Isotope* refers to a type of atom, while *nuclide* refers to the nucleus of a specific isotope characterised by its atomic number, mass number and energy. Radionuclides are unstable nuclides, which emit ionising radiation. The rates at which disintegration of radionuclides takes place are not the same for all elements. The term *half-life* is used for the time taken for the activity of a radionuclide to decay to half its original level, that is, for half the nuclei to disintegrate. The SI unit used for radioactivity is the *becquerel* (Bq), which is defined as one radioactive disintegration per second.

11.1.1 Radioisotopes

There are three different types of natural radionuclides. Some, such as uranium (U) and thorium (Th), have always been with us. They have extremely long half-lives of millions of years. A second group, daughter elements, is produced by the disintegration of members of the first group. Thus radium-226 (Ra-226) is produced by the disintegration of U-238. Ra, in its turn, produces a daughter of its own, radon (Rn-222), which is itself unstable, producing lead (Pb-210), and this, in its turn, produces a daughter, polonium (Po-210). The third group of radionuclides is constantly being formed by the action of cosmic rays on elements in the atmosphere. One of the most important of these is carbon-14, which is made from the nucleus of nitrogen. These naturally occurring radionuclides account for the 'natural background' in the presence of which we have evolved[2]. In recent years man-made additions have more than doubled the levels of radiation to which we are exposed. These have come from nuclear weapons, emissions of the nuclear industry and the increasing use of radioactive materials in medicine and industry.

When a nuclear bomb is exploded, or a nuclear reactor operated, a complex mixture of many different radionuclides is produced by fission of the fuel materials. Included are the long-lived heavy elements, such as U-235 and plutonium (Pu-239), as well as caesium (Cs-137), half-life 30 years, strontium (Sr-90), half-life 28 years, cobalt (Co-60), half-life 5.27 years, ruthenium (Ru-106), half-life one year and iodine (I-131), half-life eight days. In an explosion, many other radionuclides are also produced, not directly from the nuclear fuel, but as a result of the action of neutrons from the fission of other elements present in the bomb or reactor's casing and in the immediate environment of the explosion. The 'activation products' can include zinc-65 (half-life 245 days) and C-14 (half-life 5730 years). The radioisotopes used in medicine, industry and research are also activation products, but made by 'neutron bombardment' of selected target atoms in specially designed nuclear reactors.

11.1.2 Radioactive contamination of foods

Man-made radionuclides from a variety of sources, such as explosions of nuclear weapons, release of waste into air and water from nuclear plants, accidents in nuclear power stations and in nuclear submarines, have, during the past half century, contributed to nuclear contamination of human foods. Radionuclides released into the atmosphere will eventually fall to earth where they can be absorbed by plants and consumed by grazing animals. Radionuclides discharged into fresh or marine waters can result in contamination of fish and other aquatic organisms. Through these pathways, they can enter the human food chain. Table 11.1 lists the most important of the artificially produced radionuclides with their half-lives, the type of radiation, and the principal food pathways they can follow.

Contamination of food with radioactive metals is a twofold hazard to health, for, in addition to the chemical toxicity which might be expected from ingestion of any poisonous metal, there is the additional hazard of radiological toxicity. However, though only specific chemical forms of non-radioactive contaminants are normally toxic, any intake of radioactivity increases an individual's exposure to ionising

Table 11.1 Food pathways of radioactivity to humans.

Nuclide	Half-life	Radiation	Pathway
H-3	12.4 y	β	all foods
C-14	5730 y	β	all foods
S-35	87.4 d	β	milk, plant foods
Sr-90	28.5 y	β	milk, molluscs, meat
Tc-99	0.22×10^6 y	β	vegetables, crustaceans, offal
Co-60	5.3 y	$\beta\gamma$	molluscs
Ru-106	1 y	$\beta\gamma$	plants, molluscs, offal
I-125	60.1 d	$\beta\gamma$	milk
I-129	15.5×10^6 y	$\beta\gamma$	milk
I-131	8.0 d	$\beta\gamma$	milk
Cs-134	2.1 y	$\beta\gamma$	milk, fish, shellfish, offal
Cs-137	30.1 y	$\beta\gamma$	milk, fish, shellfish, offal
Pu-238	87.74 y	α	offal, molluscs
Pu-239	24 000 y	α	offal, molluscs
Am-241	432 y	$\alpha\gamma$	offal, molluscs
Np-237	2.2×10^6 y	$\alpha\gamma$	offal, molluscs

Adapted from Ministry of Agriculture, Fisheries and Food (1994) *Radionuclides in Food*. Food Surveillance Paper No. 43. HMSO, London.

radiation. Thus, exposure to uranium and its compounds is known to cause renal toxicity (the main chemical effect), as well increasing the risk of developing cancer (an effect of ingestion of α-particle-emitting uranium).

Both types of effect have been observed in a small number of US Gulf War veterans who were exposed to depleted uranium (DU) in armour-piercing projectiles used in that conflict[3]. Though there is concern about the chemical effects of DU in shrapnel fragments still lodged in body tissues of these soldiers[4], it is the radiological risk that normally attracts most attention from health and environmental authorities as well as from national and world bodies, especially the International Commission for Radiological Protection (ICRP)[5]. It is also the reason why permissible levels of radionuclides in foods and the environment are generally much less than those for chemical contaminants[6].

Not all radionuclides behave in the same way after they are ingested in food. Some are not absorbed from the GI tract to any significant extent and mainly pass through the gut and are excreted. These include the actinides, plutonium-238 and -239, americium (Am-241) and neptunium (Np-237). Others, such as caesium (Cs-137), are rapidly and largely absorbed. Once absorbed, the behaviour of the radionuclide will depend on its chemical form. Thus Cs-137, which is similar in chemistry to potassium, accumulates, as does potassium, in soft tissue, including muscle. The strontium nuclide Sr-90, which can substitute for calcium, to which it is chemically very similar, accumulates in the skeleton. The outcome of accumulation of any radionuclide in such ways will depend, ultimately, on what sort of radioactivity it produces, and its persistence, which in its turn depends on its physical and biological half-life.

Another factor which has to be considered in relation to radioactive contamination of food is that the behaviour of radionuclides, both in foods and in the human body, can be affected by the presence of chemical analogues. An example of this is the effect of calcium on radioactive strontium. Cows grazing on Sr-90-contaminated fodder have been shown to be able to discriminate against the strontium in favour of calcium in their milk. As a consequence, the Sr-90/calcium ratio in human bone varies, depending on whether calcium is mostly obtained from milk and dairy products or from vegetables and cereals. Another example of interaction between a radionuclide and a stable isotope is that of iodine-131 and stable iodine. In a widely used and effective countermeasure following nuclear accidents, administration of stable iodine, in the form of iodide or iodate, is used to block uptake of I-131, especially in children who have been exposed to the radionuclide[7].

In the light of the above considerations, when assessing the impact of radionuclide-contaminated food on consumers the following should be taken into account:

(1) the pathway of the radioactivity from source to the human diet
(2) the chemical form of the contaminant
(3) its uptake from ingested food
(4) its physical and biological half-lives and the type of radiation emitted
(5) the presence of competing stable analogues.

In addition, when considering the question of the health effects of consumption of a particular food contaminated with a radioactive metal, it must be recognised that the radiation dose from any such food will depend on four factors:

(1) the quantity of the food eaten over a period of time
(2) the level of radioactivity in the food consumed
(3) the type of radiation (α, β or γ) emitted by the radionuclide(s) in the food
(4) the chemical properties of the radionuclides, which determine whether they are

uniformly distributed throughout the body or are concentrated in particular tissues.

The effects of these factors are summarised in the concept of a *Dose Conversion Factor*. The DCF relates the amount of radioactivity consumed (in Bq) to the resulting *Dose Equivalent*. This is expressed in SI units as *sieverts* (Sv), and is the quantity obtained by multiplying the absorbed dose by a factor that allows for the different effectiveness of the different types of ionising radiation in harming tissues. This is usually taken as 1 for β- and γ-radiation and 20 for α-particles.

In order to determine the radiation dose that an individual will receive from a particular contaminated food, it is necessary to know the level of activity of each radionuclide it contains as well as the actual amount of the food consumed. Dietary surveys can provide the latter information. The identity and amounts of the nuclides present in the food can be determined by the use of a high-resolution gamma spectrometer or other analytical procedures. The health effects of radiation are described as either *non-stochastic*, in which the severity of the effect is related to the magnitude of the dose received, or *stochastic*, in which the probability of an effect occurring, but not its severity, is related to the dose received[9]. Non-stochastic effects, which include various degrees of radiation sickness, only occur with relatively large doses. Stochastic effects may be cancer or a genetic effect, which is not immediate and for which there is no threshold dose below which it does not occur.

11.1.2.1 Food contamination caused by the Chernobyl accident

The reality of radioactive contamination of food was brought home to the world by the Chernobyl accident, which occurred on 26 April 1986[10]. A reactor in the nuclear power station at Chernobyl, some 100 km north of Kiev in the Ukraine, then part of the USSR, was accidentally accelerated in a few seconds from low to more than a hundred times full power. The result was a massive explosion that lifted the roof off the building and spewed out debris on the surrounding area. A radioactive cloud was produced which was fed by emissions from the damaged plant for several days. The cloud was blown on the wind over much of Europe, as well as east towards Japan and China. It contained a wide range of fission products, many of them highly radioactive, the most prominent of which were iodine-131, caesium-137 and tellurium-132.

The impact of the Chernobyl nuclear explosion on radionuclides in UK foods is illustrated by the average level of Cs-137 detected in Scottish milk. In 1985 this was 0.1 Bq/litre. After Chernobyl it had risen to 7.8 Bq/litre. In the following year it had decreased to 5.4 Bq/litre, and by 1990 to 0.2 Bq/litre[11]. Similar rises, and subsequent falls, were detected in several other foodstuffs, including mutton and lamb.

Heavy rain caused deposition of large amounts of radioactivity as the cloud passed over Scandinavia and on across Europe. It crossed the English Channel six days after the explosion and moved across England, Wales and Scotland and on to Ireland before disappearing out over the Atlantic Ocean. Monitoring by health authorities in the countries over which the radioactive cloud passed found levels approaching and, in some cases, exceeding ICRP and other official intervention levels in some foods. In France, Germany, Greece and other countries, certain types of vegetables, especially spinach, were withdrawn from sale. In Finland the sale of reindeer meat was prohibited. In the UK restrictions were imposed on the slaughter of lambs produced on contaminated grass in parts of Wales, the Lake District and Scotland.

Now, more than a decade and a half later, the effects of the Chernobyl disaster still persist in the environment. Though the acute effects have appreciably decreased, authorities in many countries continue to monitor food and the environment for possible delayed effects of the nuclear accident. Cs-137, one of the main radionuclide carried by the winds to the UK from the Ukraine in April 1986, still persists on the hillsides of Wales and is detectable at more than background level in sheep reared on its upland farms[12]. Contamination of grazing sheep by Cs-137 still occurs in pasture land in southern Finland which was in the path of the radioactive cloud released at Chernobyl[12]. In Greece, which was less seriously affected than some other countries, there is still concern at its possible impact on food, and monitoring continues[14].

11.1.2.2 Food contamination and the nuclear industry

Nuclear power stations, nuclear fuel production and reprocessing, production and use of medical and industrial isotopes, and scientific research all produce radioactive waste. Its discharge to the environment, or storage in various forms and holding areas, must be carefully controlled if environmental contamination is to be avoided or at least kept to a minimum. Under government regulations in the UK, as in most other countries, only solid waste with very low levels of activity may be disposed of with other industrial or household refuse in local authority tips; if the waste has a higher level of radioactivity it may be buried in special landfill sites. Waste with the highest levels of contamination must be stored, either at the site of production or elsewhere in special storage facilities. None of these methods of disposing of solid radioactive waste should, normally, result in contamination of food. The same can be said of gaseous waste which is normally discharged from high stacks to ensure that any radioactivity emitted is diluted and dispersed before it is carried to the ground. Though this method should also minimise the likelihood, it may in some circumstances result in contamination of food crops, as can also occur when liquid waste is dispersed.

Liquid discharges from nuclear industries are normally made in the UK, and in other countries where a convenient coastal location is available, through long pipelines into the sea. This is done to ensure that any radioactivity is greatly diluted and widely dispersed. The success of this method depends on the efficiency of in-house controls of levels of radioactivity in the discharges. Where these are inadequate, or accidental spills occur, as has been the case at some nuclear installations, unacceptably high levels of contamination of marine organisms can occur.

Data published by the MAFF in the UK for levels of Cs-137 in plaice caught offshore at Sellafield in Cumbria[15] indicate that spills have occurred in the past. In 1963 levels in the fish were 19Bq/kg. Four years later they had risen to 26 Bq/kg. Over the following years they continued to rise, with a jump in 1975 to 1100 Bq/kg. Levels peaked in the following year at 1500 Bq/kg. Since then they have fallen to 940 in 1977 and 15 Bq/kg in 1991, indicating an improvement in plant hygiene and an absence of further uncontained emissions. Nevertheless, concern continues to be expressed by environmentalists and others on both sides of the Irish Sea at the possible effects of emissions from the Sellafield plant on fish and other marine foods.

Sellafield is not, in fact, the only source of radioactivity in British coastal waters. Cs-137 has been detected in marine organisms at every site examined around the whole coastline[16]. There is clearly more than one source of the radionuclide, though Sellafield is the most significant[17]. For this reason, a strict surveillance programme is in place to monitor levels of intake of radionuclides in foods derived from within close proximity of the plant where contamination by discharges of waste may have occurred. This is done under the Terrestrial Radioactivity Monitoring Programme (TRAMP) which was set up by MAFF in the early 1990s[18].

Results of the 1995 programme indicate that, in spite of waste discharge from the plant into the atmosphere, the activity of Cs-137 in the diets of both adults and children in the vicinity of the Sellafield plant was similar to that of a control group in a non-contaminated area. Moreover, levels of Cs-137 intake found in the TRAMP study were over an order of magnitude lower than those that had been recorded in Cumbria nine years earlier. However, activity of another radioactive metal, Sr-90, was relatively high in children's diets at Sellafield, compared to adults and the control group. However, overall, the TRAMP results showed that exposure from the consumption of foodstuffs collected near Sellafield did not exceed the recommended limit of 1.0 Sv/year[19].

11.2 The catalytic metals

Many countries now require that internal combustion engines of vehicles be equipped with catalytic converters to control levels of exhaust emissions. These converters usually contain 1–3 grams of two different platinum metals (mainly platinum, palladium and rhodium), supported on a ceramic honeycomb monolith[20]. Their purpose is to control emissions of nitrogen oxides, as well as carbon dioxide and hydrocarbons. During use of the converters, there is abrasion of the surface of the catalyst and emission of platinum and palladium occurs. It has been estimated that platinum is emitted at the rate of 1–3 μg/mile by cars equipped with converters[21].

11.2.1 Platinum metals in the environment

The requirement for catalytic converters in all new automobiles has had a considerable impact on platinum production and use worldwide, and, as a consequence, has raised concerns regarding possible effects on the environment and the diet. In the US, where converters have been a requirement since 1975, use of platinum metals approximately doubled over the following decade, with the automobile industry using as much of these metals as all other industries in the country combined. There appears to have been a corresponding increase in levels of platinum metals in dust on the sides of heavily-trafficked streets, far higher than the expected natural abundance of the metals[22].

Similar findings have been made in many other industrialised countries. A German study found that platinum levels in motorway soils ranged from 15 to 30 ng/g, considerably higher than background concentrations (0.09–4 ng/g)[23]. There is growing concern that some of this emission can settle on food crops and contaminate the human diet. In Italy, analyses of roadside dust in Rome for platinum, palladium and rhodium have pointed to the need to continue to monitor urban sites in view of the increasing number of cars using catalytic converters[24]. In Australia, where increasing levels of platinum metals have been detected on roadside vegetation, it has been claimed that as a result, diet is now a major pathway of platinum into body[25].

11.2.2 Platinum metals in food and diets

Table 11.2 lists the concentration of both platinum and palladium determined in UK foods in the 1994 TDS[26]. This was the first time that the platinum metals were included in a UK TDS. In addition to the two main metals of the group, iridum, rhodium and ruthenium levels in foods were also determined. Concentrations of all three fell to between < 0.0001 and 0.002 mg/kg, with one exception, a value of

Table 11.2 Platinum and palladium in UK foods.

Food group	Mean concentration (mg/kg, fresh wt) Platinum	Palladium
Bread	0.0001	0.002
Miscellaneous cereals	0.0001	0.0009
Carcass meat	0.0001	0.0004
Offal	< 0.0001	0.002
Meat products	< 0.0001	0.0006
Poultry	0.0001	< 0.0003
Fish	< 0.0001	0.002
Oils and fats	0.0002	0.0004
Eggs	< 0.0001	0.0004
Sugar and preserves	0.0001	0.0005
Green vegetables	0.0001	0.0006
Potatoes	< 0.0001	0.0005
Other vegetables	0.0001	0.0005
Canned vegetables	< 0.0001	0.0004
Fresh fruit	< 0.0001	0.0004
Fruit products	< 0.0001	0.0005
Beverages	<0.0001	0.0004
Milk	< 0.0001	< 0.0003
Dairy products	0.0001	0.0004
Nuts	0.0001	0.003

Adapted from Ysart, G., Miller, P., Crews, H. *et al.* (1999) Dietary exposure estimates of 30 elements from the UK Total Diet Study. *Food Additives and Contaminants*, **16**, 391–403.

0.004 mg/kg for ruthenium in nuts. As the authors of the report noted, the platinum group of elements are notoriously difficult to analyse, as the concentrations in foods are generally close to the limits of detection. They can also be prone to interferences when measured by ICP-MS. Thus, while concentrations were found to be very low in all the foods tested, the reported values should be regarded as 'worst case' levels.

Mean and upper level dietary exposure levels in UK adults were estimated to be 0.0002 and 0.0003 mg/day for platinum, and 0.0002 and 0.002 mg/day for palladium. Mean intakes of other platinum metals were estimated to be: iridium 0.002 (upper range 0.003), rhodium 0.0002 (0.0004), and ruthenium 0.004 (0.006) mg/day. An Australian group found that platinum intake in that country was, on average, 1.44 μg/day. Information on levels of intake of the platinum metals in diets elsewhere is not readily available, but they are likely to be of the same order, about 2 μg/day, in countries with a similar transport infrastructure to Britain. Intakes will, undoubtedly, increase in coming years as older cars without catalytic converters are phased out and increasing numbers of new cars are produced[27].

11.2.3 Health implications of platinum metals in the diet

Platinum metal compounds are a recognised hazard for workers occupationally exposed to them[28]. The metal itself appears to be biologically inert. Halogenated compounds, such as sodium tetrachloroplatinate, can cause allergenic reactions, with eczema, skin lesions, wheezing and other signs of mucous membrane inflammation.

Non-halogenated compounds as well as neutral complexes, including the anticancer drug cisplatin $Pt(Cl)_2(NH_3)_2$), are not allergenic.

There is little information about the biological effects of platinum metals in foods. Metallic platinum is not, apparently, a problem, because of its insolubility and inert nature. However, it has been shown that platinum in road dusts can be solubilised and enter the soil via rain run-off into sewage and irrigation water. Platinum in soil can be taken up by plants[29]. There is evidence that platinum in plant and animal tissues is bioavailable. However, there is no evidence that platinum compounds entering the human food chain in this way are a health hazard for the general public. This may not be the situation in the future, as emissions from catalytic converters continue to increase. Certain platinum compounds which could be dispersed into the environment in this way are known to be cytotoxic and have mutagenic and carcinogenic effects[30]. A chlorinated compound of platinum (H_2PtCl_6) has been shown to cause acute toxicity in fish, with lysis of mucosal cells of the intestine[31]. Even though the anticancer drug cisplatin is non-allergenic, its medical use is not without hazard[32]. It is for such reasons that authorities in the UK and elsewhere maintain a watching brief on this relatively new source of pollution.

11.3 The electronic metals

Every year millions of computers, ranging in size from handheld play stations through home PCs to mainframes and large pieces of telecommunication systems, are discarded throughout the world. In the UK alone, approximately six million PCs were sold in 2000. Estimates of sales elsewhere are more than ten times that figure. All of this equipment, as well as the multitudes of phones, television sets and other items which depend on their electronic brains to function, contains a wide range of metals, some very rare and costly, in their circuit boards and other parts.

Until recently much of this equipment ended up in landfills or other dumps, with little recycled or re-used. But that is now changing. A European Union Directive on Waste and Electronic Electrical Equipment (WEEE) will before long require that at least 75% of each computer and similar item must be recycled. In addition, there is growing awareness that dumping is a highly uneconomic way of disposing of these precious metal-rich items. Money can be made out of recovering their contents. Though the electronic metal recovery industry is still in its infancy, it is growing in some countries. In the UK one plant in Wales deals with about 100 tonnes of circuit boards each month. In one year the plant extracted about 500 kg of gold, one tonne of silver and 150 kg of palladium from scrapped electronic equipment[33].

Circuit boards and related items contain several other metals besides the 'old metals', gold and silver, and the more recent palladium. These 'new metals' include tantalum, gallium and several others whose electrical properties have made them ideal for the electronic age. Until quite recently these were rare metals, seldom encountered except in the metallurgy laboratory and certainly never in the normal household. Some of them were, in fact, given the name 'rare earths' by chemists, and that is what they were. Unlike the common metals such as iron and zinc, and even the less common lead and mercury, the human body never had to cope with them, except in the most minute background amounts, in the everyday environment or diet.

But now almost every household in the technologically advanced world will have equipment in it that contains one or other of these formerly rare metals. In tips and dumps, on land and sea, there is obsolete equipment and parts from which germanium, tantalum and other metals can potentially leach into the environment. They have been detected in soil and in sewage sludge from which they can enter the human food chain. How much this has occurred is only now becoming quantified, as government authorities become aware of the problems that could be caused by this new environmental challenge.

Little is, as yet, known about the full extent of this problem. Data on levels of these new electronic metals in the environment and in human diets have only recently begun to become available, and are still far from complete. Here we will look at what is known about some of these metals, not to provide a full picture but to draw attention to what could become a major health problem and a challenge for the food industry.

11.3.1 Germanium

Germanium (symbol Ge, atomic number 32, atomic weight 72.59) is a member of Group 14/IVA of the periodic table of the elements, along with carbon, silicon, tin and lead. It is a grey-white metal that retains its lustre in air and is resistant to acids and alkalis. Its chemistry has similarities to that of silicon. Its forms two oxides, GeO and GeO_2, as well as hydrides and several organometallic compounds. It has important physical properties which account for its growing industrial uses. Its high refractive index makes it a natural 50% beamsplitter. In addition, it can act as a rectifier when alloyed with small amounts of other metals. Germanium occurs widely in the Earth's crust, though at low levels. It is found in more concentrated form, along with silver, in the mineral argyrodite, as well as in sulphide ores of lead, zinc and other metals. It is obtained as a by-product of zinc refining and to a limited extent from waste products of the coal and coke industry.

Until recently the main use of germanium was in the manufacture of special optical glass used in optical filters for thermal imaging and similar applications. A rapidly growing use is in the manufacture of transistors and other electronic devices, as well as in radar. Small amounts of the metal and its organic compounds are used in the health industries, especially as dietary supplements.

Germanium can be detected, at very low levels of concentration, in all foods. A UK study has reported that levels in all food groups ranged from <0.002 to 0.004 mg/kg[34]. These are lower than levels reported in earlier studies in the US in which some foods were found to contain more than 2 mg/kg of germanium. The differences can most probably be accounted for by better analytical procedures used in the UK study (ICP-MS) than in the 1967 US investigation[35]. Dietary intakes of germanium in the UK are estimated to be 0.004 mg/day, for adults, with an upper range of 0.007 mg/day[36]. These are considerably lower than the approximately 1.5 mg/day calculated for US adults in the earlier study[37].

Little is known about the metabolism of germanium ingested in the diet. Absorption from the GI tract is believed to be rapid and complete, followed by rapid excretion, mainly in urine[38]. There is no evidence that germanium has any functional role in the body. However, in recent years it has been claimed that certain organic compounds of germanium may have some immune enhancing, anticancer and antiviral properties. The evidence for this is not clear and there are no reports of results of well designed clinical trials relating to the medical use of germanium. Nevertheless, a substance known as Ge132, as well as other organic germanium preparations, are promoted as dietary supplements by practitioners of alternative medicine on the Internet[39].

Germanium has been shown to produce toxic effects, including degenerative changes in liver and kidney, in experimental animals[40]. The element does not appear to be toxic to humans at levels of intake found in normal diets. However, consumption of doses of the order of 50 to 250 mg/day over periods of 4–18 months can cause serious harm to health and even death[41].

11.3.2 *Tantalum*

Tantalum (symbol Ta, atomic number 73, atomic weight 180.9) is a transition element, along with vanadium and niobium in Group 5/VB of the periodic table. Though it was isolated and named by Berzelius nearly two centuries ago in 1820, it has, until quite recently, remained an almost unknown rarity of the chemical world. That situation is now rapidly changing as tantalum has become recognised as a key metal in the electronics industry, especially as a component of mobile phones[42]. Tantalum is a white, ductile, malleable metal. It is resistant to attack by most strong acids. It burns in air to produce tantalum pentoxide (Ta_2O_5), which can combine with metallic oxides and hydroxides to form compounds called tantalites.

Tantalum is fairly widely distributed in the Earth's crust in small concentrations. Its main ore is tantalite ($FeTa_2O_6$), which usually occurs along with the corresponding ore of niobium, columbite ($FeNb_2O_6$). Major deposits occur in Australia and Scandanavia, with lesser amounts in the US, Brazil and South Africa. The two metals, tantalum and niobium, are extracted together from their ores by fusion with potassium fluoride and are subsequently separated by solvent extraction or selective crystallisation.

Tantalum has a number of important industrial uses. Most of the world's production is used for capacitors in electronic circuits and rectifiers in low-voltage circuits and, until recently, the most rapidly growing application was in mobile phones and such devices as Sony play stations. Tantalum steel alloys are very resistant to corrosion and for this reason are used in surgical and dental instruments, as well as to pin together broken bones. For the same reason, tantalum has largely replaced platinum in standard weights and laboratory ware. A former use of tantalum was as a 'getter' to remove gas from vacuum tubes, and in electric lamp filaments.

Tantalum levels in foods and diets are unknown. They are not likely to be high. Its fate on ingestion and absorption is probably similar to that of niobium, which has been reported to be present in low levels in human tissues. However, there appear to be no reported studies on tantalum, a gap in nutritional and toxicological knowledge that will, no doubt, be filled before long.

11.3.3 *Caesium*

Caesium, or to use its alternative spelling, cesium (symbol Cs, atomic number 55, atomic weight 132.905) is a soft, white, chemically reactive metal, which is in Group 1/IA of the periodic table. It is one of the alkali metals, along with the far more common lithium, sodium, potassium and rubidium, and the even less common francium. We have already come across the radioactive isotopes of caesium, in particular Cs-137, which is one of the more common radionuclides found in foods as a result of emissions from nuclear industries and accidents. Here we are concerned with the non-radioactive element.

Caesium compounds are very like those of potassium and rubidium, with similar formulae and properties. Metallic caesium oxidises readily in air. On exposure to light, it emits electrons, a property which makes it ideal for use in the photosensitive surface of the cathode of photoelectric cells. It is this emission of electrons that makes the metal a candidate for use in the proposed ion engine which could be used to propel spaceships in the near perfect vacuum of outer space.

Caesium is extracted from a few rare minerals, including pollucite (or pollux), $CsAlSi_2O_6$, and carnellite. Rubidium is always present along with caesium in most of these ores and commercially caesium always contains rubidium. One of the richest known deposits of pollucite, containing up to 34% of the metals, is on the Mediterranean island of Elba. It is also found in commercial quantities in the US. The use of caesium in photoelectric cells has already been mentioned. It is also used in a number of other applications in the electronic industry and as a catalyst in the chemical industry. Caesium bromide is used for optical crystals in certain types of spectrophotometers.

Most of the limited information we have on caesium levels in foods and diets and its metabolism has come from the extensive studies that were undertaken following the Chernobyl accident. Concentrations in most foods and diets appear to be very low, at less than 0.5 μg/kg, in the absence of accidental contamination. Daily intakes in the UK were reported to be about 13 μg/day[43]. On the basis of their close chemical similarity, it is generally accepted that caesium and potassium follow the same metabolic pathways in the human body. It is apparently rapidly and totally absorbed from food in the GI tract. Like potassium, caesium is distributed throughout the body and accumulates in soft tissues with a good blood supply, especially muscle. It is rapidly excreted in urine[44]. However, this is not necessarily always the case and neither potassium nor rubidium can be considered a biological analogue of caesium[45]. There is no evidence that caesium plays any important metabolic role in the body, still less that it is an essential for life. It has been found that caesium can replace the essential element potassium as a nutrient for some lower organisms, but not for higher animals[46]. It does not appear to be toxic, at least at levels normally found in foods.

References

1. Nriagu, J.O. (1980) A silent epidemic of environmental metal poisoning? *Environmental Pollution*, **50**, 139–61.
2. Enrlich, P.R., Ehrlich, A.H. & Holden, J.P. (1973) *Human Ecology. Problems and Solutions*, p. 139. Freeman, San Francisco, Ca.
3. Hartmann, H.M., Monette, F.A. & Avci, H.I. (2000) Overview of toxicity data and risk assessment methods for evaluating the chemical effects of depleted uranium compounds. *Human and Ecological Risk Assessment*, **6**, 851–74.
4. McDiarmid, M.A., Keogh, J.P., Hooper, F.J. *et al.* (2000) Health effects of depleted uranium on exposed Gulf War veterans. *Environmental Research*, **82**, 168–80.
5. International Commission for Radiological Protection (1984) *Protection of the Public in the Event of Major Radiation Accidents*, Publication No. 40. Pergamon Press/ICRP, Paris.
6. Ministry of Agriculture, Fisheries and Food (1994) *Radionuclides in Food*. Food Surveillance Paper No. 43. HMSO, London.
7. International Commission for Radiological Protection (1984) *Protection of the Public in the Event of Major Radiation Accidents*, Publication No. 40. Pergamon Press/ICR, Paris.
8. O'Flaherty, T. (1988) Reducing the risk to human health from radioactivity in the food chain. In: *Proceedings of the Conference on Pure Food Production: Implications of Residues and Contaminants*, pp. 61–78. Agricultural Institute, Dublin.
9. International Commission for Radiological Protection (1984) *Protection of the Public in the Event of Major Radiation Accicents*, Publication No. 40. Pergamon Press/ICRP, Paris.
10. Cope, D. (1988) International dimensions of the implications of the Chernobyl accident for the United Kingdom. In: *The Chernobyl Accident and its Implications for the United Kingdom* (eds. N. Worley & J. Lewins), pp. 71–3. Elsevier Applied Science, London.
11. International Commission for Radiological Protection (1984) *Protection of the Public in the Event of Major Radiation Accidents*, Publication No. 40. Pergamon Press/ICRP, Paris.

12. International Commission for Radiological Protection (1984) *Protection of the Public in the Event of Major Radiation Accidents*, Publication No. 40. Pergamon Press/ICRP, Paris.
13. Paasikallio, A. & Sormunenchristian, R. (1996) The transfer of Cs-137 through the soil–plant–sheep food chain in different pasture ecosystems. *Agriculture and Food Science in Finland*, **5**, 577–91.
14. Kritidis, P. & Florou, H. (2001) Radiological impact in Greece of the Chernobyl accident – a 10-year retrospective synopsis. *Health Physics*, **80**, 440–6.
15. International Commission for Radiological Protection (1984) *Protection of the Public in the Event of Major Radiation Accidents*, Publication No. 40. Pergamon Press/ICRP, Paris.
16. Johnson, M.S., Copplestone, D., Fox, W.M. & Jones, S.R. (2001) Annual cycle of radionuclide contamination of tide-washed pasture in the Mersey estuary, NW England. *Estuaries*, **24**, 198–203.
17. International Commission for Radiological Protection (1984) *Protection of the Public in the Event of Major Radiation Accidents*, Publication No. 40. Pergamon Press/ICRP, Paris.
18. MAFF (1995) *Terrestrial Radioactivity Monitoring Programme (TRAMP) Report for 1994. Radioactivity in Food and Agricultural Products in England and Wales*. Ministry of Agriculture, Fisheries and Food, London.
19. Sanchez, A.L., Singleton, D., Walters, B. & Cobb, J. (1997) Radionuclides in whole diets of people living near the Sellafield nuclear complex, UK. *Journal of Radioanalytical and Nuclear Chemistry*, **226**, 267–74.
20. Farago, M.E., Kavanagh, P., Blanks, R. *et al.* (1998) Platinum concentrations in urban road dust and soil, and in blood and urine in the United Kingdom. *Analyst*, **123**, 451–4.
21. Hodge, V.F. & Stallard, M.O. (1986) Platinum and palladium in roadside dust. *Environmental Science and Technology*, **20**, 1058–60.
22. Hodge, V.F. & Stallard, M.D. (1986) Platinum and palladium in roadside dust. *Environmental Science and Technology*, **20**, 1058–60.
23. Farago, M.E., Kavanagh, P., Blanks, R. *et al.* (1988) Platinum concentrations in urban road dust and soil, and in blood and urine in the United Kingdom. *Analyst*, **123**, 451–4.
24. Petrucci, F., Bocca, B., Alimonti, A. & Caroli, S. (2000) Determination of Pd, Pt and Rh in airborne particulate and road dust by high-resolution ICP-MS: a preliminary investigation of the emission from automotive catalysts in the urban area of Rome. *Journal of Analytical Atomic Spectrometry*, **15**, 525–8.
25. Vaughan, G.T. & Florence, T.M. (1992) Platinum in the diet. *Science of the Total Environment*, **111**, 47–51.
26. Ysart, G., Miller, P., Crews, H. *et al.* (1999) Dietary estimates of 30 elements from the UK Total Diet Study. *Food Additives and Contaminants*, **16**, 391–403.
27. Helmers, E. & Kummerer, K. (1999) Platinum group elements in the environment – anthropogenic impact: platinum fluxes: quantification of sources and sinks, and outlook. *Environmental Science and Pollution Research*, **6**, 29–36.
28. Rosner, G. & Merget, R. (1990) Platinum. In: *Immunotoxicity of Metals and Immunotoxicology* (eds. A.D. Dayan, R.F. Hertel, E. Heseltine, G. Kazantis, E.M. Smith & M.T. Van Der Venne), pp. 93–102. Plenum, New York.
29. Verstraete, D., Riondato, J., Vercauteren, J. *et al.* (1998) Determination of uptake of $Pt(NH_3)_4(NO_3)_2$ by grass cultivated on a sandy loam soil and by cucumber plants, grown hydroponically. *Science of the Total Environment*, **218**, 155–60.
30. International, Programme of Chemical Safety (1991) *Platinum. Environmental Health Criteria*, **125**. World Health Organization, Geneva.
31. Jouhaud, R., Biagianti-Risbourg, S. & Vernet, G. (1999) Effects of platinum on *Brachydanio rerio (Teleostei, Cyprinidae)* 1. Acute toxicity: accumulation and histological effects in the intestine. *Journal of Applied Ichthyology*, **15**, 41–8.
32. Ueda, H., Sugiyama, K., Yokota, M., Matsuno, K. & Ezaki, T. (1998) Reduction of cisplatin toxicity and lethality with sodium malate in mice. *Biological and Pharmaceutical Bulletin*, **21**, 34–43.
33. Baron, C. (2001) There's gold in those old computers. *The Independent* (Newspaper), 18 June 2001, p. 9.

34. Ysart, G., Miller, P. Crews, H. *et al.* (1999) Dietary estimates of 30 elements from the UK Total Diet Study. *Food Additives and Contaminants*, **16**, 391–403.
35. Nielsen, F.H. (1986) Other elements. In: *Trace Elements in Human and Animal Nutrition*, (ed. W. Mertz), Vol. 2, pp. 415–63. Academic Press, New York.
36. Ysart, G., Miller, P. Crews, H. *et al.* (1999) Dietary estimates of 30 elements from the UK Total Diet Study, *Food Additives and Contaminants*, **16**, 391–403.
37. Nielsen, F.H. (1986) Other elements. In: *Trace Elements in Human and Animal Nutrition*, (ed. W. Mertz), Vol. 2, pp. 415–63. Academic Press, New York.
38. Rosenfeld, G. & Wallace, E.J. (1953) Germanium toxicity. *Archives of Industrial Hygiene and Occupational Medicine*, **8**, 466–79.
39. http://members.tripod.com/jeou_shun/g2.htm
40. Van Dyke, M.S., Lee, H. & Trevors, J.T. (1989) Germanium toxicity in selected bacterial and yeast strains. *Journal of Industrial Microbiology*, **4**, 299–306.
41. Matsusaka, T., Fujii, M. & Nakano, T. (1988) Germanium-induced nephropathy: report of two cases and review of the literature. *Clinical Nephrology*, **30**, 341–5.
42. McIlwraith, J. (2001) Tantalum boom lifts mining game. *The Australian* (Newspaper), 23 April 2001.
43. Hamilton, E.I. & Minski, M.J. (1972) Abundance of the chemical elements in man's diet and possible relations with environmental factors. *Science of the Total Environment*, **1**, 375–94.
44. Guerin, M. & Wallon, G. (1979) The reversible replacement of internal potassium by caesium in isolated turtle heart. *Journal of Physiology*, **126**, 293–8.
45. Oughton, D.H., Salbu, B. & Day, J.P. (1991) Stable Cs and ^{137}Cs transfer and distribution in soil, grass and sheep. In: *Trace Elements in Man and Animals – TEMA 7* (ed. B. Momčilović), pp. 27/15–17. Institute for Medical Research and Occupational Health, University of Zagreb, Croatia.
46. Nielsen, F.H. (1986) Other elements. In: *Trace Elements in Human and Animal Nutrition* (ed. W. Mertz), Vol. 2, pp. 415–63. Academic Press, New York.

Chapter 12
Barium, beryllium, thallium and the other metals – summing up

So far, in the previous eleven chapters, only about one-third of the elements of the periodic table that can be described as metals or metalloids have been considered, in varying detail. While some of the other two-thirds are of considerable commercial and chemical interest, they are not normally of major consequence as food contaminants. However, a few have, in fact, featured in major, usually accidental, poisoning incidents. Some of the others, like the metals that have been covered in the previous chapter, could become significant food contaminants as a result of their growing industrial uses. It will be useful, therefore, to look briefly at a number of these other potential food contaminants, from the point of view of present uses and in anticipation of possible future developments.

12.1 Barium

Barium belongs to Group 2/IIA of the periodic table, along with beryllium, magnesium, calcium, strontium and radium. They are known collectively as the *alkaline earth metals*. Its widespread distribution in the environment and its commercial uses in many domestic and medicinal applications make it of particular interest as a potential food contaminant. In fact there have been well-substantiated cases of serious poisoning as a result of ingestion of barium salts.

12.1.1 Chemical and physical properties of barium

Barium (symbol Ba, atomic number 56, atomic weight 137.3) is the heaviest of the alkaline earths, with a density of 3.5. It is a soft, silvery-yellow metal that is malleable and ductile. Its chemical properties are similar to those of calcium. It is very reactive and ignites spontaneously on exposure to moist air. It forms both an oxide (BaO) and a peroxide (BaO_2). It is a divalent element and forms compounds with all ordinary anions which correspond to those of calcium. Among its compounds of industrial use is the hydroxide, known as 'caustic baryta', $Ba(OH)_2.8H_2O$. Another, the sulphate, $BaSO_4$, is one of the most insoluble compounds known. Barium titanate ($BaTiO_3$) is a crystalline substance with ferroelectric and piezoelectric properties.

12.1.2 Production and uses of barium

The main commercial barium ore is barytes or heavy spar ($BaSO_4$). It is also found in witherite ($BaCO_3$). Some 80% of the barytes extracted from deposits is used as it occurs, without any treatment apart from crushing, to make special mud used for oil and gas drilling. The remaining barytes are usually reduced with charcoal to the sulphide, which is then used to make the various chemical compounds that are required by industry. The metal itself, which is difficult to produce because of its high reactivity, can be prepared by electrolysis of its molten salts.

Barium compounds have a number of important industrial applications. A small amount is used to make special alloys with aluminium, nickel and magnesium. These were formerly much used in the construction of radio valves and other vacuum tubes because of their high thermionic electron emission. They are still used today in sparking plugs, as well as for bearings and other specialised purposes.

Barium compounds are used in the manufacture of glass, as fillers for paper, textiles and leather, and in ceramics, television tubes, bricks, pigments in paints and dyes, in lubricating oil additives and a number of domestic products. These include cosmetics, insecticides, vermicides, detergents and bleach. An important medical use of barium sulphate, which relies on its insolubility and its opaqueness to X-rays, is as the basis of the 'barium meal' in radiological investigations. Thus barium is very much part of the modern human environment and way of life, a fact of some significance when it is realised that, unlike calcium and magnesium, barium and its compounds are poisonous.

12.1.3 Barium in food and diets

Barium is found in all foods, usually at levels of about 0.05–1.0 mg/kg. Data on levels in the different food groups consumed in the UK are summarised in Table 12.1. These

Table 12.1 Barium in UK foods.

Food group	Mean concentration (μg/kg fresh weight)
Bread	1.0
Miscellaneous cereals	0.79
Carcass meat	0.05
Offal	0.07
Meat products	0.27
Poultry	0.02
Fish	0.18
Oils and fats	0.03
Eggs	0.47
Sugar and preserves	0.84
Green vegetables	0.38
Potatoes	0.16
Other vegetables	0.50
Canned vegetables	0.27
Fresh fruit	0.35
Fruit products	0.28
Beverages	0.06
Milk	0.07
Dairy products	0.31
Nuts	56

Adapted from Ysart, G., Miller, P., Crews, H. *et al.* (1999) Dietary exposure estimates of 30 elements from the UK Total Diet Study. *Food Additives and Contaminants*, **16**, 391–403.

are similar to the limited data reported in some earlier studies[1]. There is, apparently, no particularly good food source of barium, with the exception of nuts. Levels of 33 mg/kg have been found in apricot kernels, and 14 mg/kg in pecans. The Brazil nut, which, as we have already seen, is believed to be the richest food source of another metal, selenium, also has extraordinary accumulating powers for barium. Levels of up to 3000 mg/kg have been found in some kernels[2].

Concentrations of barium in drinking water have been reported to vary considerably geographically. Levels of as much as 1000–10 000 μg/litre have been reported in some places[3]. Domestic reticulated drinking water will normally have considerably lower concentrations. The US EPA's maximum contaminant level for barium in community water systems is 2 mg/litre[4].

Dietary intake levels of barium have been reported for only a few countries. In the UK the mean adult intake has been estimated to be 0.53 mg/day, with an upper range of 1.3 mg[5]. Similar intakes have been reported in the US[6]. Intake in Japan has been estimated to be 359 μg/day for the average adult[7]. WHO has estimated that total intake of barium of an adult from food and water is 300–1700 μg/day[8]. There is some concern that, in certain circumstances, dietary intakes of barium can be increased to an undesirable extent. It has been found that barium can be leached from glazed ceramic ware, resulting in contamination of their contents. Laboratory studies have shown that it is possible for a person who drinks a relatively large volume of coffee brewed in a ceramic pot and consumes orange juice from a similar container to ingest as much as 9.7 mg/day of barium. This is higher than the EPA's estimated oral reference dose (RfD) of 0.07 mg/kg body weight/day, equivalent to 4.9 mg/day for a 70 kg adult[9].

Ceramic glazes traditionally contain a variety of metals, including lead, nickel and lithium as well as barium. Because of the well-recognised potential health hazards that can be caused by leaching of lead from ceramic ware, barium salts have been recommended as a substitute for lead in the frit formulations for glazes. This may not be appropriate advice in the light of findings that potentially toxic barium can also be leached from ceramic ware[10].

12.1.4 *Absorption and metabolism of barium*

Absorption of barium from food depends on the chemical form of the element. Some of its salts, such as the sulphate, are insoluble. Most of the other forms of the element found in foods appear to be at least relatively insoluble and much of the element ingested in this way is rapidly excreted. Other compounds of barium readily dissolve in digestive fluids. The form of barium found in Brazil nuts (as yet unidentified), for instance, has been shown to be rapidly absorbed from the gut and retained in the body both in experimental animals and human volunteers[11]. In a normal meal between 1 and 30% of the barium in food will enter the blood. Some of this will be retained in bone and teeth, with much rapidly excreted. Barium has been shown to cross the placental barrier in pregnancy.

There is no convincing evidence that barium is an essential trace nutrient for humans. Early studies that seemed to show it was required for growth in rats have not been substantiated[12]. Soluble barium salts are extremely toxic. Ingestion of barium chloride and barium carbonate at doses as low as 0.2–0.5 g have caused acute poisonings in human adults, with symptoms that include vomiting, gastroenteritis and paralysis. Severe respiratory weakness and failure of the cardiovascular system has also been reported[13]. Barium compounds have caused accidental as well as suicidal death by ingestion[14]. Certain barium salts are used as rodenticides[15]. Chronic exposure to barium is often associated with effects on the cardiovascular system. There may be a connection between high levels in drinking water and high blood pressure as well as cardiac mortality[16]. Chronic barium poisoning in a Chinese community, caused by consumption of contaminated salt, appeared to be associated with increased cardiovascular disease[17].

12.2 Beryllium

The metal beryllium came to public attention in the 1940s when what seemed like an epidemic of *beryllosis*, or beryllium poisoning, occurred among industrial workers, especially in the newly established nuclear industry. There was considerable concern when what were known as 'neighbourhood cases' began to occur in people living in the vicinity of plants that used the metal. The outbreaks were relatively short-lived and ceased after control programmes were produced in beryllium-using industries. An official statement was made which asserted that beryllium was a hazard mainly for industrial workers and 'oral ingestion of beryllium compounds is of minimal concern'[18]. Unfortunately, that was not, in fact, the end of the story. As better analytical techniques have been developed for beryllium, and the dangers to public health of even small amounts of toxic substances in food and the environment are increasingly recognised, beryllium has once more come to the attention of health and environmental authorities[19]. In 1993 an official US review concluded that, for the general population, ingestion of food and drinking water was by far the most important route of beryllium exposure[20].

12.2.1 *Chemical and physical properties of beryllium*

Like barium, beryllium (symbol Be, atomic number 4, atomic weight 9.0) is one of the alkaline earth metals of Group 2/IIA. The metal is silvery-white in colour, is very strong and flexible, and has a high melting point. It is the lightest of all solid, chemically stable substances. It has very good electrical and thermal conductivities. It is corrosion-resistant and transparent to electrons, reflects neutrons and has a low coefficient of expansion. Chemically beryllium resembles aluminium. Its salts readily hydrolyse. In water the hydroxide, $Be(OH)_2$, is amphoteric. When dissolved in alkali, compounds of the type Na_2BeO_2 (sodium beryllate) are formed[21].

12.2.2 *Production and uses of beryllium*

Beryllium occurs in a number of ores, the best-known of which is the mineral beryl, beryllium aluminosilicate ($Be_3Al_2Si_6O_{18}$). It is also found in other gemstones, such as aquamarine and emerald. The metal is extracted by electrolysis. The physical properties of beryllium make it ideal for a number of industrial applications: as windows for X-ray tubes, in equipment requiring high electrical and thermal conductivity, in springs that must resist frequent vibrations. Beryllium forms a number of valuable alloys with copper, nickel, aluminium and other metals, and most of the metal is used for this purpose. The addition of as little as 2% of beryllium to copper results in an alloy six times stronger than copper, without reducing the high thermal and electrical conductivity of that metal[22]. Beryllium alloys are used for making non-sparking tools, dyes and high-strength lightweight parts of aircraft, missiles and nuclear reactors. Beryllium compounds, especially the oxide, are used in the electronics industry in ceramic chip carriers, resistor cores and laser tubes. Beryllium ceramics are also used in heat-resisting coatings, for example on rocket nozzles and re-entry cones on space vehicles and in crucibles. A former, but now abandoned, use of the oxide was in fluorescent tubes and screens.

12.2.3 *Beryllium in food and diets*

There is a surprising lack of information on levels of beryllium in foods and beverages, in spite of the considerable interest in its environmental and toxicological importance. An early study, carried out before reliable analytical procedures had been developed for the metal, estimated that daily intake in an industrial society was about 100 μg/day, with levels of about 0.1 mg/kg dry weight, in a variety of vegetables and cereals[23]. More recent data suggest that intakes are considerably lower and levels in food are in the nanogram-per-gram range, with, for example, 0.2–1.4 μg/kg in potatoes[24]. Levels in drinking water are also very low, usually less than the US EPA standard of 4.0 μg/l[25]. The range of beryllium in US drinking water has been reported to be 0.01–0.7 μg/l, with an average of about 0.2 μg/l[26]. Water in the Netherlands has been reported to have a mean of < 0.05, with a range of < 0.05–0.21 μg/litre[27].

Dietary intakes of beryllium in the US have been estimated to be approximately 0.12 μg/day[28]. An older UK study estimated that intakes in that country were < 15 μg/day[29]. An estimate of intake in Japan of 5.0 ± 5.8 μg/day has been reported[30]. Apart from these estimates, there are no other published data on beryllium intakes in different countries and population groups.

12.2.4 *Absorption and biological effects of beryllium*

Animal experiments indicate that absorption of beryllium from ingested food is low. What is absorbed is carried in blood, apparently as a colloidal phosphate bound to protein, to the liver, where it is retained initially. Some of this is subsequently deposited in bone where it behaves in a manner akin to magnesium. Most of the absorbed beryllium is eventually excreted in urine, though some may appear in milk[31].

Beryllium and its compounds are highly toxic to animal cells, even at very low levels. At high concentrations they can cause death. Beryllium is a powerful inhibitor of several enzymes, in particular alkaline phosphatase. It can also affect protein and nucleic acid metabolism and interfere with immune function. Exposure to dust, fumes and aerosols of beryllium was believed to be the cause of 'neighbourhood' beryllium disease. Direct contact with soluble beryllium compounds causes contact dermatitis and possibly conjunctivitis as well as skin ulceration. Inhaled beryllium and its compounds may cause cancer, in addition to incapacitating lung disease[32].

At present there is no clear evidence that ingested beryllium can cause health problems similar to those reported in industrial workers exposed to the metal. It is believed by some investigators that usual beryllium levels in food and beverages do not pose a serious health risk. However, this view is based almost entirely on older data, many of which were obtained by analytical methods considered unreliable today[33]. Current information on levels in foods and diets is, at best, a rough estimate and is insufficient for a reliable health risk assessment[34].

12.3 Thallium

Thallium came to prominence in the 1970s following a number of serious poisoning incidents caused by misuse of thallium-containing pesticides. An editorial in *The*

Lancet called for the utmost vigilance in the use of such products and recommended that restrictions should be placed on their availability[35]. Articles in newspapers and magazines brought the problem to the attention of the public and alerted them to the potential danger of what was, until then, a relatively unknown metal. Though that public concern has largely been replaced by other much more disturbing problems of food safety, thallium poisoning still remains a real threat under certain conditions.

12.3.1 Chemical and physical properties of thallium

Thallium's symbol is Tl and its atomic number is 81. With an atomic weight of 204.59 and specific gravity of 11.85 it is one of the heaviest of the elements. The metal is physically like lead, is white, with a bluish tinge, and is malleable, with a low melting point of 303.6°C. Thallium belongs to Group 13/IIIA of the periodic table, along with boron, aluminium and gallium. Like them, it normally is trivalent, but also exhibits other valencies. Thallium(I) compounds, such as Tl_2SO_4 and Tl_2CO_3, are soluble and resemble those of alkali metals, especially potassium. Tl(III) compounds, such as $Tl(OH)_3$, resemble those of aluminium.

12.3.2 Production and uses of thallium

Thallium is a rare metal, though widely distributed at low concentrations in rocks and soil. It is often found in the sulphides of heavy metals and is recovered as a by-product of smelting of lead and zinc ores. It has a number of industrial uses, in alloys, as a catalyst in the petroleum and other chemical industries, in the manufacture of semiconductors, electric switches, pigments, dyes, luminous paints, artificial gems and window glass. A minor use is in the manufacture of fireworks, to which it gives a green colour, and in pesticides, particularly rat poison. Formerly, thallium had a number of medical applications, including the treatment of scalp ringworm. Such use has been banned in the UK for many years. Thallium isotopes are used in clinical imaging[36].

12.3.3 Thallium in food and diets

Little information is available on levels of thallium in foods and diets. It would seem from published data that the normal level in most foods is 0.1 mg/kg or less. Measurements of thallium in foods in the UK made in the 1994 TDS found that in nearly all cases levels were at or below the analytical limit of detection, and ranged from < 0.001 to 0.003 mg/kg[37]. There is some evidence that thallium is readily taken up and accumulated from contaminated soil by certain vegetables, especially of the *Brassica* genus. Levels of up to 0.125 mg/kg in cabbage[38], with an extreme instance of 42 mg/kg in some grown on heavily polluted soil in China, have been reported[39]. Levels of 0.003–0.12 mg/kg have been found in lettuce also grown on polluted soil[40]. A German study has reported that concentrations of thallium exceeded 0.1 mg/kg in samples of kohlrabi and broccoli[41].

There is some concern that people who consume home-grown vegetables on thallium-contaminated soil are likely to have elevated intakes of thallium[42]. Thallium in soil behaves like potassium and is readily taken up by plant roots. Thallium pollution through atmospheric deposition of contaminated fumes emitted from industrial plants, such as smelting works and cement manufacturing, is known to occur[43]. While normal soils usually contain about 1 mg/kg of thallium, horticultural soils in the vicinity of a cement plant have been shown to have levels up to ten times the normal level. Local people who consumed vegetables grown on this soil were found to have a mean urinary concentration of thallium of 0.0052 mg/litre, compared to a mean of 0.0003 mg/litre in a control population, indicating that contaminated vegetables are a significant route of exposure to thallium through contaminated food[44]. The dietary intake of thallium in the UK has been estimated to be 0.002 mg/day with an upper range of 0.004 mg/day[45]. German intakes have been estimated at about 0.002 mg/day[46]. Higher levels have been reported from Japan, where it is estimated that the mean adult daily intake is 0.040 $\pm$ 0.064 mg[47].

12.3.4 *Absorption and metabolism of thallium*

Thallium is readily and totally absorbed from food. Once absorbed it is rapidly distributed throughout the vascular space and from there enters the central nervous system and other tissues. Elimination of absorbed thallium commences about 24 hours after ingestion, with about two-thirds secreted into the intestine, from which significant reabsorption occurs, and the remaining one-third excreted in urine[48].

Thallium is highly toxic. The toxicity depends on the element's ability to bind to sulphydryl groups in membranes and nerve axons, and to mimic potassium, with which it can exchange and thus disrupt fundamental cellular metabolism. Acute poisoning results rapidly in nausea, diarrhoea and abdominal pain. Delirium, convulsions and circulation problems follow and may culminate in death. A characteristic feature of the disease is hair loss, usually accompanied by skin rash. The alopecia is reported to be due to the formation of insoluble intracellular complexes between thallium and riboflavin, which may account for symptoms mimicking those of riboflavin deficiency[49]. Recovery from thallium poisoning may be incomplete, leaving mental abnormalities, especially in children.

Thallotoxicosis, or thallium poisoning, is usually the result of accidental or intentional ingestion of thallium compounds. It was the deliberate use of thallium by the man who became known as 'the St Albans poisoner' to kill his victims, in one of the most notorious murder cases in the UK, that drew public attention to thallium in the 1940s[50]. There appears to be only one published report of thallium toxicity caused by consumption of thallium-contaminated food. This occurred among villagers in Guizhou province in China who became ill after consuming vegetables grown on thallium-contaminated soil[51]. In theory a similar outcome can be expected in anyone who consumes vegetables which have accumulated ten or more times the level of thallium normally found in foods. However, to judge by the admittedly limited data available, current thallium intakes in the UK, and presumably in other countries with similar environmental conditions, should not cause concern[52].

12.4 The other metals

Little is known, from the point of view of food and human toxicology, about the other elements which will be considered briefly here. They are a disparate collection and, apart from the common chemical properties they share as metals and metalloids, have little in common except the potential of several of them to become food contaminants as they come into increasing industrial use. Unfortunately, we do not know very much about their levels in human diets or their effects on human metabolism, but it is likely that this situation will change as investigation into the health significance of metal contaminants of foods expands. What will be done here is to 'tag' these elements as potential problems and once again to leave it to future editions to expand on the details if further research indicates that fuller treatment is called for. As a glance at the table of contents of the last (2nd) edition of this book will show, this is what has been done in the present edition in the case of boron, germanium and thallium.

12.4.1 *Bismuth*

Bismuth (symbol Bi, atomic number 83, atomic weight 208.98) is a member of Group 15/VA, along with arsenic and antimony. It shares many of their chemical characteristics, though it is much more metallic in character than the other two metalloids. Bismuth is found in nature in the pure state and also occurs in combination as bismuth ochre, Bi_2O_3, and bismuth glance (Bi_2S_3), from which it can be produced by heating in air. It is also produced as a by-product of the refining of lead and copper. The metal is lustrous, hard and brittle with a reddish tint. It readily forms alloys which have low melting points. Bismuth alloys are used to make electrical fuses, safety plugs and automatic sprinkler systems, as well as in nuclear reactors. The metal has a number of other industrial applications, such as the manufacture of paint and in rubber vulcanising. Many cosmetic and pharmaceutical preparations, such as face powders, antiseptics, absorbents and antacids, use bismuth compounds. The chelate tripotassium dicitratobismuthate has relatively recently become of importance in treatment against *Helicobacter pylori* in patients with gastric or duodenal ulcers. It is also a widely used over-the-counter remedy for indigestion[53].

Levels of bismuth in food are normally low, ranging from 0.0001 to 0.001 mg/kg in all the food groups covered in the UK TDS[54]. Dietary intake in the UK has been estimated to be 0.0004, with an upper range of 0.0007 mg/day[55]. Estimated intakes in Japan are somewhat higher at 3.3 $\pm$ 2.6 µg/day[56]. Bismuth and its compounds taken orally are not readily absorbed. What is absorbed is rapidly excreted, mainly in the urine. There is no evidence that bismuth is an essential nutrient for humans. Excessive intakes of water-soluble bismuth compounds can cause renal damage and encephalopathy, as well as less serious problems of skin irritation and pigmentation[57]. There are a number of reports of bismuth poisoning in children who consumed large amounts of bismuth-containing pharmaceuticals[58]. There have also been reports of adults being treated with bismuth medications who developed bismuth poisoning when they exceeded the prescribed dose[59]. Bismuth intake in normal food does not appear to be a problem.

12.4.2 *Lithium*

Lithium (symbol Li, atomic number 3, atomic weight 6.939) is one of the alkali metals in Group 1/IA, along with sodium, potassium, rubidium and caesium. It is the lightest of the group, in fact the lightest solid substance known, with a specific gravity of 0.57. It is a silvery-white metal. Chemically it resembles sodium, though many of its salts are only sparingly soluble, unlike those of other alkali metals. Lithium has a number of industrial applications. It is used in certain alloys, in the manufacture of special glass and glazes and in electrical batteries. Lithium is probably best known, at least by reputation, for its use in psychiatric medicine to treat manic-depressive psychosis.

Lithium occurs in all foods, though at low levels. A range of 0.003–0.06 mg/kg has been reported in UK foods, with an estimated mean adult daily intake of 0.017 mg[60]. Similar intake levels of 14 $\pm$ 53 µg/day have been reported in Japan[61]. Little more information is available on food levels and dietary intakes of lithium. Lithium is absorbed efficiently in the intestine and is excreted mainly in urine. Though there is some evidence that lithium may have an essential role to play in certain animals, this has not been confirmed[62]. No complete explanation of the element's therapeutic efficacy in psychoses has yet been given. It has been suggested that it may act by altering the distribution of electrolytes within the brain[63]. Toxicity from misuse of lithium medication is well known, though the same has not been reported for lithium ingested in food. High doses of lithium may cause interference with glucose metabolism, and hypothyroidism.

12.4.3 Zirconium

Zirconium (symbol Zr, atomic number 40, atomic weight 91.22) is usually described in chemistry textbooks as one of the 'rare metals'. Yet it is, in fact, the twentieth most common element on the Earth's surface. It is even more abundant in soils than are copper and zinc. It is probably best known to many as the gemstone zircon, $ZrSiO_4$. Its chemistry is similar to that of titanium, with which it shares a place in Group 4/IVB. Zirconium has several important industrial uses, such as in alloys. Its dioxide, ZrO_2, known as zirconia, is a valuable refractory material, with a very high melting point and low coefficient of expansion. The metal is also used in industrial abrasives and in flame proofing. Various zirconates, such as Na_2ZrO_3, are used in the production of enamels and opaque glass.

Little information is available on levels of zirconium in foods and diets. It is believed to be present in almost all foods and daily intake is probably in the microgram range, rather than the 3.5 mg/day reported in an early study[64]. Little is known about absorption or metabolism of zirconium. It is believed to have a low level of toxicity when ingested by rats. It has been shown to accumulate in a variety of living cells[65]. There are no indications that zirconium in food is a cause for concern.

12.4.4 Cerium and the other rare earth elements

The term 'rare earths' refers to elements of atomic numbers 57 to 71. They all have physical and chemical properties that are very similar to those of aluminium. In fact the elements resemble each other so closely that their separation was an exceedingly difficult task before the advent of ion-exchange techniques. They are found in a number of complex silicate and phosphate minerals which usually contain a mixture of the elements. A major source is monazite and other mineral sands which are found in Australia and some other countries.

In recent years these metals have come into prominence in a number of industrial applications. An alloy of iron with a mixture of rare earth metals, of which about 25% is lanthanum, is known commercially as misch metal. It is used to manufacture the so-called 'flints' of cigarette lighters. A mixture of the fluorides of the metals forms the core of carbon arcs used in the motion picture industry for studio lighting and projection. This application uses about a quarter of the rare earth compounds produced. Another use in a related industry is in colour television tubes to improve colour rendition. Rare earths have considerable catalytic powers and are employed in the refining of petroleum.

Little is known about the biological effects of the rare earth metals or of their levels in human foods and diets. Lanthanum and its compounds have a low to moderate acute toxicity rating and their industrial use requires care[66]. There are reports that rare earth metals, especially lanthanum and cerium, occur in some fertilisers and that these may cause contamination of agricultural soils, with consequences for animal and possibly human health[67]. Wastewater from both sewage treatment and food processing plants have been found to contain a wide range of rare earth metals[68]. High levels of cerium in certain soils which are eaten by geophagists in parts of central Africa may be associated with a locally high incidence of endomyocardial fibrosis[69].

12.5 The remaining metals: summing up

Less than half the elements categorised as metals and metalloids have so far been considered here. There are many others that could warrant inclusion in a book with the title *Metal Contamination of Food*. Several of them, such as gold, indium, niobium and rubidium, have been considered appropriate for inclusion in their reviews of important metals in the environment and in biological tissues by several experts in the field, such as Nielsen[70] and Tsalev[71]. These, as well as some other metals included among the thirty determined in the official UK TDS[72], such as ruthenium and indium, could well have been discussed here. But we still know very little about them as contaminants of foods, and they are still mainly of interest in the area of occupational health and environmental toxicology. That situation will undoubtedly change as many of these metals find new and extended industrial and domestic uses. As a result they will, like lead and other metals discussed here at some length, become food contaminants of note, deserving of consideration in a study such as this.

The point was made in the second edition of this book that a major factor which would help to bring metals previously not considered as significant food contaminants to the attention of food scientists, legislators and others was improvement in analytical techniques. Advances, particularly in atomic absorption spectrophotometry and in the availability of a much wider range of certified reference standards, were to be expected over the following decades. These have, indeed, taken place and have considerably enlarged the available database on metal contamination of foods. But, while improvements in AAS have certainly contributed to knowledge, the real advances have been in ICP-MS and related multielement instruments as well as in sample preparation techniques.

As a result we now have reliable information on a wide range of the less well-known elements that occur in foods and biological materials, at levels of reliability and concentrations that not long ago were far beyond our analytical capabilities. Particularly significant has been the reduction, both in cost and in time, made possible by these new and improved instruments and techniques. It is beyond doubt that, with the use of these new instruments and procedures, data will continue to be obtained to fill the lacunae in our knowledge of metals in food.

There was another prediction in the previous edition of this book, namely that another important step was about to be taken in understanding metal contamination of foods and that rapid progress could be expected. This was in relation to the speciation of elements in food, and the need to go beyond the simple investigation of total concentrations to determining their chemical forms. Unfortunately, unlike the situation regarding improvements in analytical techniques, speciation studies are still very much in their infancy and the predicted progress has not been made.

Food composition databases, even in recently published official documents such as those produced in the UK by the Food Standards Agency, still continue to provide information mainly, if not exclusively, in terms of total metal contents. There are a few exceptions, such as the separate listing on occasion of organic and inorganic arsenic and mercury in fish and other foods, but in most cases only total quantities appear in the tables. The same situation occurs in food legislation, both national and international. FAO/WHO, for instance, has established maximum limits for certain metals in some foods and beverages, but, apart from mercury where the limits refer to its methyl form, limits are generally given in terms of total metals. Similar regulations, with reference to total lead, mercury and cadmium in food, have been established by the European Union. Even in the case of essential minerals permitted in certain foods, European legislation pays little attention to the chemical forms of the elements, though these can have significant effects on their biological efficacies.

As the authors of a recent review have commented, this situation results in the development of a vicious circle. Because legislators continue to build almost exclusively on expressing the content of trace elements in a matrix only by the total content, without regard to chemical species, there is no challenge to produce simple and fast analytical methods to provide more relevant speciation data[73]. It is up to analysts to break out of that vicious circle and to develop the methods to provide these missing data.

As has been noted by one of the UK's leaders in the field, the concept of speciation applies not only to metals of nutritional importance, but also to potentially toxic elements, such as arsenic, cadmium, mercury and lead. A better understanding of the predominant chemical forms of such contaminants in foods would, therefore, seem to be essential for those who have to make decisions concerning dietary requirements and related legislation[74]. It is difficult not to agree with this author's conclusion that:

> *the study of human nutrition and health requires more information about trace element species in food. Proper dissemination of accurate information gives consumers informed choice, it allows industry to make the best use of foods, supplements and processing (which might affect chemical species), and it provides governments with the basis for good advice and legislation. It may be that one day essential trace element species will be listed on food composition labels or that legislation for potentially toxic elements will be based on the chemical species present rather than the total concentration*[75].

References

1. Reeves, A.L. (1979) Barium. In: *Handbook on the Toxicology of Metals* (ed. L. Friburg, G.F. Nordbergy & V.B. Vouk), pp. 62–77. Elsevier/North Holland, Amsterdam.
2. Lisk, D.J., Bache, C.A., Essick, L.A., Reid, C.M., Rutzke, M. & Crown, K. (1988) Absorption and excretion of selenium and barium from consumption of Brazil nuts. *Nutrition Reports International*, **38**, 183–91.
3. WHO (1990) *Environmental Health Criteria 107*, Barium. World Health Organization, Geneva.
4. EPA (1990) *Barium, IRIS: Integrated Risk Information System*. Environmental Protection Agency, Washington, DC.
5. Ysart, G., Miller, P., Crews, H. *et al.* (1999) Dietary exposure estimates of 30 elements from the UK Total Diet Study. *Food Additives and Contaminants*, **16**, 391–403.
6. Nielsen, F.H. (1986) Other elements. In: *Trace Elements in Human and Animal Nutrition* (ed. W. Mertz), p. 449. Academic Press, New York.
7. Shimbo, S., Hayase, A., Murakami, M. *et al.* (1996) Use of a food composition database to estimate daily dietary intake of nutrient or trace elements in Japan, with reference to its limitations. *Food Additives and Contaminants*, **13**, 775–86.
8. Ysart, G., Miller, P., Crews, H., *et al.* (1999) Dietary exposure estimates of 30 elements from the UK Total Diet Study. *Food Additives and Contaminants*, **16**, 391–403.
9. Assimon, S.A., Adams, M.A., Jacobs, R.M. & Bolger, P.M. (1997) Preliminary assessment of potential health hazards associated with barium leached from glazed ceramic ware. *Food Additives and Contaminants*, **14**, 483–90.

10. Rossitol, M. (1979) Ceramic glazes may poison foods. *Art Hazards News*, December.
11. Stoewsand, G.S., Anderson, J.L., Rutzke, M. & Lisk, D.J. (1988) Deposition of barium in the skeleton of rats fed Brazil nuts. *Nutrition Reports International*, **38**, 259–62.
12. Edel, J., Bahbouth, E., Barassi, V., Di Nucci, A., Gregotti, C., Manzo, L. & Sabbioni, E. (1991) Metabolic behaviour of barium in laboratory animals. In: *Trace Elements in Man and Animals – TEMA* 7 (ed. Momčilović), pp. 30/9–10. Institute for Medical Research and Occupational Health, Zagreb, Croatia.
13. Agency for Toxic Substances and Disease Registry (ATSDR) (1992) *Toxicological Profile for Barium and its Compounds*. US Department of Health and Human Services, Public Health Service, TP-91/03. DHHS, Washington, DC.
14. Downs, J.C.U., Milling, D. & Nichols, C.A. (1995) Suicidal ingestion of barium-sulfide-containing shaving powder. *American Journal of Forensic Medicine and Pathology*, **16**, 56–61.
15. Lisk, D.J., Bache, C.A., Essick, L.A., Reid, C.M., Rutzke, M. & Crown, K. (1998) Absorption and excretion of selenium and barium in humans from consumption of Brazil nuts. *Nutrition Reports International*, **38**, 183–91.
16. Brenniman, G.R., Kojola, W.H., Levy, P.S., Carnow, B.W. & Namekata, T. (1981) High barium levels in public drinking water and its association with elevated blood pressure. *Archives of Environmental Health*, **36**, 28–32.
17. Rosa, O. & Berman, L.B. (1971) Barium poisoning due to consumption of contaminated salt. *Journal of Pharmacology and Experimental Therapy*, **177**, 433–9.
18. Skilleter, D.N. (1990) To Be or not to Be – the story of beryllium toxicity. *Chemistry in Britain*, **26**, 26–30.
19. Meehan, W.R. & Smythe, L.E. (1967) Occurrence of beryllium as a trace element in environmental materials. *Environmental Science and Technology*, **1**, 839–44.
20. Agency for Toxic Substances and Disease Registry (1993) *Toxicology Profile for Beryllium*. US Department of Health and Human Services, Public Health Services. Report TP-92/04. DHHS, Washington, DC.
21. Reeves, A.L. (1986) Beryllium. In: *Handbook on the Toxicology of Metals*, Vol. II: Specific Metals (eds. L. Frieberg, G.F. Norberg & V.B. Vouk), 2nd edn., pp. 95–116. Elsevier, Amsterdam.
22. Lang, L. (1994) Beryllium: a chronic problem. *Environmental Health Perspectives*, **102**, 526–31.
23. Eisenbud, M. (1949) Beryllium in the environment. *Journal of Industrial Hygiene and Toxicology*, **31**, 282–94.
24. Hofele, J., Sietz, M., Grote, M. & Janssen, E. (1994) Untersuchungen zu Berylliumgehalten von Kartoffeln and Böden. *Agribiological Research*, **47**, 273–9.
25. US EPA (1992) National Primary and Secondary Drinking Water Regulations: synthetic organic chemicals and inorganic chemicals, Final Rule. 40 CFR Parts 141 and 142. *Federal Register*, **57**, 138. United States Environmental Protection Agency, Washington, DC.
26. APHA (1991) Section 3500 – Be beryllium, *Standard Methods for the Examination of Water and Wastewater*, 18th edn., pp. 3–53. American Public Health Association, Washington, DC.
27. Vaessen, H.A.M.G. & Szteke, B. (2000) Beryllium in food and drinking water – a summary of available knowledge. *Food Additives and Contaminants*, **17**, 149–59.
28. Agency for Toxic Substances and Disease Registry (ATSDR) (1992) *Toxicological Profile for Barium and its Compounds*. US Departments of Health and Human Services, Public Health Service, TP-91/03. DHHS, Washington, DC.
29. Hamilton, E.I. & Minski, M.J. (1972) Abundance of the chemical elements in man's diet and possible relation with environmental factors. *Science of the Total Environment*, **1**, 375–94.
30. Shimbo, S., Hayase, A., Murakami, M. *et al.* (1996) Use of food composition database to estimate daily dietary intake of nutrient or trace elements in Japan, with reference to limitation. *Food Additives and Contaminants*, **13**, 775–86.
31. Reeves, A.L. (1979) Barium. In: *Handbook on the Toxicology of Metals* (ed. L. Friburg, G.F. Nordbergy & V.B. Vouk), pp. 62–77. Elsevier/North Holland, Amsterdam.

32. Skilleter, D.N. (1990) To Be or not to Be – the story of beryllium toxicity. *Chemistry in Britain*, **26**, 26–30.
33. Vaessen, H.A.M.G. & Szteke, B. (2000) Beryllium in food and drinking water – a summary of available knowledge. *Food Additives and Contaminants*, **17**, 149–59.
34. Vaessen, H.A.M.G. & Szteke, B. (2000) Beryllium in food and drinking water – a summary of available knowledge. *Food Additives and Contaminants*, **17**, 149–59.
35. Anonymous (1970) Thallium poisoning. *Lancet*, **ii**, 564–5.
36. Baldwin, D.R. & Marshall, W.J. (1999) Heavy metal poisoning and its laboratory investigation. *Annals of Clinical Biochemistry*, **36**, 267–300.
37. Ysart, G., Miller, P., Crews, H. *et al.* (1999) Dietary exposure estimates of 30 elements from the UK Total Diet Study. *Food Additives and Contaminants*, **16**, 391–403.
38. Geilmann, W., Beyerman, K., Neeb, K. & Neeb, R. (1960) Thallium ein Regelmassig Vorhandenes Spurenelement in Tierischen und Pflanzlichen organismus. *Biochemische Zeitschrift*, **333**, 62–70.
39. Zhoud-X. & Liu, D-N. (1985) Chronic thallium poisoning in a rural area of Ghizhou Province, China. *Journal of Environmental Health*, **48**, 14–18.
40. Hislop, J.S., Collier, T.R., Pickford, C.J. & Norvell, W.E. (1982) *An Assessment of Heavy Metal Pollution of Vegetables Grown near a Secondary Lead Smelter*. Report no. AERE 2383. Atomic Energy Research Establishment, Harwell.
41. Richter, U. (1999) Thallium in food. *Ernahrungs-Umschau*, **46**, 360–73.
42. Sherlock, J.C. & Smart, G.A. (1986) Thallium in food and diets. *Food Additives and Contaminants*, **3**, 363–70.
43. Brockhaus, A., Dolgner, R., Ewers, U., Kramer, U., Soddermann, H. & Wiegand, H. (1981) Intake and health effects of thallium among a population living in the vicinity of a cement plant emitting thallium containing dust. *International Archives of Occupational and Environmental Health*, **48**, 375–89.
44. Richter, U. (1999) Thallium in food. *Ernahrungs-Umschau*, **46**, 360–73.
45. Ysart, G., Miller, P., Crews, H. *et al.* (1999) Dietary exposure estimates of 30 elements from the UK Total Diet Study. *Food Additives and Contaminants*, **16**, 391–403.
46. Richter, U. (1999) Thallium in food. *Ernahrungs-Umschau*, **46**, 360–373.
47. Shimbo, S., Hayase, A., Murakami, M. *et al.* (1996) Use of a food composition database to estimate daily dietary intake of nutrient or trace elements in Japan, with reference to its limitations. *Food Additives and Contaminants*, **13**, 775–86.
48. Moore, D., House, I. & Dixon, A. (1993) Thallium poisoning. Diagnosis may be elusive but alopecia is the clue. *British Medical Journal*, **306**, 1527–9.
49. Cavagh, J.B. (1991) What have we learned from Graham Frederick Young? Reflections on the mechanisms of thallium neurotoxicity. *Neuropathology and Applied Neurobiology*, **17**, 3–9.
50. Holden, A. (1995) *St Albans Poisoner: Life and Crimes of Graham Young*. Black Swan/ Transworld, London.
51. Richter, U. (1999) Thallium in food. *Ernahrungs-Umschau*, **46**, 360–73.
52. Richter, U. (1999) Thallium in food. *Ernahrungs-Umschau*, **46**, 360–73.
53. Baldwin, D.R. & Marshall, W.J. (1999) Heavy metal poisoning and its laboratory investigations. *Annals of Clinical Biochemistry*, **36**, 267–300.
54. Ysart, G., Miller, P., Crews, H., *et al.* (1999) Dietary exposure estimates of 30 elements from the UK Total Diet Study. *Food Additives and Contaminants*, **16**, 391–403.
55. Ysart, G., Miller, P., Crews, H. *et al.* (1999) Dietary exposure estimates of 30 elements from the UK Total Diet Study. *Food Additives and Contaminants*, **16**, 391–403.
56. Shimbo, S., Hayase, A., Murakami, M. *et al.* (1996) Use of a food composition database to estimate daily dietary intake of nutrient or trace elements in Japan, with reference to its limitations. *Food Additives and Contaminants*, **13**, 775–86.
57. Tsalev, D.L. (1983) *Atomic Absorption Spectrometry in Occupational and Environmental Health Practice*, Vol. 1, p. 96. CRC Press, Boca Raton, Fa.
58. Boyette, D.P. (1946) Bismuth poisoning of children by pharmaceutical compounds. *Journal of Paediatrics*, **28**, 193–7.
59. Playford, R.J., Matthews, C.H., Campbell, M.J., Delves, H.T., Hla, K.K. & Hodgson, H.J.

(1990) Bismuth induced encephalopathy caused by tripotassium dicitratobismuthate in a patient with chronic renal failure. *Gut*, **31**, 359–60.
60. Ysart, G., Miller, P., Crews, H. *et al.* (1999) Dietary exposure estimates of 30 elements from the UK Total Diet Study. *Food Additives and Contaminants*, **16**, 391–403.
61. Shimbo, S., Hayase, A., Murakami, M. *et al.* (1996) Use of a food composition database to estimate daily dietary intake of nutrient or trace elements in Japan, with reference to its limitations. *Food Additives and Contaminants*, **13**, 775–86.
62. Mertz, W. (1993) The history of the discovery of the essential trace elements. In: *Trace Elements in Man and Animals – TEMA 8* (eds. M. Anke, D. Meissner & C.F. Mills), pp. 22–8. Verlag Media Touristik, Gersdorf, Germany.
63. Mertz, W. (1986) Lithium. In: *Trace Elements in Human and Animal Nutrition*, Vol. 2, pp. 391–7. Academic Press, London.
64. Schroeder, H.A. & Balassa, J.J. (1966) Trace elements in food: zirconium. *Journal of Chronic Diseases*, **19**, 537–42.
65. Pais, I. Novak-Fodor, M., Jantzso, B. & Suhajda, A. (1993) New results in the research of hardly known trace elements (titanium and zirconium). In: *Trace Elements in Man and Animals – TEMA 8* (eds. M. Anke, D. Meissner & C.F. Mills), pp. 735–6. Verlag Media Touristik, Gersdorf, Germany.
66. http://www.corrosionsource.com/handbook/periodic/57.htm
67. Avtsyn, A.P., Rish, M.A. & Yagodin, B.A. (1991) Trace element research in the USSR. In: *Trace Elements in Man and Animals – TEMA 7* (ed. B. Momčilović), pp. 12/1–5. IMI, Zagreb, Croatia.
68. Kawasaki, A., Kimura, R. & Arai, S. (1998) Rare earth elements and other trace elements in wastewater treatment sludges. *Soil Science and Plant Nutrition*, **44**, 433–41.
69. Reilly, C. & Henry, J. (2000) Geophagia: why do humans consume soil? *BNF Nutrition Bulletin*, **25**, 141–4.
70. Nielsen, F.H. (1986) Other elements. In: *Trace Elements in Human and Animal Nutrition* (ed. W. Mertz), p. 449. Academic Press, New York.
71. Tsalev, D.L. (1983) *Atomic Absorption Spectrometry in Occupational and Environmental Health Practice*, Vol. 1, p. 96. CRC Press, Boca Raton, Fa.
72. Ysart, G., Miller, P., Crews, H. *et al.* (1999) Dietary exposure estimates of 30 elements from the UK Total Diet Study. *Food Additives and Contaminants*, **16**, 391–403.
73. Berg, T. & Larsen, E.H. (1999) Speciation and legislation – where are we today and what do we need for tomorrow? *Fresenius' Journal of Analytical Chemistry*, **363**, 431–4.
74. Crews, H.M. (1998) Speciation of trace elements in foods, with special reference to cadmium and selenium: is it necessary? *Spectrochimica Acta*, **B53**, 213–19.
75. Crews, H.M. (1998) Speciation of trace elements in foods, with special reference to cadmium and selenium: is it necessary? *Spectrochimica Acta*, **B53**, 213–19.

Glossary

ADI	acceptable daily intake
AFS	atomic fluorescence spectrometry
ANZFA	Australia New Zealand Food Authority
AAS	atomic absorption spectrophotometry
CFI	continuous flow injection
CV	coefficient of variation
CVAAS	cold vapour atomic absorption spectrometry
DAN	diaminonaphthalene
DCP	direct current discharges
DCV	dose conversion value
DRI	dietary reference intake
DRV	dietary reference value
EAR	estimated average requirement
ETAAS	electrothermally heated graphite tube atomic absorption spectrophotometry
FAAS	flame atomic absorption spectrometry
FAO	Food and Agriculture Organization of the United Nations
FDA	Food and Drug Administration
FI	flow injection
FNB	Food and Nutrition Board
GFAAS	graphite furnace atomic absorption spectrometry
GRAS	generally recognised as safe
HGAAS	hydride generation atomic absorption spectrometry
HG-ICP-MS	hydride generation–inductively coupled plasma–mass spectrometry
HPLC	high-performance liquid chromatography
ICAP	inductively coupled argon plasma spectroscopy
ICP-AES	inductively coupled plasma atomic emission spectrometry
ICP-OES	inductively coupled plasma optical emission spectrometry
ICP-MS	inductively coupled plasma mass spectrometry
IDMS	isotope dilution mass spectrometry
INAA	instrumental neutron activation analysis
IUPAC	International Union of Pure and Applied Chemistry
JEFCA	Joint Expert Committee on Food Additives
JFSSG	Joint Food Safety and Standards Group
LOAEL	lowest-observed-adverse-effect level
LNRI	lower reference nutrient intake
MBS	Market Basket Survey
MF	modifying factor
NAA	neutron activation analysis
NOAEL	no-observable-adverse-effect level

NSL	nutrient safety level
PMTDI	provisional maximum tolerable daily intake
PRI	population reference intake
PTWI	provisional tolerable weekly intake
RDA	recommended dietary allowance
RfD	reference dose
RIA	radioimmunological assay
RM	reference material
RNI	recommended nutrient intake
SDS-PAGE	polyacrylamide gel electrophoresis
SF	safety factor
SI	safe intake
TDS	Total Diet Study
TPN	total parenteral nutrition
TVP	textured vegetable protein
UF	uncertainty factor
UL	tolerable upper intake level
XRF	X-ray fluorescence

Index